ALLE · ZEIT · WACH · 1842

THE INTERNATIONAL INSTITUTE FOR APPLIED SYSTEMS ANALYSIS

is a nongovernmental research institution, bringing together scientists from around the world to work on problems of common concern. Situated in Laxenburg, Austria, IIASA was founded in October 1972 by the academies of science and equivalent organizations of twelve countries. Its founders gave IIASA a unique position outside national, disciplinary, and institutional boundaries so that it might take the broadest possible view in pursuing its objectives:

To promote international cooperation in solving problems arising from social, economic, technological, and environmental change

To create a network of institutions in the national member organization countries and elsewhere for joint scientific research

To develop and formalize systems analysis and the sciences contributing to it, and promote the use of analytical techniques needed to evaluate and address complex problems

To inform policy advisors and decision makers about the potential application of the Institute's work to such problems

The Institute now has national member organizations in the following countries:

Austria
The Austrian Academy of Sciences

Bulgaria
The National Committee for Applied Systems Analysis and Management

Canada
The Canadian Committee for IIASA

Czechoslovakia
The Committee for IIASA of the Czechoslovak Socialist Republic

Finland
The Finnish Committee for IIASA

France
The French Association for the Development of Systems Analysis

German Democratic Republic
The Academy of Sciences of the German Democratic Republic

Federal Republic of Germany
Association for the Advancement of IIASA

Hungary
The Hungarian Committee for Applied Systems Analysis

Italy
The National Research Council

Japan
The Japan Committee for IIASA

Netherlands
The Foundation IIASA–Netherlands

Poland
The Polish Academy of Sciences

Sweden
The Swedish Council for Planning and Coordination of Research

Union of Soviet Socialist Republics
The Academy of Sciences of the Union of Soviet Socialist Republics

United States of America
The American Academy of Arts and Sciences

Insuring and Managing Hazardous Risks: From Seveso to Bhopal and Beyond

Editors:
Paul R. Kleindorfer · Howard C. Kunreuther

Springer-Verlag Berlin Heidelberg NewYork
London Paris Toyko

Paul R. KLEINDORFER
Professor of Decision Sciences and Economics
University of Pennsylvania
Philadelphia, PA 19104
USA

Howard C. KUNREUTHER
Professor of Decision Sciences and Public Policy and Management
Director of the Center for Risk and Decision Processes
University of Pennsylvania
Philadelphia, PA 19104
USA

IIASA
International Institute for Applied
Systems Analysis
A-2361 Laxenburg
Austria

ISBN-13: 978-3-642-83076-1 e-ISBN-13: 978-3-642-83074-7
DOI: 10.1007/978-3-642-83074-7

Softcover reprint of the hardcover 1st edition 1987

2161/3020-543210

Preface

This book is the *Proceedings* of the International Conference on Transportation, Storage, and Disposal of Hazardous Materials, which was held at the International Institute for Applied Systems Analysis (IIASA), 1–5 July 1985. The Conference brought together representatives of academia, business, and government from East and West to discuss the nature of current problems in the area of hazardous materials. An important objective of the Conference was to suggest steps that could be undertaken by industrial firms, the insurance industry, and government agencies to improve the safety and efficiency with which hazardous materials are produced and controlled in industrialized societies.

Conference sponsors were IIASA, the Geneva Association, and the Center for Risk and Decision Processes of the University of Pennsylvania. Additional financial support was received from the US Environmental Protection Agency, the Monsanto Corporation, the Center for Organizational Innovation at the University of Pennsylvania, and the Canadian IIASA Committee. We are grateful to all of these institutions for their generous support of this Conference.

Within IIASA, a long history of research in risk activities is evident. This owes much to the vision of IIASA's founding Director, Howard Raiffa, and program leaders who have promoted risk research at IIASA. The present Conference continued this tradition with the strong support of IIASA's current Director, Thomas H. Lee, and Deputy Director, Vitali Kaftanov.

We wish to express our appreciation of Vivien Landauer's able coordination of the Conference and the assistance of Janice Malseed in putting together these *Proceedings*.

Paul R. Kleindorfer
Howard C. Kunreuther
Philadelphia, August 1986

Contents

Editors' Introduction

In July 1984, the Geneva Association convened a meeting for the purpose of planning an international conference on transportation, storage, and disposal of hazardous materials, with special focus on the role of compensation, regulation, and insurance. The Geneva Workshop recommended that the Conference be held at the International Institute for Applied Systems Analysis (IIASA) in Vienna, which has been an important focus for international risk research. The recommendations of the Geneva Workshop were strongly supported by the IIASA Directorate. Thus was launched the International Conference on Hazardous Materials, the *Proceedings* of which follow.

The focus of the Conference was on petrochemical industry problems, with emphasis on regulation and insurance. The Conference had as its major objective a research agenda for the next five to ten years for hazardous materials research in these areas. Given the international nature and scope of hazardous materials problems, participants at the Conference included broad representation from the international community and a rich mixture of practitioners and scholars [1].

The basic themes for the Conference were laid out at the Geneva Planning Workshop. They came to be structured under the following headings:

(1) *Historical Background*. This topic was intended to provide perspectives on the nature and magnitude of accidents and losses from previous technological disasters, notably Seveso and Bhopal.

(2) *Problem Context.* We were concerned with hazardous materials problems in the following contexts: production, transportation, handling, storage, and disposal of hazardous materials. These contexts were meant to include both the dangerous goods sector (e.g. chlorine and sulfuric acid) and the hazardous waste area.
(3) *Risk Analysis.* Here we planned to discuss the traditional problems of hazard identification, risk estimation, risk evaluation, and related perception and communication problems. We were specifically concerned with linking risk analysis to available policy instruments for managing hazardous materials risks.
(4) *Risk Management and Insurance.* Finally, the prescriptive focus of the Conference was principally on risk management and insurance measures. We were interested in determining what policy instruments could be used to mitigate risks, to reduce or eliminate risks, to spread risks, and to absorb the financial and other loss potential of risks in socially and financially acceptable ways.

Figure P.1 summarizes the above areas and shows the principal stakeholders associated with the hazardous materials problem. The overriding theme of the Conference, as it evolved, was to link theory and practice in the use of policy instruments and legal institutions for resolving conflicts among these stakeholders for the problem contexts depicted in *Figure P.1*.

We now provide a brief overview of the Conference papers to give the reader a foretaste of the contents of this volume. The chapters are organized under the four headings listed above.

I.1. Historical Background

The chapters in Part One were commissioned to provide historical perspectives on the nature and magnitude of the hazardous materials problem, with particular attention being paid to the nature of serious accidents in this area. We were also concerned with providing an overview of the magnitude of losses suffered through man-made environmental disasters.

In the first chapter, Lagadec discusses several case studies of considerable interest in the risk management area. These include the release of dioxin in Seveso, the explosion of a liquefied propane gas tank in Mexico City, and the leakage of toxic gas from a pesticide plant in Bhopal, India. He uses these case studies to derive a framework for describing how organizations and public authorities have reacted to crisis situations. Lagadec also uses these case studies to describe the means for coping with crisis, including better emergency planning,

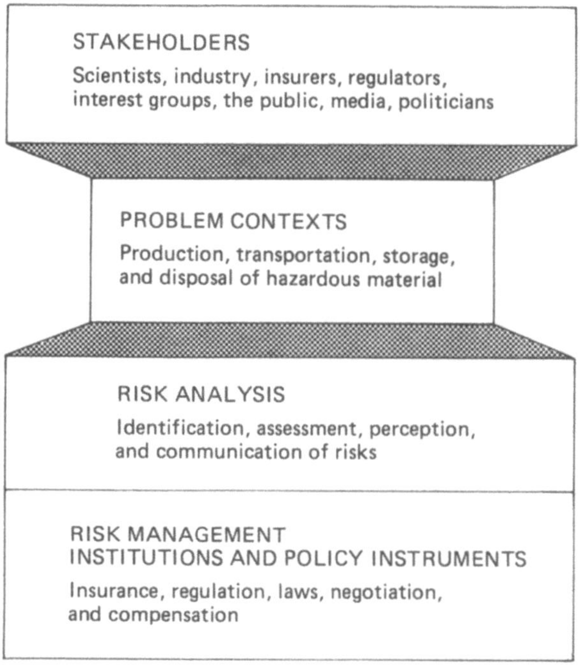

Figure P.1. Basic themes of the Conference.

organizational and institutional design considerations, and approaches to the management of crisis.

This chapter is followed with a description by Naschi of the engineering aspects of severe accidents, with specific reference to Seveso, Mexico City, and Bhopal. Naschi's major focus is on errors in technical design and/or management of the operation which lead to catastrophic failure. He describes some of the safety devices in place for the three cases and indicates why they did not function properly at the time of the accidents. Naschi summarizes his argument by suggesting that the major causes of Seveso, Mexico City, and Bhopal were a combination of "design deficiencies, operating errors, and managerial mistakes."

In their chapter Pocchiari *et al.* describe Seveso and its aftermath in detail from a public health point of view. This original paper was presented in a colorful fashion by Professor Pocchiari at the Conference and described both the Seveso incident and the evacuation procedures following the Seveso accident. Professor Pocchiari uses Seveso as a case study to discuss the uncertainties surrounding the problem of whether such evacuations should be ordered to safeguard public health.

In Chapter 4 Smets addresses the following question: What is the magnitude of environmental damage caused by industrial activities? And, further, are such damages insurable? Smets reviews damage due to oil pollution at sea, accidents involving dams, and air, water, and noise pollution, as well as environmental impairment resulting from radioactive pollution and hazardous wastes. In each of these cases, Smets looks at the best estimates available on actual damage and the proportion of losses covered by insurance and other payments to victims. He concludes that environmental damage is not, by nature of its magnitude and occurrence levels, an uninsurable event. The magnitude and occurrence levels appear to be no higher, for example, than those associated with commercial airline accidents. This original paper set the stage for a lively discussion at the Conference as to what distinguishes environmental impairment accidents, for which insurance is largely unavailable in the USA today, from accidents of other types, for which insurance is clearly available. This discussion is reflected in the comments following Smets' chapter, as well as in several other chapters reviewed below.

I.2. Problem Context

As noted, the theme of the Conference was the use of the policy instruments of insurance, compensation, regulation, and negotiation to promote safe and efficient practices in the hazardous materials area. Several Conference papers explored the relationship between these policy instruments and the specific problem contexts of *Figure P.1*, namely the production, transport, storage, and disposal of hazardous materials.

In Chapter 5 Kleindorfer and Kunreuther investigate the use of insurance and compensation as policy instruments in the context of hazardous waste management. First, they review the nature of hazardous waste management activities, from decisions by firms as to how much waste to generate, through the decisions related to the transport and disposal of this waste. They discuss the complex interweaving of liability and insurance considerations with these decisions. In theory, there is an opportunity to utilize insurance as a policy instrument for encouraging industrial firms to engage in risk reduction measures. However, recent court rulings in the USA and elsewhere on the nature of liability for health effects associated with hazardous materials have created problems for industry and insurance firms with regard to the implementation of such a plan of action. Kleindorfer and Kunreuther then describe the current stalemate in siting new hazardous waste facilities. They recommend the use of insurance and compensation as policy instruments for sharing the benefits to a region from locating a hazardous waste facility with those stakeholders who have to bear the risks associated with such facilities.

In Chapter 6 O'Hare describes the importance of bargaining and negotiation in risk management in the context of hazardous materials transportation. He points out the tremendous importance of negotiation both in striking deals to appropriately spread risks and benefits, and as a means of communication to arrive at an informed consensus about the facts associated with a particular hazard. O'Hare describes the negotiation problem for hazardous materials transportation by considering first the negotiable issues (e.g. classification of substances, handling procedures, and emergency response measures). He then describes the various stakeholders to the hazardous materials transportation negotiation process and the impediments in bringing all of them together in attempting to negotiate the issues involved. These frequently turn out to be very serious problems since the issues are technically complex and there are many stakeholders involved. O'Hare argues, however, that there are also significant opportunities for negotiations to improve the regulation and management of hazardous materials transportation.

Chapter 7 by Kasperson, has as its problem context the siting of hazardous waste facilities, both for radioactive and for chemical wastes. Kasperson points out the very contentious nature of the current stalemate among the stakeholders depicted in *Figure P.1*. By now everyone is clear on the nature of the NIMBY ("not in my backyard") and LULU ("local unwanted land use") syndromes. These acronyms reflect the difficult dilemma facing society when there is substantial benefit to the general population from the production of goods and potential risk to a much smaller set of individuals who are exposed to the risks of having waste products transported and stored from industry in their backyard. Kasperson argues for the increased use of policy instruments such as public communication, benefit sharing or compensation, and public participation in resolving these conflicts. However, given the prevailing scientific uncertainty associated with the consequences of hazardous waste, Kasperson suggests that we have a very rocky road ahead of us in siting hazardous waste facilities. He proposes a set of ethical and/or equity principles as guiding principles for winning and maintaining public trust in the regulation of hazardous materials and the siting of new facilities.

I.3. Risk Analysis

The third group of Conference papers focus on risk analysis, encompassing the traditional areas of hazard identification, risk estimation, and risk assessment. We were particularly concerned with the way in which people and firms perceive and evaluate risks.

In Chapter 8 Covello and Merkhofer describe and evaluate prominent methods for hazardous risk assessment for determining the nature of different chemicals. These methods are organized according to the component of the risk assessment process they are meant to address. Methods designed for characterizing a source of risk are discussed initially, followed by methods for assessing exposures, dose–response assessments, and lastly methods for risk estimation. Covello and Merkhofer explore in detail the complex relationships between the various phases of chemical risk analysis and management illustrated in *Figure P.1*.

In the next chapter Lind surveys current methods of risk analysis and recent advances in this area. Lind describes, through a set of examples, current methods for assessing both the probabilities of failures and their consequences through fault trees and event trees. He also emphasizes the crucial influence of data and uncertainty in risk analysis. At the Conference this paper triggered an animated discussion on the limitations of risk analysis, which is taken up at length in both of the discussants' comments on this paper. These comments describe the practical use and promise, as well as the pitfalls, of risk analysis for insurers and industry.

In Chapter 10 Slovic discusses the problem of communicating risk to the public. The objective of informing and educating citizens about risk issues has triggered a concern among policymakers as to just how to present information to the public. This means among other things finding ways of making technical and scientific uncertainties comprehensible, as well as understanding the public's concerns and anxieties about the risks caused by complex hazards. Slovic describes the current state of research and its possible uses for overcoming these obstacles.

In Chapter 11 von Winterfeldt describes a new methodology, value tree analysis, for understanding the values which various stakeholders may have in relation to policies affecting the risks associated with hazardous materials. Professor von Winterfeldt describes the historical and disciplinary roots of value tree analysis in decision analysis and multi-attribute utility techniques. He also describes its use in various applications to date. Value tree analysis allows a hierarchical description of the value structure of various stakeholders associated with a particular hazardous materials problem. Such an analysis may help in diagnosing and resolving conflicts and in evaluating alternative policy options from the different stakeholder perspectives. He illustrates this methodology in an extended case analysis of options for offshore oil development in southern California.

I.4. Risk Management and Insurance

The final group of papers at the Conference consider the institutional arrangements which society has developed for coping with risks. The key areas of interest involved regulations, both by government agencies and by industry, legal institutions, and, primarily, insurance.

In Chapter 12 O'Riordan and Wynne compare regulatory styles for hazardous waste management in various countries. The authors ask whether there are fundamentally different or convergent regulatory styles in each of these countries induced by the nature of the hazardous waste problem itself. The essential differences across countries relate to centralization versus decentralization of control. However, fundamental similarities are found to be induced by the common problems of scientific uncertainty and technical complexity in regulating hazardous wastes.

The next chapter, by Baram, considers the legal background of liability insurance and risk analysis for chemical industry hazards. Certainly, the key institution for adjudicating and resolving conflicts among parties will be the legal system. For this reason, Baram reviews recent developments in toxic tort law and insurance law. The impact of new developments in the USA and in other countries significantly affects the economic vulnerability of industry and insurers. Baram also suggests that defensive strategies on the part of firms or insurers in restricting their liability or curtailing insurance will do little to prevent risks or satisfy the public. Indeed, they may lead to further risk regulation which may impose costs on industry and insurers in excess of their own private initiatives. Professor Baram concludes that insurers should join with industry, government, and academia in promoting the development of active strategies, based on risk analysis and risk management, to protect both their own interests and societal well-being.

One of the high points of the Conference was the extended discussion elucidating the role of insurance for environmental impairment liability (EIL). This discussion ranged from theoretical explanations of insurance and regulation to the realities of insurance in practice.

In Chapter 14 Klaus provides an introductory discussion of EIL for land-based incidents. He indicates that, from the insurer's viewpoint, surprisingly low losses in relation to fire claims have resulted from the well-known recent environmental disasters. This raises a natural question as to why the insurance industry is so concerned about offering EIL coverage. Klaus suggests that, with the possible exception of the USA, where court settlements are prohibitive, EIL is insurable. However, he also argues that pollution of the environment entails very complex management and insurance issues. He recommends strong adherence to risk assessment and risk management practices on a cooperative basis between insurance firms and companies to increase the likelihood that EIL insurance will be offered on a broad basis in the future.

In the next chapter Orlando discusses recent developments concerning the transportation of hazardous materials by sea. He points out that marine insurance has a very long tradition: marine underwriters have been willing to provide collision liability protection for ship, cargo, and freight. In the area of EIL associated with ocean transport of oil and oil products, various pooling and fund agreements provide workable arrangements for insuring such liabilities. Orlando indicates, however, that attempts to arrive at an international agreement covering transport of other hazardous substances besides oil have not been successful for several reasons that he outlines.

The above contributions highlight a perplexing dilemma. Smets indicates that the nature and magnitude of losses in the environmental area is not extraordinarily high compared to other areas for which insurance is currently available. Both Klaus and Orlando propose some workable arrangements for insuring environmental liabilities which might be profitable for the insurance industry. Nonetheless, several participants pointed to the stark reality that many firms increasingly were going "naked", unable to purchase EIL coverage at any reasonable price due to limited worldwide capacity in these lines. The institutional arrangements and decision processes of the insurance industry itself come under closer scrutiny in an attempt to explain this state of affairs.

Aicken describes the basic logic of risk spreading in the insurance industry in Chapter 16. He indicates several features which make a set of risks insurable and then comments on aspects of EIL which make it a very risky business for insurers. These issues include the well-known problems of latent effects and gradual occurrence, as well as uncertainties in establishing causality for toxic effects and large court settlements negotiated by toxic tort lawyers.

In the final chapter, Pfennigstorf considers these insurance-specific issues in more detail. He discusses the manner in which the catastrophic character of certain environmental risks affects the insurability and coverage of these risks. After a detailed comparison of EIL insurance in the USA and Europe, he considers the outlook, challenge, and market prospects for insurance in the hazardous materials area. This paper stimulated a very active interchange among Conference participants on the reasons for lack of available coverage against environmental risks. Some of the flavor of this interchange is contained in the two discussant comments following the Pfennigstorf contribution. One of these, by Cowell, suggests that the lack of insurance in risk spreading for technological and environmental hazards is one of the key problems of our times.

I.5. Small Group Sessions and Research Planning Agenda

A major objective of this Conference was to plan a research agenda for the next decade in the hazardous materials area. The last third of the Conference was spent in doing just that. An extended panel discussion, followed by small group meetings, developed an agenda for future research. The research proposals generated by the small group meetings were presented at a concluding plenary session and were based on ideas from the papers included in this volume, and the discussant and participant comments which followed. We summarize these research recommendations at the end of this book in an Epilogue, which we hope will serve as a prologue for significant future work in the area of hazardous materials management.

Note

[1] A full list of participants at the IIASA Conference is included at the end of this volume.

PART ONE

Historical Background

CHAPTER 1

From Seveso to Mexico and Bhopal: Learning to Cope with Crises

P. Lagadec

1.1. Introduction

A well-defined failure not unknown to statistical series, codified emergency procedures, a limited number of people involved, a breakdown brought rapidly under control, press releases drafted with no great difficulty by the press offices concerned, and relatively easy coverage by insurance: these are the features of the accident, the province of safety specialists. Major technological hazards explode this system of reference (Lagadec, 1981a,b).

The very large-scale event, extremely serious in its immediate effects and disturbing in its long-term consequences, causes sudden immersion in a universe quite different from that of the "conventional" emergency. Enormous and unexpected difficulties that defeat or wrong-foot the operational arrangements in force; agonizing and paralyzing uncertainty; a critical phase that goes on and on and is therefore wearing on mechanisms, men, and organizations, and an extraordinary increase in the number of people involved: these are some of the features of the post-accident dynamic following a major accident.

The logic is scaled up from that of the "ordinary" accident to that of the crisis. Disproportion, hypercomplexity, and strongly destabilizing tendencies are the hallmarks of the crisis phenomenon which we must now learn to understand better and bring under control.

The chemicals sector can no longer ignore this problem in all its different dimensions – technology, organization, decision-making, and social policy.

Some organizations have been violently confronted with the problem in recent years and have often had to invent – on the spot and with the storm at its height – new tools, new behaviors, and new policies. It is possible to learn lessons from these (often painful) experiences, to shed a little light on the problem, and to carry out research that may help all concerned (industry, the authorities, the organizations involved, and the general public) to enhance their skills in this respect.

Events show there is no time to be lost. During the 1970s Flixborough, Seveso, and Mississauga were so many "warnings" whose cost, fortunately, was not too high, but 1984 brought large-scale disasters:

(1) Cubatao, Brazil, 25 February: oil spillage and fire in the middle of a shanty town – 500 deaths.
(2) Mexico City, 19 November: gas explosions with a domino effect in an industrial site plumb in the center of a densely populated area – 452 deaths according to official sources; 1000–2000 according to the press; perhaps more.
(3) Bhopal, 2–3 December: release of poisonous gas affecting one quarter of the population of the capital of Madhya Pradesh (800 000 inhabitants) – over 2000 immediate deaths.

These three events overtook all the statistics compiled since the Second World War. The accident in India rocked America's third largest chemical company to its foundations. The problem of the major hazard – the structural vulnerability of our industrial systems – has become an urgent strategic question.

Our purpose here is simply to signpost the field of study and clarify the many different aspects of the problem to be defined. This will be done under three headings, in this order:

(1) Reminders of actual cases to illustrate the complexity and severity of the subject.
(2) Points of reference to facilitate the approach to and understanding of the crisis phenomenon.
(3) Pointers to help formulate responses or make them more effective.

1.2. The Shock of the Events

The first aspect of a crisis is the ordeal of the "black-out" – a stunning and oversudden change of state, the unthinkable event that overwhelms and destabilizes. The functions, relations, and missions of the system or systems it strikes seem to have no relevance. Language itself seems incapable of naming the ordeal that has begun. The organizations concerned are thrown into an unfamiliar universe. If the presence of these phenomena is only faint the crisis may be described as "incipient", but if they fill the stage and structure the course of events then that is a crisis situation.

But let us not go too fast. For a start, the crisis does not take the form of an ordered series of isolatable difficulties. Rather it is a global phenomenon where the usual analytical approaches find no purchase. The nature of the problem with which everyone is suddenly faced – compact, elusive, and all-embracing – is a powerful destabilizing factor.

Before we go on to a more analytical study of the crisis phenomenon, therefore, we need to dwell for a moment on this immediate and inescapable challenge: the shock of the event viewed overall. A few significant cases will take us straight into this world of crisis.

1.2.1. Seveso

Today, some comfort is rightly derived from the fact that the consequences of the accident on 10 July 1976 were limited, but we should not forget the ordeal that those responsible and the populations concerned went through in the weeks and months following the dioxin leak.

Let us return to the peak of the crisis when uncertainty was so high, as can be seen from the following calendar of statements (D = day of the accident):

$D + 1$: The manufacturer releases the information that a product used in making herbicide has been accidentally discharged and that precautions are advisable (Pecorella, 1977, p 106).

$D + 3$: The public health authorities write to the mayors of Meda and Seveso as follows: "According to the inquiries that have been made there is no fear of any danger to the people living in the areas surrounding the plant" (Pecorella, 1977, p 106).

$D + 12$: The prefecture is reassuring: "At this time there is no cloud of toxic gas" (Conti, 1977, p 15).

$D + 13$: The prefecture again: "Other health measures should not be considered necessary or urgent" (Conti, 1977, p 16).

$D + 13$: Speaking on television, the Regional Health Director claims: "Everything is under control" (Cerrutti, 1977, p 13).

$D + 13$: The Medical Research Board Director (G. Reggiani) of the industrial group concerned (Hoffmann–La Roche) declares: "The situation is very serious and drastic measures are called for. 20 cm of earth needs to be removed, the works buried and the houses destroyed" (Cerrutti, 1977, p 13).

D + 14: The Regional Health Director replies: "This man has been parachuted in: nobody was expecting him. It does not follow that he is an official spokesman or speaks on behalf of the firm. I have confronted him with the seriousness of his statements and I have the impression he is bluffing. He will have to answer for what he has said" (Conti, 1977, p 15).

D + 14: Change of tone. Official communiqué: "179 people will have to leave their homes within 24 hours" (Cerrutti, 1977, p 13).

During the six months that followed, many of the people in authority remained caught up in this confusion. Uncertainty about the effects of the contamination and the inability to find effective decontamination methods were combined with confrontations of a sociopolitical nature (central versus regional government, Milan versus Seveso, Christian Democrats versus Italian Communists, the Church versus advocates of abortion, the authorities versus the manufacturer, etc.). All this produced a situation of complete helplessness.

Whence the conclusion of the Regional Health Director in Milan: "If the steps taken do not produce positive results within three months, we will let nature take its course" (Conti, 1977, p 100). A new reality governed the situation: the major chemical risk.

1.2.2. Mississauga–Toronto

Here too the ordeal was severe because it was general. No one knew what was in the inferno: the train's manifest was illegible, the freightcars were unapproachable, the information given by railway officials was incorrect (they said there was no chlorine), the presence of polychlorinated biphenyls (PCBs) was rumored, and there was a succession of explosions, one tank being blown a distance of nearly 700 m. Finally, it was decided to act on the assumption that there was a chlorine tankcar in the blaze.

Arrangements had to be made to evacuate on an unprecedented scale – 220 000 people – and, more particularly (an underrated detail), for over 24 hours, so that the social fabric was broken apart with consequences of many kinds. The problems included the hospitals. They did, of course, have emergency plans enabling them to receive an inflow of victims but what they were asked to do was completely different, i.e. evacuate as well – and they had no plan for that. They could be given 20 minutes' warning whereas they needed over 4 hours to evacuate their patients. At the site of the accident the propane cars needed spraying with water, whereas water had to be kept away from the chlorine car, also in the fire.

And, apart from the specific problems to which answers had to be found, there were policy questions such as whether the police emergency plan should be applied or that of the region. There was more experience of the former (for lower-level incidents at least) but the latter, with more involvement of policymakers, was reportedly "safer" in terms of responsibility in the case of serious problems (Burton *et al.*, 1983).

1.2.3. Taft, Louisiana

At 11 PM on 10 December 1982, the management of Union Carbide ordered part of its workforce to leave. There was a temperature surge problem in one of the acrolein tanks. The tank exploded, 17000 people were evacuated and river traffic on the Mississippi was brought to a halt. The case would be of little interest except it is an illustration of how crisis conditions can be created by a blatant lack of communication between those responsible.

Neighboring towns had very well equipped and experienced emergency centers but the information reaching them about the seriousness of the event was too late and too indirect. Hence their surprise, for example, at receiving calls from residents in the zone concerned asking them what evacuation routes to take (what evacuation?) and at suddenly being asked by the factory to shut off all roads up to 8 or 10 km out (when it was only a question of a minor incident representing "no danger"). "No-one told us anything," said a public official.

Sophisticated emergency systems were in place (e.g. direct telephone lines between crisis centers and the dangerous plants in the area). Specialized emergency teams were available, but when they arrived in the plant they were taken under the wing of the public relations people and not allowed to attend the technical meetings. The whole structure was cut off at the base by a single factor: mistrust (Quarantelli, 1983).

1.2.4. The case of the 41 drums of waste from Seveso

This is an example of another type: the "media accident". The affair had been brewing since September 1982 and came to a head in March 1983 with the publication of an article asking the question: where have the 41 drums of waste from Seveso gone? A symbolic word "dioxin", a very well documented press article, unwise management by several businessmen, and assurances given and accepted without double-checking: all this set the scene for a situation of acute social turbulence that kept the whole of Europe in suspense for two long months.

The Italian authorities asserted that the 41 drums had left Italy under official guard (as far as the French frontier) for a destination "somewhere" in the north. France stated that the cargo had also left French territory and let it be understood that Germany was the country of destination. The FRG issued a denial but started inquiries. The GDR denied information given in Rome. The Swiss firm (Hoffmann–La Roche) claimed that the cargo had already been buried in a controlled dump by the authorities of the host country who should, therefore, be fully informed. Suspicion grew and every government or agency with some responsibility quickly came up against formidable problems of credibility. Fingers were pointed at France, Germany, Belgium, the UK, the North Sea, Italy, and even ... the USSR.

In April, Hoffmann–La Roche learnt that the documents on which its information was based were false. The assurances it had given governments were therefore worthless, as were the statements that governments themselves had issued.

The questions became an obsession. Where were the drums? Who knew? Claims, insinuations, denials, and corrections from Milan, Rome, and elsewhere kept the excitement at fever pitch. A continent-wide hunt for the drums and for the "liars and dissemblers" went on at a vigorous pace, with extensive coverage in all the European dailies. The FRG organized a full-scale search in a suspect waste dump as a spectacular operation calculated to satisfy those with the strongest suspicions but also to set a precedent that was both of limited effectiveness and industrially suicidal – "break in everything, everywhere" could not be the most rational instruction.

The issue was serious. A generalized suspicion of governments developed but also, and above all, of the chemical firms. A strict boycott was mounted against the Swiss group, which said it knew nothing. Dumps of waste chemicals were systematically incriminated. In the end, suspect drums were being seen everywhere and the checks made (though always negative) regularly brought to light situations reflecting little credit on the dumps themselves or the firms concerned.

Seven countries and governments and over 40 organizations had the searchlights of the media thrown on them, for the case held the front page or another prime position throughout. The officials and businessmen concerned wondered just how far this unstoppable flood would spread ... (Lagadec, 1984b).

1.2.5. San Juan Ixhuatepec, Mexico City

This was not just a plant exploding (like Flixborough) but a whole industrial site going up in flames. The dreaded domino effect came into action. How far would it go? Large masses of flying metal prompted fears of disastrous chain reaction effects. What is more, the site was in the middle, not of a country area (like Flixborough), but of a densely populated urban zone.

Pemex, the leading national oil and gas company, was confronted with one of the most serious industrial disasters in history. A searching light was thrown on everything that could have been a contributory factor. Industrial safety was remorselessly exposed to the probing of the press whose list of questions included such things as:

(1) Problems of design: proximity of the different installations in the zone to each other and closeness of the industrial site to the conurbation.
(2) Inadequate prevention measures: absence of any plans at all for the installation (*Proceso*, 26 November 1984) and highly inadequate maintenance, also mentioned, incidentally, in a report of the Pemex Health and Safety Committee dated 17 September 1984 (*Excelsior*, 23 December 1984).
(3) No awareness of precursor accidents: "Everything is under control," as one municipal delegation was told that had expressed concern after an accident a few months previously (*Alarma*, No 1127).
(4) Highly inadequate government supervision: laxity and incoherence (*Proceso*, 26 November 1984; *Por Esto*, 19 December 1984.
(5) Corruption: installation of a bypass so that gas could be supplied to private distributors from the storage site without what was done being recorded (*Proceso*, 26 November 1984).

Apart from all this there were also the social problems – rural exodus, poverty belts, uncontrolled land use and speculation – which explain the worst effects of the disaster because of their partial responsibility for the settlement of populations in the immediate proximity of so dangerous a gas storage area. The questions seemed to be too serious: no (accurate) final accounts of the disaster were published (Lagadec, 1985).

1.2.6. Bhopal

The shock for Union Carbide (a foreign company for the country concerned, unlike Pemex in Mexico) was on the same scale as the event. It had to cope with immediate problems and at the same time safeguard the future. It had to give a great deal of information to save what it could of its image but it had only limited access to the Indian data. Lastly, each of its statements was likely to impair its case in the courts. The crisis turned every question into a trap:

(1) Were the safety measures at Bhopal the same as at the other Union Carbide methyl isocyanate (MIC) plant at Institute (West Virginia)? If the answer were "no", that opened the door to charges of exploiting the Third World. If the answer were "yes", it could generate a panic or at least serious trouble at the United States (US) location.

(2) Was the firm intending to take immediate steps? To mitigate the effect of an affirmative answer to the preceding question, all MIC production could be halted until what happened at Bhopal was fully understood, but could such a decision take the place of policy, the collection of information being difficult and lengthy?

(3) Was Union Carbide's safety policy on a level with what was required for such hazards? The reply could only be "yes". But then, how could one account for the avalanche of problems uncovered – "revealed" – at Bhopal: design faults, maintenance deficiencies, inadequate preventive measures and insufficient staff training? In its inquiry, the *New York Times* (28 January 1985) identified ten violations of rules that ought to have been followed. While it was right to point out that the Indians were responsible for the operation of the plant, it could not be pretended that headquarters at Danbury (Connecticut) were not keeping serious watch on these problems which Union Carbide said were a top priority. Nor could there be any question of laying everything at the door of the Indians. Interests in India (and elsewhere), now and in the future, ruled that out.

(4) Was the company in a position to pay? Here, too, the answer had to be "yes", but the path to be trodden was a hairline. Overassurance could tempt applicants (and their lawyers of which there were plenty) to step up their claims, which could change the group's financial situation. The big question was that of the basis of compensation. If North American standards were used, that could raise some doubts about the firm's ability to pay. Taking a yardstick with more affinity to the country concerned could again spark off the polemic about multinationals and the Third World, strategically a rather dangerous question. A further point was that thoe firm had also to contend with attacks from within: its own shareholders had filed a court action against the management that had jeopardized their profits in this way.

Vicious circles and perverse effects colored the scene which – it has to be said – hardly favored nuanced declarations, even though everything was done to avoid the simplified logic that emotion and the media both tended to demand.

On the Indian side, the situation was not easy either. The responsibility of the local subsidiary was unquestionable, but the Americans could hardly be blamed for urban planning (except on the score of not giving enough information about the product). Certain efforts to inform the public would have helped to save a very large number of lives (H'rouda, 1985). Links between managerial staff and senior local political officials were also rather embarrassing: the officials were in the same party as the Prime Minister and the elections were in the offing.

The various illustrations – from Seveso to Mexico City and Bhopal – clearly show the general and variform nature of present and potential crisis situations. They present a challenge to customary Cartesian logic because the dynamic of crisis does not lend itself to breakdown into independent subproblems. That said, we shall now take a more analytical approach to the crisis phenomenon and endeavour to define its principal dimensions.

1.3. The Crisis Dynamic

For the sake of simplicity we can identify three dimensions to the extreme turbulence characteristic of the crisis situation:

(1) Crisis has the features of an *unfurling wave*. It overwhelms the usual instruments of management, rendering them useless and even counterproductive. It strips bare and leaves its stamp, that of incapacity.

(2) Crisis throws things *out of order*, reducing operational mechanisms to uselessness. Worse, the mechanisms help to aggravate the situation. The result is helplessness.

(3) Crisis causes a complete *break*. The missions and goals of the system have to be reconsidered, too. The break – a fault line through which many different eruptive manifestations may break out – calls for revisions that are not simply tactical or organizational but more fundamental, i.e. strategic and "political".

In combination, these three factors not only produce difficulties rather more serious than the norm but generate a very special phenomenon – the crisis dynamic.

1.3.1. Crisis as a tactical breakdown: when tools no longer work

The ordinary tools of routine management are basically characterized by their frame of reference, which is confined to the usual rules and patterns. The unexpected, the improbable, and, more still, the abnormal are not generally included (and it is best that this should be so for the satisfactory running of stable systems). Before the event, questions that might call for radical changes to the system are not put. Of themselves, these facts explain why a crisis – consisting largely of the unexpected and the abnormal – will find the system at a loss. Operationally, these structural limits express themselves as serious constraints. Straining the analysis a little (though this is justified because processes become rigid in crisis situations), the conventional response systems may be said to be capable – but only capable – of:

(1) Dealing with a limited number of difficulties at one and the same time.
(2) Working at overload within fairly narrow margins and for a limited span of time.
(3) Coping with relatively slow developments, not with complete breaks.
(4) Acting within the framework of predefined regulations.
(5) Operating within stable homogeneous units, not in the looser framework of networks whose contours and own dynamics are in a process of rapid change.
(6) Mobilizing a limited potential of resources.
(7) Processing information that is relatively accurate, reliable, and verifiable.
(8) Applying to a specific part of a system, the latter – overall – being stable and well under control, the ceateris paribus condition here being an essential reference.
(9) Dealing with a limited number of persons and representatives.
(10) Dealing with difficulties in the framework of a trial and error process of which irreversibility and the critical gravity of the induced effects are not a feature.
(11) Dealing with difficulties which are not immediately exposed to the glare of publicity, etc.

On all these points, a crisis is the exact opposite: difficulties pile up, the struggle is long term, the usual frameworks are in malfunction, action has to be taken at high speed, the basic rules have to be changed, the whole system starts resonating, the general aims are no longer known and no one knows how to formulate strategies and decide objectives, or with whom. The unfurling wave effect is not the only problem: things are out of order.

1.3.2. Crisis as an organizational breakdown: when regulations no longer operate

Crisis is to be recognized "not only by the growth of uncertainties and unknowns but also by the breakdown in regulations, i.e. the unfurling of antagonisms and uncontrolled processes that are self-accelerating and self-amplifying" (Morin, 1981, p 16).

These basic mechanisms, acting in a time of considerable tension, can cause the headlong plunge into crisis. The following processes need to be considered in a crisis situation:

(1) The system will, if anything, function less effectively than ordinarily. The corrective mechanisms cease to work and the pace of events heightens contradictions in contrast with the generally held notion that when there is a problem everyone rallies round and shows extraordinary dedication.

(2) Indeed, mobilization is the exception rather than the rule. Instead of a general drive energizing a common action, one often witnesses the prudent withdrawal of a large number of potential sources of help. There are many individual organizations that see a crisis as a major threat to their position.

(3) Neither will there be any mobilization of teams. Investigations show that, in unprepared structure, only individuals – wholly on their own in most cases – face up to problems ... watched in fascination, hypnosis, anxiety, or irony by those around them.

(4) Nor will the very many channels of communication be set up that are necessary to link together the large number of organizations confronted with the problem. If there has been no advance preparation, the different parties involved are more likely to form separate islands. The situation of extreme tension and vulnerability is hardly conducive to the formation of these essential links.

(5) Where there is a vital need for trust, the opposite is what usually materializes. It grows into conflict unless all concerned make serious preparation to fight it immediately. So, instead of shoulder-to-shoulder unity in distress, what dominates is latent or open conflict or even the sizing-up of opportunities for the settling of old scores.

(6) The propensity to believe in myth is particularly acute because, in combination, the worries, uncertainties, rumors, and mysteries that are nursed increase the attraction of generalizing many people and organizations (this was particularly clear in the case of the disappearance of the Seveso drums). Instead each person, each entity, tends to rely on some item of information dredged out of the grey areas always present in a crisis situation for the assurance it gives (in terms of self-assurance and the conferring of some petty power that he, she, or it holds the key to the interpretation of the crisis and its solution).

Other avenues need to be studied and the analysis needs to be deepened but a central lesson is already to be learned from these observations. It is that, given all these forms of disorder, what is wanted is not to counter all failures one by one but to understand that crisis is disorder and calls for other means of action.

1.3.3. Crisis as a "political" breakdown: when missions and goals no longer operate

A crisis is not simply the result of unsuitable tools or inadequate organizational capacities. It is the evidence of a deeper failing in the general context which structured the life of the system concerned up to that point. It is not the accidental failure of a particular element that is most to be feared but the vulnerability of a general sociotechnical architecture.

Tactical incapacity, of course, makes organizational inadequacies that much worse, and both together increase the exposure to basic vulnerability. Crisis results from the interaction of these three fault lines. Some of the examples already discussed clearly illustrate the point.

Seveso

The problem here was the helplessness of science, technology, organizations, governments, and states in the face of certain hazards of industrial society. During the summer of 1976 it was shown that extreme insecurity (irreversibility and incapacity) could arise from the very heart of technological development, so promising in other respects; the warning shot came from the chemical sector whereas many had expected it to come from the civil nuclear industry.

The Seveso Drums

Admittedly, tactical problems, such as customs procedures, were considered for a time, and more serious problems like waste management were raised, but the real issue related to more fundamental questions of waste production and industrial policy. Clearly anyone who perceived, in this situation, no more than a problem of customs papers or waste management would be incapable of understanding the essential factors of the crisis dynamic.

San Juan Ixhuatepec, Mexico City

The gas industry and its hazards were the first targets of post-accident activity but the tremor spread in many other directions as well. Questions probed deeper and reached the crucial problems to which the disaster had given new force.

Searching light was thrown on the whole question of safety in Mexico. There were symbols: the big Azcapotzalco refinery located in the very heart of the capital in a district numbering a million inhabitants, the gas pipelines in poor condition threatening the whole of the north of the country and the airport, ringed by urban districts. More generally, there was the reality of an extremely fragile urban system. But there was some wavering. Where was one to begin? How much room was there to maneuver in? Would it be enough if the case were taken up?

There was a general clamor for spectacular measures. The President of the Republic set up a working group to study the problem of major industrial hazards that could be threats to Mexican conurbations.

But doubts were voiced. The difficulties of defusing the metropolitan powder keg of 17–18 million inhabitants are enormous, wrote *The News* on 26 November 1984. The newspaper recalled the editorial it had published a month and a half before the disaster, entitled "Exodus or urban Hara-Kiri", in which it had reviewed the reasons for successive failures in the efforts to deconcentrate the capital – the key to improving the safety situation in Mexico. Others smothered the immediate crisis in despair. "As usual, the government will do nothing" wrote *Proceso* on 26 November giving other and yet worse disasters as the only prospect.

Bhopal

The disruption went deep and took many forms. The headline in *Business Week* on 24 December 1984 was "Union Carbide fights for its life"; rarely was an accident to have so severe an impact on so powerful a firm.

In the wider context, the multinationals, and their relation with the Third World, were again potential targets. The chemical industry itself was faced with a frightening possibility: the fears attaching to the nuclear industry in particular might suddenly shift (this had already happened – at regional level – at Mississauga). The protection offered by statistical argument had gone and was now reversed. One quarter of a regional capital was hit. The images of warfare (chemical warfare) invaded the world of industrial hazards (the exact number of deaths in Mexico City could also, perhaps, have warranted a similar transposition).

Another unresolved question was what could happen if there were a repeat of this kind of rout or even a minor accident in an industrialized country (and it happened, in August 1985, in the Union Carbide MIC plant at Institute – the "cost" of this second incident being perhaps higher for the image of the firm). Cumulative phenomena are to be considered. Perhaps Bhopal has laid a minefield for the future. Hence the acute nature of the subject which will leave its mark for many years.

To sum up:

(1) A grave technical failure can affect a system in its key equilibria. A major accident may explode into a crisis, i.e. a process of extreme social disruption.
(2) Several fault lines (tactical tools, organizational capacities, basic "political" factors) crisscross the post-accident stage. The very high number of factors clash with each other and the outcome depends on their interaction.
(3) Everything may suddenly crystallize around an event of secondary rank which will overturn what has previously been taken for granted and trigger off the crisis in this or that direction. Attention to apparently harmless events is therefore essential without, however, sight ever being lost of the basic structuring of the setting on which the crisis plays itself out. In the famous words of Montesquieu, if the chance of one particular incident causes vast repercussions it is also because the general conditions existed to give it that destabilizing power.

1.4. Coping with Crisis

1.4.1. Tactical capacities for stabilizing an emergency situation

Speed is the essence of emergency action. The rule of thumb with the fire brigades is one minute for a glass of water, ten minutes for a tender, and one hour for a complete brigade. Chemical accidents and major hazards, however, set higher requirements: technical expertise for the direct handling of the accident, high-performance emergency organization resources, and overall planning fully capable of dealing with situations on a very large scale and lasting a considerable length of time.

Basic Arrangements

Several components have to be put together for all the necessary skills to be available. Briefly, these include (Cashman, 1983; Cumberland, 1982; Lagadec, 1983b): hazardous materials response teams; communications and advice centers; on-site emergency plans; off-site emergency plans; mutual aid systems.

But the important thing today is to make sure that they are still really operational instruments and that their scale is sufficient to cope with a major situation.

Emergency Arrangements Scaled to Match the Major Chemical Hazard

The gravity of the hazards entailed demands nothing less than excellence in the quality of the response systems. The following points need particularly careful attention.

Knowledge of effects and their range. Very special attention needs to be drawn to this essential point in the case of major hazards. There are numerous manuals and data sheets on the dangers associated with chemicals but the information they give relates more often than not to the action to be taken in the event of an accident at work affecting a limited number of staff. What are needed now are appropriate documents for major releases that may affect large populations.

Without this knowledge of the range of effects, emergency action can only be tentative. This is obviously a "burning" issue. Investigation would show that in many cases the distance separating a plant from an urban area (a few dozen meters, except in the explosives industry where the lessons of the many accidents that happened during the 19th century have been heeded) is too small. Apart from anything else, this raises big legal questions: if expropriation proves necessary, who foots the bill? So far, no one wants to grasp the nettle.

Homogeneity and coherence of arrangements and measures. It is vital to consider a given installation in the setting of the broader system of which it is part. Two points call for special attention: overall knowledge of the hazards in the zone where the dangerous installation is located (Health and Safety Executive, 1978, 1980; Rijnmond Authority, 1982); comprehensive emergency planning (Gray and Quarantelli, 1981; Quarantelli, 1981).

Informing the population (and the workforce). A population informed about the hazards of the place where it lives and knowing what to do in the event of an emergency is that much less vulnerable, as many examples show. But this social awareness of hazards is largely lacking, and the reflex action ensuring safety and protection even more so. It is therefore to be feared, for instance, that, in a situation where the population has to be under strict instructions to stay at home, there will quickly be an uncontrollable flood of people onto the roads.

The European Economic Community (European Communities Council, 1982) Seveso Directive requires that this information be given to populations. In many cases, the leeway to be made up will demand a considerable effort. There is one prior condition, however, that is not always met. Information within the firm itself is sometimes inadequate.

Systems that are alive: planning rather than plans. Emergency arrangements work only if they are alive, i.e. if their quality is constantly being improved. The purpose of a plan is not to buy cheap reassurance. The authorities that bore the heavy responsibility of bringing the Mississauga accident under control warn against the traditional inadequacies of emergency plans. Three words –"No paper plans"– sum up their message. As Inspector Silverberg put it (Silverberg, 1983):

(1) "Emergency planning is a continuous activity requiring participation and understanding of all government departments, agencies, voluntary groups, private sectors" (p 18).
(2) "Paper plans that have been developed without consultation with all interested departments are of little use" (p 18).
(3) "The heart of emergency planning is: an active process of review, consultation, exercises and training to develop teamwork and preparedness" (p 18).
(4) "An emergency plan must accurately reflect existing operational capabilities and resources. Many written plans reflect more co-ordinated planning than actually exists. Actual participation is a must by all concerned in the planning effort" (p 19).
(5) "'Compliance' plans or 'generic' plans represent little or no real planning activity at local level and have often been produced to satisfy government requirements" (p 20).

Even if a serious effort of preparation is made, doubt or at least caution, must constantly be kept alive in the minds of those concerned. The major event is not controlled that easily. Very relevantly, the same Canadian official notes:

(1) "Warning stage is important – but remember – a disaster can occur so fast that there may be no time for warning" (p 28).
(2) "Don't assume your communication system is going to work" (p 28).
(3) "Remember: the reality of impact totally changes the environment" (p 28).
(4) "No exercise will fully reproduce the actual atmosphere of a disaster" (p 22).
(5) "Co-ordinated effort among technical sectors that may not work together during normal times is a vital aspect" (p 23).

(6) "There are no such things as purely technical decisions: political factors are more prominent than ever during aftermath of disaster. Prepare policy makers and administrators to deal with the problems they will face" (p 23).

1.4.2. Organizational capacities able to bring the runaway dynamic under control

A major accident causes more than just an emergency situation and the mobilization of relief teams. It sets off a turbulence which puts vast systems to the test and penetrates far deeper than just the outer fringe of the organizations concerned. The major hazard creates the need for a real "organizational defence in depth" allowing the general momentum generated by the event to be brought under control. Often the problem is not really grasped because, here too, there is an implicit clinging to the "accident" concept for which the front-line action of the emergency services will suffice.

Crisis is different. It is a situation in which a large number of organizations, wrestling with critical problems, exposed to intense external pressures and acute internal tensions, and placed in mutually conflicting positions and stances, are suddenly and for a long period of time thrust to the front of the public stage in a society of mass communications – in other words, they are "live" – and are guaranteed an unfailing place in the headlines of press, radio, and television (Lagadec, 1984a).

An Organizational Culture that often Excludes the Possibility of a Major Accident

The problem would be simple if organizations merely had to put together response plans for the major accident eventuality. But there are many other prior needs at a more fundamental level. The real basis on which organizations' response capacity is founded consists of their customary references, their standards – in other words, their "culture". Before considering what tools to provide or rules to follow, therefore, we need to explore these basic springs of action which regularly inhibit the preparation, initiation and control of the necessary response.

Mindset. The case studies that have been made teach one very first lesson: if problems of major hazards, crisis, and the exceptional are not part of the culture of an organization, it will be incapable of responding with the rapidity, skill, capability, and perseverance required. Worse still perhaps, the unexpected and the abnormal will cause paralysis and reactions that will aggravate the situation.

It has to be understood that here we are up against deep-lying difficulties: the morale of the organization and the stability necessary for daily routine do not welcome any admission of the exceptional. It is only recently that the reality of major accidents has become clearly evident. The subject itself seems fraught with menace: would not even recognizing it be a kind of acceptance of "defeat"?

These reasons explain the reluctance there often is to tackle the problem frankly. That being so, organizations find themselves at a serious loss if the major accident happens. The whole of their system of representation is caught with its flank exposed, which sets off a chain reaction of difficulties.

Helplessness in the face of the "unexpected". Many recent cases yield a disturbing diagnosis. Unwilling to consider even the idea of an exceptional disruption, organizations tend to enter into crises awkwardly, "backwards" so to speak. "Too late and too little" would seem to be a fair description of all their attempts at response.

Exaggeratedly sometimes, the organization goes through a chain of difficulties:

(1) It shuts its eyes to the many signs that precede the crisis in most cases.

(2) It deciphers the start of the crisis a long time after many other key actors.

(3) Hypnotized by the unexpected, the "unthinkable", its first reaction is to shrink back into itself when, on the contrary, it ought to be multiplying its relations with the environment. This is the period of (suspect) silences, the "no comment" statements (particularly dangerous for the organization's credibility), the denials that mislead no one (except perhaps their authors, which is serious), and hasty statements of the kind "everything is under control", immediately interpreted as meaning that the situation is completely out of hand.

(4) The organization puts up the shutters, cultivates the "fortress under siege" spirit, breaks up into a multitude of islands eyeing each other watchfully, and soon offers itself as an easy victim to the crisis and those who know how to benefit from it. When (for example) too many denials have been disproved by the facts and television pictures, the organization's room for maneuver is seriously curtailed. It is then likely to harden its attitude still further and worsen its position unless those in charge are able to intervene and get different strategies adopted – a change of heading it is very difficult to obtain when the storm is at its height.

We would stress one capital point: these processes can happen at breathtaking speed. Take the example of the *Mont-Louis*, the French ship which sank off the Belgian coast in August 1984 with drums of uranium hexafluoride in its holds.

In less than 24 hours, after Greenpeace had sounded an immediate alarm and following the silences and labored denials of the organization in charge of the cargo, the Belgian Environment Minister accused the French Government, on radio and television, of concealing the truth. Amusingly, the Belgian Minister, while violently denouncing the fact that his country had been unable to obtain the slightest piece of information about the ship's cargo and the dangers incurred by his country, gave assurances in his statement that the country was at no risk. In 24 hours the custom of the immediate "reassuring" denial had thus placed not one, but two, governments (and certain firms or agencies in the nuclear industry) in a very uncomfortable position vis-a-vis the ecologist organization. The latter – again following habit – then threw away its advantage by making wild exaggerations, thus enabling those responsible to extricate themselves from a very difficult position. Although the risk, in the opinion of the specialists, was minimal, the culture of the organizations involved had very nearly converted a minor accident into a "media disaster".

The point to remember, therefore, is that even before the rules for responding to an emergency situation could be brought out of the safe, the "culture" of the organization had had the time to do considerable harm. So what purpose do writing clear press releases, holding press conferences, showing one knows the facts, and so on, serve in such circumstances? It is extremely difficult for even the best tactical measures to rescue an organization from a strategic defeat.

So the first requirement to help an organization respond more successfully to a crisis is not to provide it with a list of instructions. The essential need is to look carefully at the deep-lying culture from which its reactions spring. Once again, it is difficult to sidestep this deep-lying culture. There is every likelihood it will surface (and violently) at a time of crisis. In communications, for example, a professedly outgoing policy pasted artificially over a culture of secrecy will not last long, and the idea that things are being concealed will quickly gain ground. So mere recipes will not do.

Crisis Management Aptitudes

A great deal of work would be necessary to draw up organizational rules for crisis management. The case studies that have been produced, however, tell us that the ability to develop a proactive attitude both inside the organization and towards the outside world is an important precondition.

The following points need to be considered.

(1) *Recognition of the reality of the major hazard.* The traditional reaction rejecting any review of this subject on the grounds of the need to show "optimism" has to be scrapped. "Clear thinking" is not necessarily synonymous with "catastrophism" or "anti-industrial attitudes".

(2) *Knowledge of the major hazards that could concern the organization.* Here, other references than just statistics and habits need to be brought in. The rare event may in future carry just as much weight as all accumulated experience. Marginal logic is no longer the only desirable intellectual reference.

(3) *The ability to recognize, quickly, the onset of crisis conditions and to manage the information collected without delay.* The point is well illustrated by the case of the *Mont-Louis* already referred to. It may be analyzed briefly as follows. A business enterprise has to perform an ongoing mission. It therefore tends to concentrate on the large masses, the regular event. The role and interest of the press and critical groups is to highlight the exceptional. These are two opposite cultures which have developed different response modes and tools. The latter set of actors possesses extraordinary capabilities in this respect: gathering of information, the swift reporting of that information, and immediate distribution. Public and private officials have duties, of course, which oblige them to exercise more prudence – which explains some delay in their response.

(4) *The ability to bring swiftly to life a network of actors relevant to the crisis.* The collection of information, the analyses that have to be made, and the decisions that have to be taken imply that a large number of relationships – often new ones – have to be woven immediately within the organization and with the outside environment. Existing conflicts, the absence of previous links, and differences in "culture" between the organizations concerned greatly impede the development of the necessary cooperation. Very often, a new mode of operation will become established only after the higher levels of authority, recognizing the seriousness of the situation, demand the necessary changes. But it is still difficult for them to modify the culture of their organization at a time when it is dangerously exposed and therefore on the defensive. The capability of proactive response has to be developed beforehand.

(5) *The ability to work with the media.* Here again the basic culture of the enterprise is deeply involved. Particularly in Europe, the rule of secrecy is often all-powerful. There are reasons for it, but the dangers of too uncommunicative an attitude have to be

carefully weighed. Journalists do not simply ask for information and wait for those responsible to be kind enough to give it to them. They are professional information hunters and not easily put off. To dispel the illusions that have long been held on this subject, it is sufficient to remember the ingenuity of the journalists at Three Mile Island (Scanlon and Alldred, 1982; Sandman and Paden, 1979). Donald R. Stephenson (Director, Corporate Communications, Dow Chemical, Canada) sets out very clearly the attitude needed in this field where the proactive approach is often tragically absent:

> 1. The public must be fully informed frequently and accurately through the media, from the outset. This must be done by one or two highly credible senior spokesmen who understand the situation and can explain it calmly and clearly in lay language. The first 24 hours of a crisis are critical.
> 2. If this is not done, a public information vacuum probably will develop rapidly – and be filled by rumors or alarms far worse than the real situation.
> 3. Silence in the midst of a crisis implies guilt, whether justified or not.
> 4. It is not enough merely to assure the public that everything is O.K. and there's no reason for alarm. To be credible, we must provide details of how that conclusion is drawn.
> 5. It is vital to realize that reporters face deadlines hour by hour. Information must always be correct, consistent and current, even if all the answers aren't immediately available.
>
> These are principles that many top executives and operating managers find hard to adopt. They fail to understand the urgency of the situation or the implication of delayed response. They are inclined to try to smother bad news rather than air it. (Stephenson, 1984, p 3)

With J. Scanlon, one must stress the crucial importance of this capacity to manage information in a crisis. In an open Information Society there can be no division drawn between operations and communication. Anyone who cannot control information problems can have no control over the operational conduct of the emergency situation.

(1) "An emergency, among other things, is an information crisis and must be treated as such" (Scanlon *et al.*, 1982, p 31).

(2) "to a considerable extent whoever controls the access to information, whoever is the source of information becomes the center of operations and control; and if you don't have communications systems operational, if you can't disseminate it, then you also lose

the power to have operational control and it will shift to whoever has that" (Scanlon, 1982, p 17).

(3) "communications are so important in the aftermath of disaster that the centers of communication may well be the centers of operational control as well" (Scanlon, 1975, p 429).

Tactical capacities and organizational aptitudes form a whole, but another dimension of response is needed to complete the picture: the ability to run the system caught in the turbulence of a crisis.

1.4.3. Management capacities

The major accident and the crisis dynamic – where weighty issues are at stake – are problems for top management. Here, too, basic references have to be completely changed. The "incident" used to have only marginal effects and could be handled by the relevant specialized technical service. A major accident can derail or even fatally damage the systems concerned. Today, managements of both public and private organizations have a duty to recognize and explore this "new frontier" and bring it under control. With brutal suddenness, Bhopal revealed the critical importance of these matters for the heads of firms and public agencies.

The men in top management must arm themselves with the specific management capabilities necessary for controlling crisis situations and prepare their organization as a whole to withstand such conditions. Their task is also, of course, to take every step to prevent major accidents and to make their constant concern the anticipation of problems, without which their field of maneuver could be extremely narrow.

Managements in the Eye of the Storm:
Controlling Systems in a Crisis Situation

From a study of recent cases, it is possible to identify a number of important requirements for the managements of systems faced with a crisis dynamic:

(1) *Put the organization on a "crisis mode" footing.* This is a question of swiftly actuating the patterns of thought, arrangements, and behaviors that the organization (and not just the emergency services) needs to adopt to be in a position to cope with the situation in every respect instead of waiting for the crisis to attack each of the subsystems concerned one by one.

(2) *Initiate an ad hoc data collection and analysis system.* Crises require the capture and verification of information received via infinitely more diversified channels than in ordinary circumstances. They also demand a continuous effort of interpretation on the basis of unusual and very exposed criteria and models. In particular, special attention has to be paid to the gross intellectual mistakes that lead to fundamental errors of judgment and persistence in those errors, each fresh item of information being forced (to the point of absurdity) into the frame of logic adopted. In spite of the absolute need for resolute and immediate action, the functioning of the organization must leave room for caution about the hypotheses that are formulated and for the active and critical clarification of the implicit hypotheses behind the reasoning. In particular, the system that management perceives may not be the real system: the gap between the two needs to be studied without delay.

(3) *Make sure no gross mistakes and blunders are made at the very outset.* It is not unusual for some subsystem in the organization to act on its own initiative, without coordination, obeying reflexes wholly inappropriate to crisis conditions, e.g. the overhasty issue of a press release claiming "nothing has happened" or "everything is under control". Foolishness of this kind can gravely compromise the position of the management (particularly if committed in its name) or necessitate disclaimers which would be highly inopportune, this not really being the time for an outbreak of internal strife. It is vital to identify these untimely initiatives that the organization might be tempted to take at the earliest moment – and prevent them happening.

(4) *Strive to maintain the organization's internal coherence and capability.* As we have seen, the crisis dynamic sends tremors through the organization, causes cracks to appear in its structure, sows doubt about its objectives and fundamental missions, weakens allegiances, resurrects bitter conflicts, etc. Soon the system has broken up into so many islands all behaving like little besieged fortresses. Immediate steps have to be taken to counter these destructive tendencies by reclarifying the key dimensions of the life of the organization: its missions, policies and strategies, the rules of communication, the rules for the settlement of disputes, etc. No imprecision must be allowed to prevail about these very foundations of the life of the system. The beginnings of any internal contributory factors – conflict, rumor, or inertia – must be identified and dealt with immediately. Here again, the management must be constantly questioning itself on the gap between its perception of the system and the system's real state,

and constantly checking to see that its decisions are effectively put into practice and produce the effect desired.

(5) *Maintain and develop the organization's external capacity.* Whereas the natural tendency is for the organization to shrink into its shell, it must – on the contrary – strive to increase relations with its environment. This rapid deployment of relational systems is necessary in order to be able to receive and give out information and to implement actions with hitherto unfamiliar organizations. Efforts need to focus on two main areas: the operational and decisional system and the media. One of top management's tasks in this respect is to check the quality of the relationships that are established, in particular the level (position in the hierarchy, authority, and power) of the persons with whom links are made.

(6) *Informing the public.* It is a critical question in a crisis situation and this is therefore one of the important functions of which management must take direct charge (which implies new patterns of internal operation in this respect if the public relations departments are not of the highest status in the company).

(7) *Manage the time factor.* A constant, questioning watch must be kept on the development of the crisis. Every time a decision is taken the question must be faced: What's next? The purpose is to prevent incoherence over time and, more fundamentally, to ensure that the response to the crisis dynamic escalates as time goes by.

(8) *Be on the alert for any possibility of incipient crises breaking out on other fronts.* Frequently the scale of a crisis will grow as a result of certain details that the organization, paying little heed to questions that seem to be of secondary importance, fails to attend to quickly enough. Managements need to keep a careful and constant watch for the outbreak of any subsidiary crisis on a secondary front. The typical example is the case of the 41 dioxin drums from Seveso. Every "suspect" waste dump could become the main factor of the crisis within 24 hours. Managing a crisis is managing a kaleidoscope: a slight shift can change the scene completely.

(9) *Strive constantly to enlarge the organization's room for maneuver.* The reference here is to the need to combat a regular effect of crises which is the overhasty and heavy-handed closing off of many possibilities.

(10) *In addition to running the organization concerned, work for the development of the overall system affected by the crisis.* While the tendency is for the organization to turn inward on its own problems and plans, top managements should take their part in the more general moves. They should consider what initiatives might be taken and give the help that outside actors may need.

Possibly they should lay the basis for a new general configuration of the system concerned and then define new places in it for the organizations involved. Managements' tasks in a period of crisis extend far beyond merely defending the immediate and specific interests of the organizations for which they are responsible.

Preventing the Major Accident and Preparing the Organization for Crisis Situations

The precondition: mindset. Again, there can be no decisive progress in the control of major hazards and crisis situations without a clear awareness and recognition of the reality of the challenge and what is at stake. This is necessary in order that the required stimuli come from the top and so that the example of the upper echelons consolidates the progress to be made in the organization as a whole. The rule is simple and obvious, but putting it into effect is certainly more difficult. After years and decades of not doing enough in this respect, changes of attitude cannot be brought about without a great effort (Mitroff and Kilmann, 1984). And yet it is here that the key to all the actions to be launched or developed to prevent major hazards and crisis situations and secure greater control over them is to be found.

A decision-making system reorganized to include the major hazard question. One primary objective is essential: safety has to be made a goal, field of decision, and problem for top management. To that end, there are several operational rules that senior executives need to institute or strengthen:

(1) *There has to be a clear perception of the safety options.* For this to be so, safety options have to be perceived as decisions (not simply "technical" provisions) at all levels of the organization and the choices made with regard to location, design, maintenance and management need to be perceived as safety decisions.

(2) *Safety problems must be given specific expression.* The danger is that technical, economic, and administrative considerations will overshadow safety questions. To avoid this trap, organizational systems need to be designed in which specific account can be taken of safety. For example, the head of a plant – whose duty is to develop his establishment – should not be the only architect of safety options and the only channel for conveying those options to top management. An internal critique on safety matters, via recognized organizational machinery, must be possible and must have every opportunity to express itself at top management level. There is no point in hiding the fact: the management of safety is another industrial bargaining situation. If certain equilibria are

not secured, there is little likelihood of safety questions being given proper consideration.

(3) *Decision-making levels have to match the importance of the issues.* A major question for top management is whether hazard problems are not covered up at the lower echelons, the danger being that there may be undivulged trade-offs at the intermediate levels based on considerations that are too narrow in the light of what is at stake. Here again, it is important to ensure that the organizational system guarantees the upward flow of information, including proposed decisions. A point worth noting is that if every organization dealt with safety questions at the right level, communication between levels would become immediately easier. In short, the question for the senior manager in the private or public sector is whether he commands an internal organization that enables him to exercise his decision-making powers effectively in safety matters.

Specific prevention efforts. In view of the surprises that came out on the occasion of the alerts and accidents discussed earlier, a first step for managements in the public or private sector could be to produce a diagnosis of the situation regarding safety problems in the systems they are responsible for. A number of audits would make it possible to identify any major technical or organizational shortcomings.

The object would be to ensure there was no chance of suddenly finding situations like those there were at Flixborough or Canvey Island:

> It was clear that no-one concerned in the design or construction of the plant envisaged the possibility of a major disaster happening instantaneously. (Department of Employment, 1975, p 36.)

> During the preliminary visits to the premises selected for detailed assessment, the investigating teams were reassured to find that, where hazardous materials were being processed, handled or stored, the managements were generally very responsive to matters of operational safety. Where relevant codes of practice were available, these had been taken into account in the design and construction of plant and facilities. However, these visits also established that none of the companies concerned had made a systematic attempt to examine and document those few potentially serious events which might cause accidents among people in the surrounding community. (Health and Safety Executive, 1978, p 8.)

Similarly, the safety assessment which the French authorities requested be made at the Union Carbide plant at Béziers, after the Bhopal disaster, revealed certain problems at the works but, more particularly, some very serious shortcomings in the system for transporting MIC between Fos (near Marseilles) and Béziers. Studies like these – which need to cover both software and hardware – are essential for decision-makers in public and private sectors. They supply a snapshot of the safety of their system, an information base enabling them to determine priorities, and an opportunity for displaying, internally and externally, the importance they attach to accident prevention.

Thus the exercise can be not only a tool of investigation but, at the same time, (1) an effective lever for propagating that internal culture of the organization, making it more sensitive to major hazard questions, and (2) a valuable talking point in the social discussion which unfailingly develops on the subject of major hazards.

This last point is worth stressing. It is vital to have accurate references in these discussions that always threaten to get out of hand when fear, suspicion, and the imagination rule the mind. It was one of the lessons of the studies carried out at Canvey Island. The work done by the Health and Safety Executive and the publications it produced helped not only to reduce appreciably the hazards in the area but also to ensure that the social debate had a solid basis (Lagadec, 1979; 1983a). The British Government had every right to say in its second report on the zone in 1980:

> We regard the report as a watershed – a unique, pioneering exercise which will prove to be a major turning point in risk assessment work. The report aroused extensive interest in the U.K. and around the world, far beyond the sphere of those having a direct local concern. In its range and depth the report provided, and still provides, a potent stimulus to the debate about risk assessment techniques and the practical decisions which have to be made about the relation between potentially hazardous industry and people who live and work nearby. (Health and Safety Executive, 1980, p iv.)

Decompartmentation and Opening Up to the Outside

The "ordinary" accident was something that stayed within the confines of the firm, but the major accident makes its effects felt far beyond and strikes at the surrounding communities. The consequence of this new state of affairs is immediate: the outside world demands an explanation from the company and even the right to look behind the scenes in industry. Clearly, such demands are unprecedented and a shock to the industrial culture previously sheltered by its protective walls.

Facilitating adaptation to these new social requirements is a matter for senior management. It becomes important to engineer new strategic positions defined in terms of credibility and legitimacy. It becomes necessary to be able to supply accurate information, to embark on programs of internal transformation, etc. Authority or invoking "Science" and "Progress" no longer carry enough weight. They are even suspect, and therefore counterproductive. The major hazard requires the company (but it is also true for the public agencies concerned) to be able to justify its activity by far more open yardsticks than those previously used.

Several initiatives can be taken along these lines. One is to accept (as is tried in France) that safety assessments be reviewed by external experts reporting to the public supervisory authority and another is to make safety reports public (not including certain passages which would otherwise reveal industrial secrets but whose absence does not affect the understanding of the text). A few years ago, this kind of possibility would have seemed unthinkable. The major hazard has made readiness for this more open type of policy advisable or even essential. But there are two possible approaches, one reactive and the other proactive. The latter offers valuable strategic advantages to a company in terms of consolidating its position in the event of a crisis (which always remains a possibility even when the maximum is done in the way of prevention).

A case of considerable significance for the future: Union Carbide's strategic decisions following the Bhopal disaster. On 20 March 1985, Mr Warren Anderson, Chairman of Union Carbide, announced the following decisions (Anderson, 1985, pp 2–3):

1. There'll be intensified sampling procedures, training and retraining sessions, process review and countless administrative and physical changes.
2. Our overseas locations will face three times as many company safety audits this year, compared to 1984. And there'll be a significant increase in such audits at our US sites as well.
3. A new committee of the Board of Directors is meeting semi-monthly to handle health, safety and environmental affairs. Head of the committee is Union Carbide director Russel Train, first administrator of the Environmental Protection Administration and presently president of the National Wildlife Foundation. Mr. Train has been a director of Union Carbide for the past eight years.
4. Another significant change established a new committee of top-level management on risk assessment, reporting directly to Union Carbide President Alec Flamm. This brings to the highest level of the corporation compliance reviews of our facilities that handle hazardous materials.

These key decisions are evidence of a qualitative change in an industrial group's policy towards safety matters. They deserve most careful attention by everyone concerned and, first and foremost of course, the whole of the chemical

These key decisions are evidence of a qualitative change in an industrial group's policy towards safety matters. They deserve most careful attention by everyone concerned and, first and foremost of course, the whole of the chemical industry.

A Third Path of Action: Anticipating Risks and Vulnerabilities

Working with the future in mind and preparing for changes of direction a long time in advance are increasingly necessary as systems become more complex and the bearings that are set become more difficult to alter. If hazards and vulnerabilities are not anticipated, efforts to prevent and, all the more so, efforts to control crisis situations will be severely handicapped. Two main directions need to be investigated although our reference to them here will be very brief.

Internal development: hazards "in ovo" in the chemical sector. Several different starting points can be taken, including hazards associated with: expected new products; new manufacturing, transport and storage technologies; new forms of business organization (computerization, early retirement with the loss of long-experienced staff); new strategies such as the development of biotechnologies, the relocation of the basic chemical industry in the Third World, and the development of certain fine chemical sectors.

The more general setting: underlying trends in the environment of the chemical industry. The question of major hazards associated with dangerous substances arises in a setting in which conditions are becoming significantly more fragile. Some aspects are:

(1) *The geographical interspersal of industrial and urban activities.* This is already raising the problem of the distance separating industrial plants from urban areas. New legal provisions need to be developed. A vital subject is compensation.

(2) *The development of networks and structures of activity through which havoc-creating chain effects are possible.* The safety of dangerous substances, for example, might be imperiled by destabilizing accidents affecting wide-ranging geographical areas. In November 1981, the Lyons region (1 million inhabitants) was cut off from the outside world following a fire in a telephone exchange. Even key government links were severed because they, too, went through the same exchange. What would have happened if some kind of major accident had happened demanding trunk communications on a massive scale?

(3) *The economic crisis.* This may have serious repercussions on firms' safety in two ways: risk-taking increases and resources allocated to control and safety (primarily maintenance and supervision) are reduced.

(4) *Trends and changes in the social "demand" for safety.* It is clear that what until recently was regarded as "acceptable" will no longer be so in the future. More than that, any major accident may completely change perceptions and requirements. This was brought out in Ontario, following the Mississauga accident. Hazards became a far more prominent subject in people's minds and the subjective rankings of various hazards were radically reordered (Burton *et al.*, 1983). The subject of many opinion polls was the general perception of science and technology; it would be wise to look into the far-reaching consequences in this field of disasters like Bhopal. Such a survey, based on these considerations, was conducted in France in March 1985 (SOFRES, 1985). Here are some of the findings:

   (a) Public opinion is concerned about major accidents. Asked whether they thought accidents like those in Bhopal and Mexico City could happen in France, 3% said it was inevitable, 55% likely, 32% unlikely, and 2% impossible. Conclusion: the theory of the perfect control of technological systems prevalent in the 1960s and 1970s can no longer claim credibility.

   (b) The subject of major hazards attracts much attention. Asked if there were too much or too little said about the technological risks that existed in France, 21% answered "too much" and 67% "too little" (12% said they didn't know). Conclusion: there is a definite demand for information on the subject.

   (c) Opinion is divided about how seriously the problem is being tackled. 44% said everything was being done to obviate technological hazards and 39% said not everything was being done (no answer: 17%). These figures give food for thought on the position that the main actors would be in were a major accident to occur.

   (d) The public do not seem to have much confidence in those with primary responsibility – heads of businesses in particular. Asked in whom they would put their trust to take effective action to reduce technological hazards, 30% replied "central government", 5% "local authorities", 9% "heads of businesses", 14% "staff of the firms concerned", 45% "accident prevention specialists", 4% "the general public", and 2% "no one" (8% did not reply). Conclusion: here again the position of business managements would be problematic in the event of a major accident.

(5) *The problem of sabotage and terrorism.* For obvious reasons this point will not be discussed here, but the brevity of the reference should not be misinterpreted. This is one of the gravest questions that those in charge at the highest level need to go into.

On these latter points, as on those referred to earlier, what remains to be done is to set up groups to exchange views and formulate proposals. Their purpose would be to go more thoroughly into the problems arising and to identify the strategic innovations that need to be introduced before the room for maneuver is seriously curtailed by the absence of any response and before, above all, a major event – but in a highly developed country this time – overturns the technical, economic, and cultural framework in which it is still possible to think out and handle problems of major hazards and crisis situations.

1.5. Conclusion

Tactical tools, organizational flexibility, strategic capability: at each step along the way research requirements are apparent. They concern a very large number of disciplines, and the problem, no doubt, is less a question of how to identify the many different demands to be met than of how to prepare the ground for fruitful investigation.

Certainly a first need is to bring together all the approaches and disciplines. The crisis dynamic is a general movement and approaching it in too compartmentalized a fashion – the classic temptation – would be dangerous. The second need is to observe, very firmly, the requirement for research to stay very close to reality in all its complexity and keep a healthy distance from models more satisfying to the intellect than in their relevance. This presupposes that the research scientist has access to data, which is, as yet, extremely difficult: the crisis situation is, by definition, critical for those involved, and that makes the quest for information particularly arduous.

Here, industrialists, senior public officials, and research scientists have to invent, as a result of discussion and experiment, new rules for work and for the exchange of information.

Many of these skills, incidentally, could be found among specialists in international relations who have been exploring the field of crises in that area for many years. But let there be no illusions: even in that field which is so crucial for mankind, the crisis question is far from being under control:

> Crisis management is an overly polite description of U.S. activities. What we really have is crisis coping and adaptation

> (Richard Beal, a senior director for crisis management systems and planning at the White House, *Science*, August 1984, p 907).

> According to a host of current and former NSC [National Security Council] staff members, much of the information available in a crisis is useless or incorrect; decision-makers have little or no crisis experience; careful planning is inadequate (Richard Beal, *Science*, August 1984, p 907).

No one holds the key to the problem. Dangerous substances, major hazards, vulnerability, crisis dynamics: the foundations necessary for fruitful work and collaboration urgently need to be laid, with determination and humility and bearing in mind the words of Warren Anderson: "We're learning the lessons of Bhopal and we'll be doing so for a long time to come" (Anderson, 1985, p 3).

References

Anderson, W.M. (1985), *Concluding Remarks*, Union Carbide Corporation Press Conference, 20 March.

Burton, I., Victor, P., and White, A. (1983), *The Mississauga Evacuation*, Final Report to the Ontario Ministry of the Solicitor General (The Institute for Environmental Studies, University of Toronto).

Cashmann, J.R. (1983), *Hazardous Materials, Emergencies – Response and Control* (Technomic Publishing Co., Lancaster, PA).

Cerrutti, G. (1977), Cent jours à la dioxine, *Survivre à Seveso* (Maspéro-Presses universitaires de Grenoble).

Conti, L. (1977), *Visto da Seveso* (Feltrinelli, Milan).

Cumberland, R.F. (1982), The control of hazardous chemical spills in the United Kingdom, *Journal of Hazardous Materials*, **6**, 277–287.

Department of Employment (1975), *The Flixborough Disaster*, Report of the Court of Inquiry (HMSO, London).

European Communities Council (1982), Council Directive of 24 June 1982 on the major-accident hazards of certain industrial activities, *Official Journal of the European Communities*, 5 August.

Gray, J., and Quarantelli, E.L. (1981), Social aspects of acute chemical emergencies, *Journal of Hazardous Materials*, **4**, 309–397.

Health and Safety Executive (1978), *Canvey, an Investigation of Potential Hazards from Operations in the Canvey Island – Thurrock Area* (Health and Safety Commission, HMSO, London).

Health and Safety Executive (1980), *Canvey, an Investigation of Potential Hazards from Operations in the Canvey Island – Thurrock Area* (Health and Safety Commission, HMSO, London).

H'rouda, P. (1985), *Fuite de gaz dans une fabrique de pesticides en Inde. Bhopal, lundi 3 décembre 1984*, Mission d'observation Médecins Sans Frontiéres; Société Française de Médecine de Catastrophe/Convergences Médicales (SAMU-94 Hôpital Henri Mondor, Créteil).

Lagadec, P. (1979), *Le probléme de la sûreté d'un grand complexe industriel: Le cas de Canvey Island* (Laboratoire d'Econométrie de l'Ecole Polytechnique/Service de l'Environnement Industriel du Ministére de l'Environnement, Paris).

Lagadec, P. (1981a), *Le Risque technologique majeur – Politique, risque et processus de développement*, Collection Futuribles [in English: *Major Technological Risk – An Assessment of Industrial Disasters* (Pergamon Press, Oxford, 1982].

Lagadec, P. (1981b), *La civilisation du risque – Catastrophes technologique set responsabilité sociale* (Editions du Seuil, Collection "Science ouverte", Paris). In Spanish: *La civilizacion del riesgo – Catastrofes technologicas y responsabilitad social* (Editorial Mapfre, Madrid, 1984).

Lagadec, P. (1983a), *Le probléme de la sûreté d'un grand complexe industriel. Le cas de Canvey Island. 1978–1982* (Laboratoire d'Econométrie de l'Ecole Polytechnique; Service de l'Environnement Industriel du Ministére de l'Environnement, Paris).

Lagadec, P. (1983b), *Dispositifs de gestion de crise* (Laboratoire d'Econométrie de l'Ecole Polytechnique; Service de l'Environnement Industriel du Ministère de l'Environnement, Paris).

Lagadec, P. (1984a), Le Risque technologique majeur et les situations de crise, *Annales des Mines*, **8**, 41–53.

Lagadec, P. (1984b), Risques technologiques et stratégies de communication – L'affaire des 41 fûts de déchets de Seveso (septembre 1982–juin 1983). Laboratoire d'Econométrie de l'Ecole Polytechnique – Service de l'Environnement Industriel du Ministére de l'Environnement.

Lagadec, P. (1985), *Défaillances technologiques et situations de crise. La catastrophe de San Juan Ixhuatepec – Mexico, 19 novembre 1984* (Laboratoire d'Econométrie de l'Ecole Polytechnique, Paris).

Mitroff, I, and Kilmann, R.H. (1984), *Corporate tragedies – Product Tampering, Sabotage, and Other Catastrophes* (Praeger, New York).

Morin, E. (1981), Preface, in M. Salomon, *L'avenir de la vie* (Editions Seghers, Paris).

Pecorella, G. (1977), Qui va payer? *Survivre à Seveso* (Maspéro-Presses universitaires de Grenoble).

Quarantelli, E. L. (1981), *Socio-Behavioral Responses to Chemical Hazards. Preparation for and Responses to acute Chemical Emergencies at the Local Community Level* (Disaster Research Center, Ohio State University, Columbus, OH).

Quarantelli, E. L. (1983), *Evacuation Behavior. Case study of the Taft, Louisiana, Chemical Tank Explosion Incident*, Final Report for the Federal Emergency Management Agency (Disaster Research Center, Ohio State University, Columbus, OH).

Rijnmond Authority (1982), *Risk Analysis of Six Potentially Hazardous Industrial Objects in the Rijnmond Area*, A report to the Rijnmond Authority (D. Reidel, Dordrecht).

Sandman, P. and Paden, M. (1979), At Three Mile Island, *Columbia Journalism Review*, **18**(7–8),43–58.

Scanlon, J. (1975), Crisis Communications in Canada, in B.D. Singes (ed), *Communications in Canadian Society*, 429–443 (Copp Clark, Toronto).

Scanlon, J. (1982), *Crisis Communications: The Ever Present Gremlins* (Emergency Communication Unit, Reference to COMCON '82, Arnprior, Ontario).

Scanlon, J. and Alldred, S. (1982), Media coverage of disasters. The same old story, *Emergency Planning Digest, Emergency Canada*, October–December, 15–22.

Scanlon, J., Dixon, K., and McClellan, S., (1982), *The Miramichi Earthquakes. The Media Response to an Invisible Emergency*, Emergency Communication Unit, Report 8211 (School of Journalism, Carleton University, Ottawa).

Silverberg, C. (1983), Conference at Royaumont Centre (France).

SOFRES (1985), Sondage Association des Industriels de France – Préventique, Les Français face aux risques technologiques, *Préventique*, **2**, 15–19.

Stephenson, D. (1984), Are you making the most of your crises? *Emergency Planning Digest, Emergency Canada*, October–December, 2–5.

CHAPTER 2

Engineering Aspects of Severe Accidents, with Reference to the Seveso, Mexico City, and Bhopal Cases

G. Naschi

2.1. The Problem of Severe Accidents

Modern technological processes may present a relevant threat to the safety of employees and the public or to the integrity of the environment from adverse events such as catastrophic fires, explosions, release of toxic substances, and so on, even if the probability of occurrence is in some cases low. The possibility of accidents with severe consequences is also related to the development of integrated industrial systems: the combination of the complexity of the processes, the extent of recent large-scale installations, and the high density of various activities in industrialized countries has led to the rather high rate of very dangerous accidents, which unfortunately still occur.

Recently, a large number of events have occurred with such immediate and delayed consequences as to receive a great deal of public attention. An as yet unpublished report by the European Economic Community (EEC) contains an inventory of severe accidents in the last decade in the European industry: about 100 events are reported, most of which involved fatalities and injuries. All these accidents were caused by errors in the design and/or by bad management of the plants: analytical diagnosis of the actual causes of an adverse event and a detailed study of the sequence of incidents leading to the event might give some insights that would be useful in identifying weak points in the installations.

In this context an examination of the most significant recent adverse events is presented from the point of view of the engineering aspects: the examples are related to the events at Seveso in 1976, and in Mexico City and Bhopal at the end of 1984. It should be pointed out that we can only mention the most probable hypotheses about the last two accidents since at present the inquiries are still in progress and the investigation data are in the hands of the pertinent authorities.

2.2. Engineering Analysis of the Case Histories

2.2.1. The accident at Seveso, Italy

This accident occurred on 10 July 1976, at ICMESA, a chemical plant for the manufacture of 2,4,5-trichlorophenol (TCP).

A brief description of the process might be a useful aid in understanding the evolution of the event. The main stages of the process are (*Figure 2.1*):

(1) Alkaline hydrolysis of tetrachlorobenzene (TCB) in a solution of ethylene glycol and xylene, at atmospheric pressure and a temperature of 135–150 °C, to obtain sodium trichlorophenate (STCP).
(2) Removal of water and recovery of the solvent by distillation at a temperature of 155–175 °C and a pressure of 1–0.026 bar.

The first two phases take place in the alkaline hydrolysis reactor.

(3) Mixing with cooling water and acidification of STCP with hydrochloric acid, to obtain TCP. This operation takes place in the acidification reactor.
(4) Settling, separation, and washing of TCP with water.
(5) Distillation of TCP in a batch still.

The accident occurred in the alkaline hydrolysis reactor.

One reasonable explanation of the event is the following (presented here in a very simplified form). After the reactor was shut down, the stirring of the liquid mass was conducted for about 15 min, then the contents of the reactor remained inside for several hours. The long stagnation of the liquid mass in the reactor, whose not submerged wall was at a temperature of about 300 °C (practically without heat transfer to the outside), resulted in a thin hot liquid stratum, within the uppermost layers of the contents, which rapidly reached a temperature of about 200 °C. At this temperature, maintained for several hours, at least two exothermic slow reactions were activated in

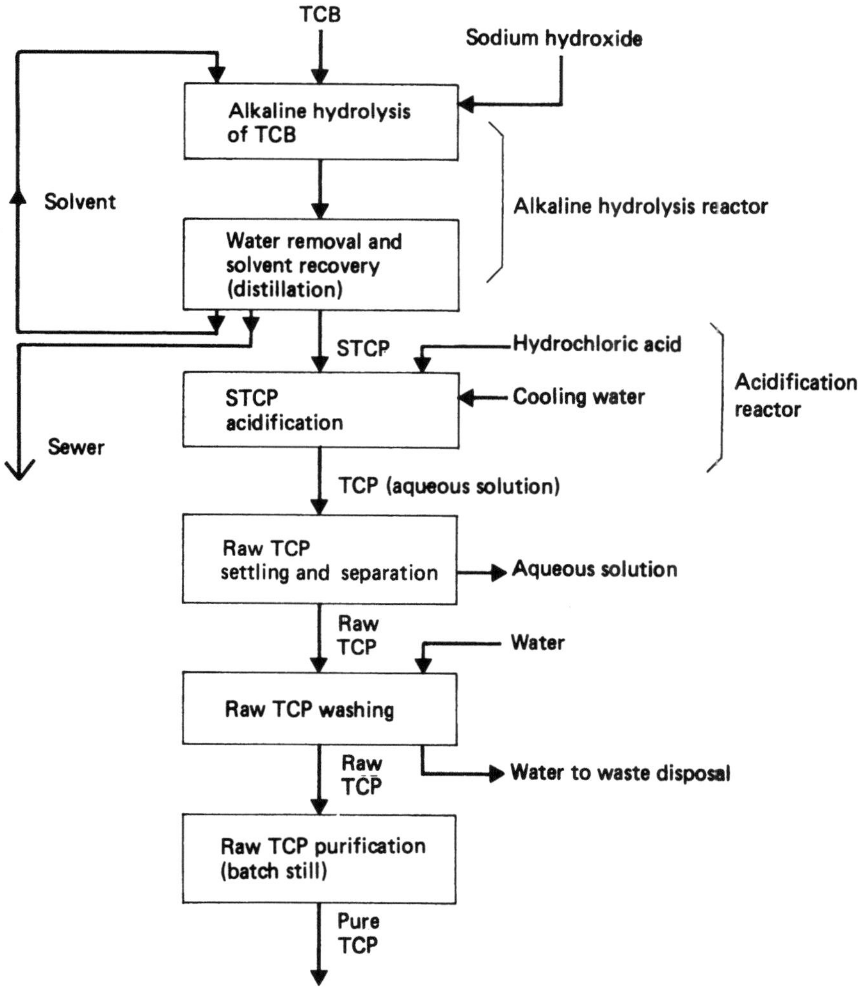

Figure 2.1. The TCP manufacturing process.

the layer, with a further temperature increase above the threshold temperature of an explosive exothermic runaway reaction yielding gaseous products and highly toxic 2,3,7,8-tetrachlorodibenzo-*p*-dioxin (TCDD). The consequent internal pressure increase caused the breaking of a rupture-disk (set-point at about 4 kgf/cm^2) and a dispersion in the air of a cloud of mixed products such as ethylene glycol, STCP, caustic soda, and TCDD.

This scenario leads to the following considerations:

(1) When in a chemical process there is a highly toxic product, it is necessary to study not only the process but also the physical parameters under normal and abnormal conditions and to provide adequate safety measures with proper procedures.

(2) The use of controlled relief systems to avoid releases of toxic materials to the open air is necessary.

2.2.2. The accident at San Juan Ixhuatepec, a suburb of Mexico City

This accident occurred on 19 November 1984, at a liquefied petroleum gas (LPG) storage and distribution center of the Petroleos Mexicanos State Company (Pemex). The total gas (propane and butane) storage capacity consisted of four 10000 bbl and two 15000 bbl spheres and of 48 horizontal cylindrical bullet tanks for a total of 2210 bbl stored. The layout of the facility is shown in *Figure 2.2*, including a square storage area (estimated 100 m × 100 m), two flare pits, a tanker lorry filling station, a gas bottling plant, a laboratory and a measurement room, a control room, and a fire protection system. Adjacent to the boundary of the Pemex center were located the privately owned distribution compounds of Unigas and Gasomatico. The installation was located in the middle of a highly populated area. At the time of the event it is thought that about 80000 bbl were stored.

Probably the disaster could have been avoided if the management of Pemex, the Executive Committee of the Petroleum Workers' Trade Union, and the Labour and Social Prevention Committee had taken into account the pressing advice which had been given by the Health and Safety Committee after an inquiry carried out two months before: many deficiencies were pointed out in the state of equipments located in the pipeline discharge sector and in the pumping and tank-filling area. For instance, the automatic fire protection system was found to be inoperative and 80% of the safety relief valves, which should operate in case of rupture or pipe disconnections, were in poor condition.

The Pemex company had asserted that periodic inspections were scheduled for the systems in the storage area and for the fire protection system. However, an assessment of the accidents that had occurred at other plants owned by Pemex during the previous year would suggest insufficient safety management by the company.

The results of the official investigation have identified, as probable initial cause, a gas escape from a cylindrical tank or from a faulty pipe in the storage area. It seems likely that there had been an extensive spread of vapor before ignition took place. At present we have no clear information as to the nature and location of the igniting source.

The 48 cylindrical tanks, packed near one another and supported on concrete stools, were set in such a way as to give rise to the possibility of combustion and explosion of any wide vapor cloud of gas in the storage area, being under partially confined conditions. The consequent over pressures would probably have been sufficient to cause

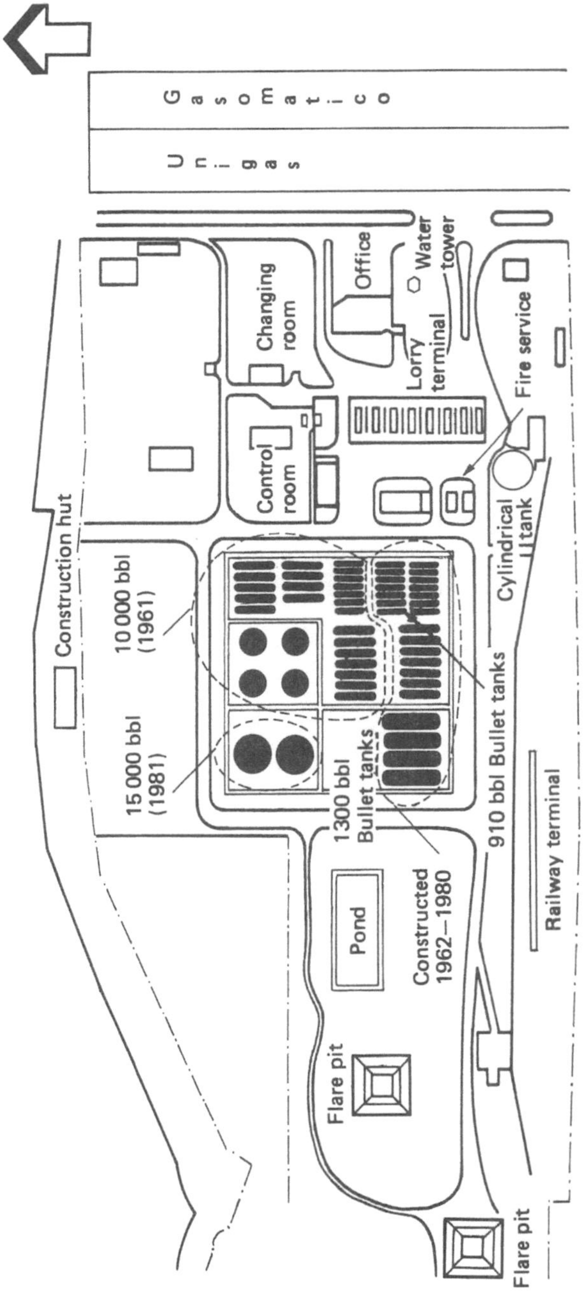

Figure 2.2. **The Pemex LPG storage installation of San Juan Ixhuatepec, Mexico City.** (***Source:*** **SIGTTO, 1985.)**

some dislodgement of tanks and adjacent pipelines, resulting in further large gas spill followed by progressive fires and explosions and by devastation of the area and its surroundings. In the very short period of emergency, the water spray system was not activated (the pumps required manual starting). The fire fighting services reached the plant in a short time, but by then devastation had already occurred.

The fire was brought under control by fire fighters more than seven hours after the beginning of the accident and it was finally extinguished after burning for 36 hours. At least 20 city blocks are thought to have been destroyed by the fires, in an area of radius about 2.4 km around the complex.

The accident was thus characterized by two main phenomena:

(1) A great extension of gas spread, which behaved like a dense cloud.
(2) The "domino" or knock-on effect: this refers to a loss-of-containment accident which interferes with the operation of other nearby systems so that further loss of containment occurs. For example, a small leak of flammable gas may ignite or explode, producing an impingement on adjacent large vessels and a consequent large spill of hazardous substances. In the case under consideration, the progressive disintegration and explosion of the cylindrical tanks produced a gradual, but rapid, sequence of ruptures. Of course the "domino" effect produces the worst consequences when an accident on one site affects the operations on nearby industrial sites.

The accident leads to the following conclusions:

(1) The arrangement of a sufficient separation distance around each part of the plant is very important for the protection of possible housings on the site.
(2) The presence of large quantities of highly flammable and explosive substances such as LPG requires the large-scale introduction of adequate, reliable, and efficient safety systems. In addition, systematic control of the plants requires up-to-date procedures for maintenance and inspection.
(3) The location of hazardous installations in the vicinity of highly populated areas must be avoided. Local planning is needed for the use of adjoining land.

2.2.3. The accident at Bhopal, India

This accident occurred on 3 December 1984 at a chemical plant for the production of pesticides owned by Union Carbide India Ltd.

Methyl isocyanate (MIC), a highly volatile chemical product, was used to manufacture some pesticides. In the factory MIC was produced by a process using monomethylamine and phosgene as raw materials and chloroform as a solvent. The refined MIC was later stored in type 304 stainless steel tanks, from which it was transferred to the derivatives units as needed.

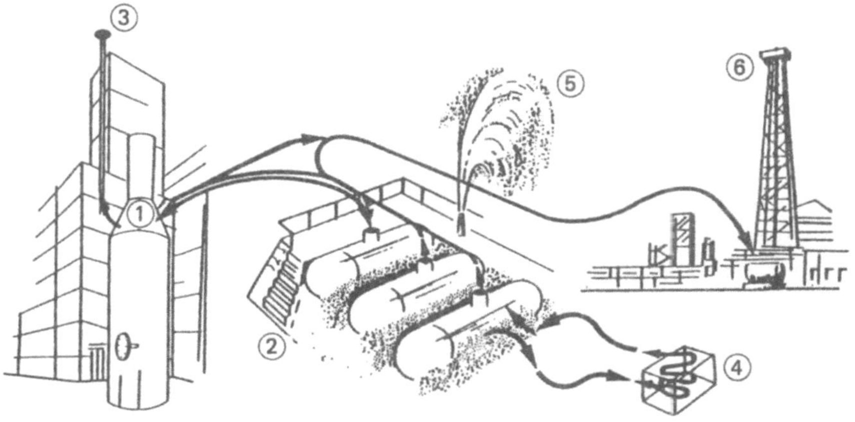

Figure 2.3. The equipment involved in the Bhopal accident: 1, ***storage tanks***, the third tank is usually empty and used for emergency and temporary off-specification MIC storage; 2, ***vent gas scrubber***, MIC escaping from tanks is passed through circulating caustic soda solution and then destroyed; 3, ***vent line***, MIC vapors are discharged to the atmosphere or to the flare tower; 4, ***refrigeration system***, this keeps MIC at 0 °C, which slows down the rate of undesired reactions; 5, ***water curtain and fire water monitors***, these can reduce part of the escaping MIC vapors; 6, ***flare tower***, this is used to burn vent gases from different plant units.

To follow the accident sequence it is necessary to outline the equipments associated with the event: the MIC storage system, the vent gas scrubber, and the flare tower (see *Figure 2.3*). MIC was stored in two of three horizontal 60 t tanks. The third tank was used for emergency storage of MIC and for temporary hold of off-specification MIC; the content of the tanks was circulated through heat exchangers cooled by a refrigeration system to maintain MIC at a temperature of about 0 °C. This system was provided to maintain the stored material at low temperature, which slows down the reaction rate and allows time for reprocessing or destruction in the case of contaminated material (i.e. off-specification MIC). It had been made nonoperational some months before, probably because it was not considered essential for the normal operation of the plant.

In case of off-specification MIC or opening of the tank safety valves the MIC could be transferred to the vent gas scrubber where all entering gases were contacted with circulating caustic soda solution and then destroyed. The vent from the scrubber could either go to the atmospheric vent or to the flare tower. The flare tower was used to burn vent gases, including normal ones from the MIC storage tanks and the vent gas scrubber: prior to the accident the flare tower had been removed from service for maintenance work and it was not operating at the time of the accident.

We can now give a description of the most probable scenario, taking into account the hypotheses about the causes. The primary cause was a direct introduction of water (estimated 500–1000 l) into an MIC tank through the piping system connected with the tank. The water with MIC led to an exothermic chain of reactions with the production of, inter alia, carbon dioxide. The increase in temperature was not signaled by the tank high-temperature alarm since it had not been previously reset.

The Union Carbide Co. Bhopal Incident Investigation Team pointed out the following additional phenomena. MIC containing a high concentration of chloroform was sent to the tank instead of to the adjacent tank used for storage of off-specification MIC. As the temperature in the tank increased, the corrosion rate increased because of the presence in the tank of an abnormally high concentration of chloroform. The corrosion products catalyzed a concurrent exothermic trimerization of MIC. Runaway reactions (MIC–water and trimerization of MIC) accelerated as the temperature increased. The pressure in the tank, due also to carbon dioxide, rose so quickly that the tank rupture-disk and the downstream safety valve opened and remained open for approximately two hours.

During this period about 25 t of MIC in vapor and liquid form were discharged through the safety valve to the vent gas scrubber. This equipment had to return to an operating mode when the operators activated the circulation pump for caustic soda: the vent gas scrubber proved inadequate to neutralize such a high flow rate of high-pressure MIC. As the flare tower was out of order, MIC contained in the tank was emitted from the vent gas scrubber stack to the atmosphere. The operators turned on the fire water monitors and water curtains and directed them to the stack and the MIC process area to knock down MIC vapor partially.

A heavy gas cloud containing toxic MIC was dispersed in the direction of high-density population areas. When the alarm system was activated, a crowd of people went towards the factory: they did not realize the magnitude of the tragedy until it was too late (*Figure 2.4*).

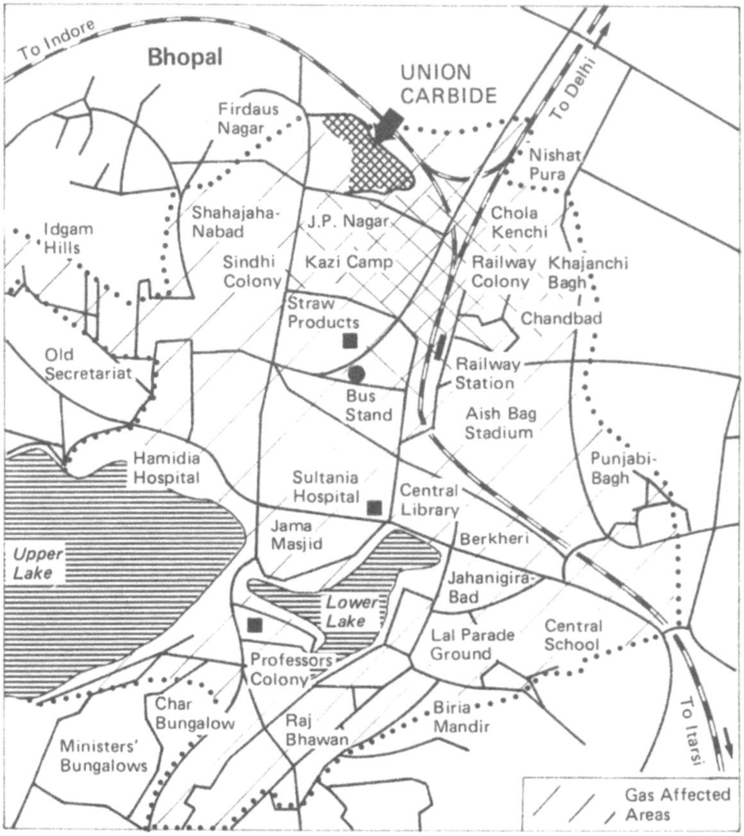

Figure 2.4. Map of Bhopal, showing the gas-affected area. (*Source*: *India Today*, 31 December 1984.)

From the above description we can draw the following conclusions:

(1) It is necessary to maintain auxiliary systems and instrumentation devices in a continuous operability condition. In this case, the refrigeration system, out of order for some months, would have decreased the exothermic reaction rate; the temperature alarm, which should have alerted workers, had not been properly set.

(2) It is important to have adequate design of protection systems when a chemical process converts or produces toxic substances. We refer to the efficiency of the vent gas scrubber, water curtain, and monitoring systems. The scrubber was not turned on until after the reaction had gone out of control.

(3) Redundant and/or alternative safety-related equipment must be available to guarantee a high level of safety during maintenance of safety-related equipment. We refer to the flare tower, which could have destroyed some of the escaping gas, but was out of order.
(4) Cooperation between manufacturers and local authorities is necessary to ensure the soundness and implementability of emergency and evacuation plans.
(5) Adequate information should be disseminated to the public and should include:

    (a) The nature of the hazards which might affect them.
    (b) The emergency arrangements.
    (c) What the public should do in a major accident situation.

2.2.4. Lessons learned

As a conclusion of the available information related to the above case histories we might summarize the causes of catastrophic accidents with the following statement: "a combination of design deficiencies, operating errors, managerial mistakes".

From the experience of the Seveso accident, we can recommend adoption of the following items:

(1) Deeper examination of the chemical process during plant design to identify the complete range of variations of the physical parameters, all the chemical products involved, and the relative properties.
(2) Development of the design along the lines of a "defence in depth" principle.
(3) A controlled relief system to a contained atmosphere, to avoid releases of toxic substances to the environment.

From the experience of the Mexico City accident we can recommend the following items:

(1) A detailed study of plant layout to avoid the possibility of the "domino" effect due to confined spaces.
(2) A preventive analysis of the effects on surrounding facilities following the occurrence of an adverse event in a high-risk installation and the adoption of adequate territorial planning in order to minimize the consequences.

(3) The adoption of specific maintenance and inspection programs to guarantee the efficiency of safety-related systems.

The experience of the Bhopal accident indicates the need for the following:

(1) Redundancy of instrumentation devices and safety provisions.
(2) The adoption of maintenance and inspection programs as mentioned for the Mexico City accident.
(3) Careful management of the plant with the adoption of written procedures following a proper quality assurance scheme.
(4) The implementation of emergency plans, taking into account information for the public.

In all the cases we have examined, plant management deficiencies had an important effect. To some extent, this is unavoidable in a production-oriented organization, which suggests the necessity of surveillance action by institutional bodies that are properly safety oriented and have adequate enforcement powers.

2.3. Conclusions

At the present stage of modern industrialization a policy of balance between workers' health protection and environmental defence on one side and economic and social development on the other seems necessary. Therefore adequate attention must be devoted to research into reasonable and effective technical solutions to safety problems related to high-risk industrial installations.

The normal practice of "trial and error" has been adopted up to now to cope with the problems of adverse events. The experience gained from the analysis of accidents that have already happened and the identification of weak points in areas or systems of the plant has led to the adoption of corrective safety measures. We think that this method in itself is not sufficient as a general approach to safety, because it implies case-by-case correction after the occurrence of an accident.

Nowadays, safety experts agree in recommending preventive global analysis as the most effective approach to safety. This means systematic analysis, at the design stage, of installations and of their modes of failure, and assessment of the consequences associated with potential accident sequences. All this implies the study of the relationship between the plant and the environment as well. The main objectives of these studies should be:

(1) A balanced level of plant protection against different malfunctions and accident sequences.
(2) A full understanding of plant response that allows proper management and operator practices and procedures.
(3) A correct perspective on the relative importance of plant equipment and components that leads to correct maintenance procedures.

It is evident that such a global approach responds to the needs that emerge from the lessons learned from the above-mentioned case histories; in addition, it promotes adequate management and prevents air and water pollution.

The overall approach to safety assessment has been most widely applied both to nuclear and space activities. In the nuclear industry the methodology of systematic safety assessment has been adopted with an "anticipatory" attitude. In 1975 the *Reactor Safety Study* or WASH 1400 Report (NRC, 1975) offered a comparison of the risk levels involved in nuclear and other human activities. After this, the progressive introduction of probabilistic methodologies in the safety analysis of plants took place; these methodologies are presently well developed and they have reached a sufficient degree of maturity to suggest their application to other fields, especially in relation to the most technologically sophisticated plants or complex situations. In fact a significant application of this methodology has been adopted in two particular situations: the Canvey Island area and, more recently, the Rijnmond area. Both the Canvey Island and Rijnmond sites, respectively located in the estuary of the Thames and in the vicinity of the Rhine delta, are examples of concentration of industrial installations involving many major hazards: the final reports of two different studies contain a set of recommendations and proposals that were presented to public inquiries and submitted for approval to the authorities (Health and Safety Executive, 1978, 1981; Rijnmond Authority, 1982).

As a final note we can point out that the Directive 82/501 of the EEC "on the major-accident hazards of certain industrial activities" (European Communities Council, 1982) in force in the European countries is in accordance with the implementation of a rational approach to the safety of industrial plants. It has been called the "Seveso Directive" because it arose from the Seveso accident. The principal objectives of the EEC Directive are:

(1) The prevention of major accidents arising from industrial activities.

(2) The limitation of the effects of such accidents both on man and on the environment.
(3) The harmonization of control measures to prevent and limit major accidents in the EEC.

The requirements of the Directive can be divided as follows:

(1) General requirements, which can be more widely applied. These require that manufacturers adopt the necessary precautions to prevent major accidents, report those that do arise, and take steps to limit their consequences.
(2) Specific requirements, to be applied only to the most potentially hazardous activities. These require that information be provided to the public that could be subjected to a major accident. Through a notification procedure, the manufacturer must submit to the competent authorities a detailed study, containing information about:

    (a) Installations and dangerous substances.
    (b) Hazards and their control.
    (c) Possible major accident situations, on-site and off-site emergency plans, alarm systems, and resources for dealing with the accidents.

References

European Communities Council (1982), Council Directive of 24 June 1982 on the major-accident hazards of certain industrial activities, *Official Journal of the European Communities*, 5 August.

Health and Safety Executive (1978), *Canvey, an Investigation of Potential Hazards from Operations in the Canvey Island–Thurrock Area* (Health and Safety Commission, HMSO, London).

Health and Safety Executive (1981), *Canvey, an Investigation of Potential Hazards from Operations in the Canvey Island–Thurrock Area* (Health and Safety Commission, HMSO, London).

NRC (1975), *Reactor Safety Study*, WASH 1400 (US Nuclear Regulatory Commission, Washington, DC).

Rijnmond Authority (1982), *Risk Analysis of Six Potentially Hazardous Industrial Objects in the Rijnmond Area*, A report to the Rijnmond Authority (D. Reidel, Dordrecht).

SIGTTO (1985), *Interim Info Review*, January (Society of International Gas Tanker and Terminal Operators Ltd.).

CHAPTER 3

The Seveso Accident and Its Aftermath

F. Pocchiari, V. Silano, and G. Zapponi

3.1. Introduction

This chapter considers the accident at the ICMESA plant at Seveso, Italy, an accident which has had profound effects on public policy relating to chemical safety within the European Economic Community.

On 10 July 1976 an exothermic reaction raised the temperature and pressure inside a reactor used at the ICMESA plant for trichlorophenol (TCP) production beyond limits, thereby causing a safety device to blow out. At the moment the safety device gave way, the principal volatile compound (*ca.* 1000 kg) inside the reactor was ethylene glycol (boiling point 198 °C). It can be assumed that, at the moment of blowout, batch temperature was well over 200 °C and that the temperature decreased somewhat (*ca.* 200 °C) while glycol was being expelled into the open air. Once glycol had been exhausted, the temperature most probably rose to a higher level during the elimination of *ca.* 1300 kg of diethylene glycol (boiling point 245 °C). Reactor temperatures are thought to have increased further (>300 °C), thus causing extensive mineralization of residual organic substances. Emission gradually dropped during this phase until it ceased altogether.

Almost certainly the violent blowout following the valve's giving way caused the escaping vapors and entrained particles to leave the reactor at a speed of some hundreds of meters per second. Smaller particles may have been carried out by the glycol vapor during its violent initial boiling phase. Mixed with various particles and chemicals, the 2,3,7,8-tetrachlorodibenzo-*p*-dioxin (TCDD) settled to the ground in accordance with prevailing wind directions at the time of blowout.

3.2. Organizational Aspects

The accident directly involved about 1800 ha of a densely populated area (Brianza) of Seveso (*Figure 3.1*) and five municipalities of the Lombardy region (province of Milan). However, as a preventive measure the administrators decided to place the inhabitants of 11 municipalities (about 220 000 persons) under medical and epidemiological surveillance (*Figure 3.2*).

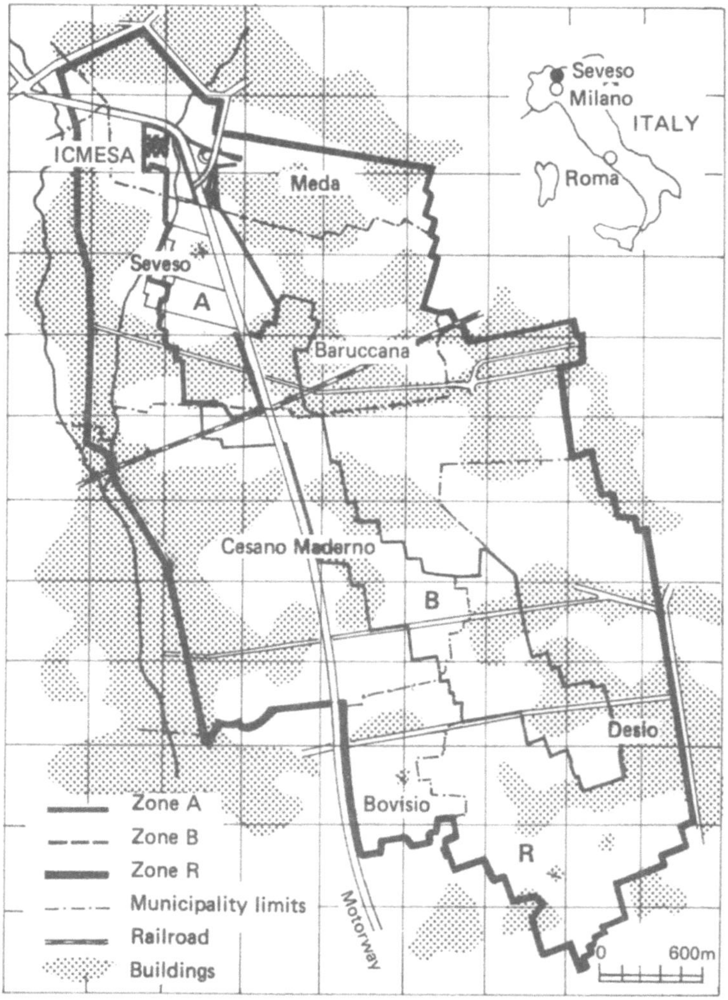

Figure 3.1. The Seveso area: zones A, B, and R at their maximum extension, showing major built-up areas (stippled) and surrounding farmlands. (*Source*: Pocchiari *et al.*, 1983.)

Figure 3.2. Map of the 11 municipalities involved in the health monitoring program by pollution zones. (*Source*: Abate *et al.*, 1982.)

At the very beginning, the Seveso emergency was dealt with by the general emergency services. The alert system was from industry to police, and immediately afterwards the mayors of Meda and Seveso were involved as heads of the local authorities responsible for contingency plan enforcement. At this early stage, coordination and expert advice were mainly provided by the province, while regional authorities were kept constantly informed. A significant feature of the Seveso emergency was the lack of an immediate understanding of the nature of the chemicals involved and of the extent of the affected area. This only became clear with the progressive appearance of adverse effects on human beings and the progress of environmental monitoring.

At this more advanced stage, it became clear that a number of municipalities had been affected and that the focal point of the accident response system had to be at a regional level. Within 16 days after the accident the Regional Minister of Health was invested with full powers by the Lombardy Regional Council. Since then the Lombardy region has been responsible for all relevant actions. A series of measures to simplify administrative procedures was adopted so as to enable the region and other local authorities to utilize more rapidly the funds provided by the Italian Government and to overcome some problems of coordination which had become evident in the meantime. On 17 January 1977 the Lombardy regional authorities issued a law specifying the norms for utilizing the government allocation of US $150 million and the procedure to be followed for the approval of operational programs. On 2 June 1977 the programs were approved by the Regional Council, but instead of passing a law, the Council passed an administrative act which facilitated timely modifications in case new contingencies should require change in one or more points of the plan. The efficacy of the governmental action was further strengthened with Act No. 27 of 17 July 1977, which established a special office headed by an ad hoc commissioner. The commissioner was responsible directly to the regional president, who, in turn, answered to the Regional Council. The powers conferred upon this commissioner were many and far-reaching, including the power to coordinate, direct technical matters and manage operations through pre-established operational plans. *Figure 3.3* shows the organizational outline of the authorities and consultant bodies involved at this stage.

The role of the Italian Government was mainly to provide sufficient funds for handling the emergency and to supply expert advice, personnel and materials (particularly for TCDD assays). For instance, an ad hoc section of the Istituto Superiore di Sanità (ISS) was established at Seveso to help local laboratories to carry out the environmental monitoring of TCDD. The government authorities, in conjunction with regional authorities, were also responsible for relations with international agencies, organizations, and experts.

3.3. The Emergency Phase

The various phases of the emergency day by day are shown in *Table 3.1*.

Courtyard animals were severely affected, many dying within a few days of the accident. At the same time, dermal lesions among human beings who had been exposed to the toxic cloud began to appear. About ten days after the accident, it was clear that TCDD had been formed in the ICMESA reactor and released into the environment with the toxic cloud.

Preliminary analytical findings concerning TCDD in soil and vegetation and available information on the sites of toxic and pathological

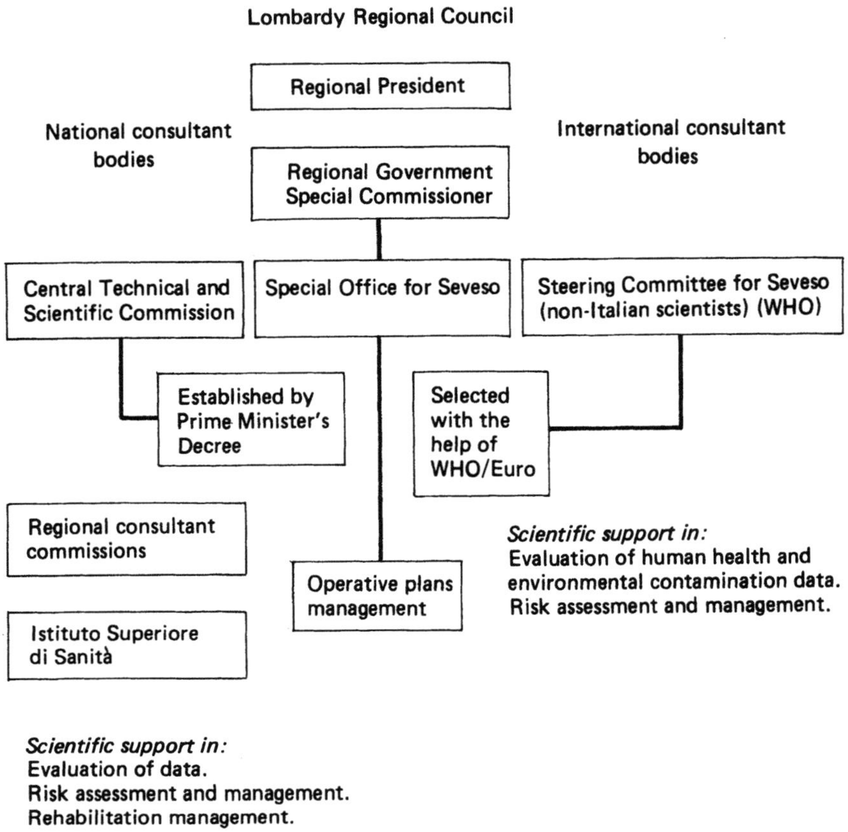

Figure 3.3. Organizational outline of the authorities and consultant bodies charged with medium- and long-term activities management (WHO/Euro refers to the World Health Organization/Regional Office for Europe, Copenhagen, Denmark). (*Source*: Pocchiari *et al.*, 1986.)

events, and on air movements at the time of blowout, were used to approximate the contaminated area. As a first step, on 26 July 1976, the Italian authorities evacuated 225 people from a 15 ha area immediately southeast of the plant. A few days later, further analytical findings concerning TCDD contents of soil and vegetation samples prompted the Italian authorities to evacuate all the inhabitants (733 people) of a wider area (coded as zone A) extending for about 2 km southeast from the ICMESA plant (approximately 110 ha). The TCDD levels detected on vegetation indicated a low-rate exponential decrease of TCDD environmental contamination with the distance from the plant. Data concerning TCDD levels in soil were the major indicators of environmental contamination, particularly during the first two years after the accident.

Table 3.1. The emergency day by day.

10 July 1976 (Day 1): At 12.40 hours, an exothermic reaction of unknown cause raised the temperature and pressure inside reactor A101 at the ICMESA plant in Meda beyond limits, causing the safety device to blow up.

11 July 1976 (Day 2): Samples of leaves, that appeared yellow after contact with the toxic cloud, were gathered by ICMESA personnel and sent to the Givaudan laboratories in Switzerland. At 17.45 hours (29 hours after the accident) two ICMESA firm representatives informed the military police station as well as the Mayor and the Public Health Officer at Meda that a cloud of herbicide vapors, containing trichlophenol (TCP) had been released (TCDD was not mentioned).

12 July 1976 (Day 3): ICMESA officials informed the Health and Hygiene Office of Seveso municipality that they had notified the people living in areas close to the plant to abstain from eating vegetables, owing to the toxicological hazard caused by herbicidal substances.

14 July 1976 (Day 5): The first manifestations of skin eruptions appeared in exposed children. In areas adjacent to the ICMESA plant, the first animals (cats, rabbits, and chickens) died.

15 July 1976 (Day 6): Other cases of edema and skin and eye irritations occurred among the people of Meda and Seveso. The mayors of Meda and Seveso declared the San Pietro area (close to the ICMESA plant) to be polluted by toxic substances. The area was marked off with pickets and the use of vegetables and fruits within it forbidden.

16 July 1976 (Day 7): Thirteen children were admitted to hospital with toxic dermatitis. Local public health officers sent samples of local vegetation to the Provincial Laboratory for Public Hygiene and Prophylaxis (PLPHP) in Milan for analysis.

17 July 1976 (Day 8): The mayors of Seveso and Meda ordered that the plants, vegetables, and crops, as well as dead animals, of the polluted area be burnt.

18 July 1976 (Day 9): The mayor of Meda issued an ordinance to close down the production sections of the ICMESA plant. The gates to plant section B were locked and sealed by law.

19 July 1976 (Day 10): The Provincial Inspectorate for Labor Conditions was requested to intervene. The Director of Givaudan chemistry laboratory arrived in Italy and admitted that TCDD had been found in the samples analyzed.

20 July 1976 (Day 11): The chief chemist from the Provincial Inspectorate for Labour Conditions issued a writ accusing ICMESA directors of having violated several articles of Italian regulations for prevention of labor accidents.

(cont.)

Table 3.1. (*cont.*) The emergency day by day.

---

20 July 1976 (Day 11) (*cont.*): All local doctors were requested to notify the authorities of all gastroenteritis and skin condition cases.
An ordinance was issued, forbidding the consumption of food products from animals of the polluted zone.
Analytical tests confirmed the presence of TCDD in the toxic cloud. These early tests, carried out at Givaudan laboratories, showed relatively low levels.

21 July 1976 (Day 12): A meeting was held at Seveso Town Hall, attended by the Mayor, the Regional Public Health Minister, and other authorities. Givaudan results showing the presence of TCDD were presented. Further safety measures were decided. A college of experts was appointed by the Public Health Minister.
Health monitoring was decreed for all the people living in the affected area. The National Ministry of Health and the ISS (National Public Health Institute) were informed. The ISS was placed at the region's disposal by the Public Health Ministry.

22 July 1976 (Day 13): Eighty children were relocated in a resort area. The National Ministry of Health supplied the regional authorities with data on TCDD from international literature and sent technical officials to the polluted zone. An urgent meeting of the Provincial Council for Public Health was called.

23 July 1976 (Day 14): The Provincial Council for Public Health met in special session with experts from the national Ministry of Health, the ISS, Milan University, and the Mario Negri Institute. They confirmed the measures taken by the Regional Public Health Ministry as follows:

(1) Vegetation was to be tested by the Institute for Plant Pathology to establish the borders of the affected zone.
(2) Dead animals were to be tested by the regional veterinary officer and the Institute for Animal Prophylaxis of Brescia.
(3) A clinic was to be established at Seveso to serve people affected by toxicant-associated symptoms.
(4) Chemical monitoring was to be carried out by the PLPHP.

The national Ministry of Health arranged for a gas chromatography research team from the ISS. ISS and PLPHP technicians commenced systematic sampling of vegetation. That night, the extraction method was worked out and results from the first analyses were obtained, using low-resolution gas–liquid chromatography in combination with MID/mass spectrometry (mass fragmentography).

24 and 25 July 1976 (Days 15 and 16): Results of the laboratory tests completed by the ISS and PLPHP during the night of 23–24 July were reported at a meeting held by the regional Public Health Minister. Officials and technicians from the Lombardy Region, the national Ministry of Health, and the ISS (*cont.*)

Table 3.1. (*cont.*) The emergency day by day.

were present, as well as the director of Givaudan laboratories. The presence of high levels of TCDD was reported and confirmed.
The boundaries of the most contaminated area zone were traced (extending southwards for approximately 750 m and covering approximately 15 ha).
Evacuation of population from this zone (coded as zone A) was decided.
Sampling and chemical analyses continued throughout that day and night to define contamination levels around zone A.

26 July 1976 (Day 17): The regional Public Health Minister was made responsible for supervision, management and coordination of the various emergency measures at Seveso (focal point). The Lombardy regional resources were put at his disposal. Two hundred and twenty-five people (170 from Seveso and 55 from Meda) were evacuated by the local authorities.

27 July 1976 (Day 18): Results from laboratory tests over the next few days caused the regional health authorities to extend zone A approximately 1600 m southwards from the plant. A further evacuation was decided.

28 July 1976 (Day 19): The regional Public Health Minister set up four technical scientific commissions for:

(1) Medical and epidemiological aspects.
(2) Gathering data on environmental pollution.
(3) Land reclamation.

29 July 1976 (Day 20): The health problem commission assessed the situation. Skin tests carried out on more than 500 people established that:

(1) Unusual skin morphological disorders were observed.
(2) None of the patients exhibited obvious general symptoms.
(3) Some subjective complaints (e.g. nausea, gastralgia, itching) were temporary.

All cases were kept under medical surveillance. Close surveillance was considered advisable for all pregnant women in contaminated areas. Prevention of new pregnancies was also considered advisable in such areas.

30 July 1976 (Day 21): The Lombardy regional government decided to declare the area a disaster zone and to appoint a special government commissioner.

2 August 1976 (Day 24): Further laboratory tests caused the zone A to be extended southwards to a distance of up to 2200 m from the source and the local population to be evacuated. Zone A now covered some 108 ha, with more than 730 people having been evacuated. A lower contamination level zone, coded as zone B, was defined, including two municipalities and covering about 260 ha.
The regional authorities ordered that children younger than one year old and women up to three months pregnant should be kept away from the area during daytime.

(*cont.*)

Table 3.1. (*cont.*) The emergency day by day.

4 August 1976 (Day 26): The Prime Minister instituted a central technical and scientific commission to study decontamination measures.

5 August 1976 (Day 27): To streamline financial administration, a coordinating committee was formed, including the regional Minister for Public Health, the mayors of involved municipalities, representatives from the trade unions, commerce, industry, craftsmen, agricultural organizations, and a national government representative.

10 August 1976 (Day 32): A decree was issued by national government, which included provisions for emergency action (40 billion lire).
The regional medico-epidemiological commission recommended that voluntary therapeutic abortion be available for pregnant women who had been exposed to TCDD within the first three months of their pregnancy.
A special office of the national Ministry of Health was established in Milan, to liaise between central and regional administrations.
The constitution of an ISS task force (already operating) was formalized.

12 August 1976 (Day 34): A medical commission was formed at the University of Milan to help and advise pregnant women wishing to abort.

14 August 1976 (Day 36): The Central Technical and Scientific Commission recommended the establishment of a third zone (zone R), around zone A and zone B. TCDD levels in zone R were below 5 $\mu g/m^2$. Zones A, B, and R now covered a 1430 ha area. Health precautions in zone R included prohibition of use of local crops and local animal products as food.

15 August 1976 (Day 37): The first complete pollution map of the area affected was released by the regional Ministry for Public Health.

In *Figure 3.4*, four contamination maps of zone A, carried out in different periods, are shown. Soil data indicated large variations of TCDD environmental levels even in limited areas: this was interpretable as the consequence of the "semidiscrete" TCDD environmental dispersion process (in the form of TCDD-containing particles of various dimensions and contamination levels) as well as due to air turbulence and obstacles during the deposition. The comparison of TCDD levels, detected in replicated soil samples gathered in the contaminated areas, indicated a lognormal distribution of replicated measurements, whose standard deviation in log-units corresponded to a factor of about 2.5 for original data. These findings showed the opportunity of an adequate caution level in data evaluation for risk assessment (e.g. reference to confidence limits of estimates) and the need for extensive TCDD monitoring of the affected environment.

Zone B (270 ha) was the natural extension of zone A along the main TCDD diffusion pathway and exhibited lower dioxin contents. Both

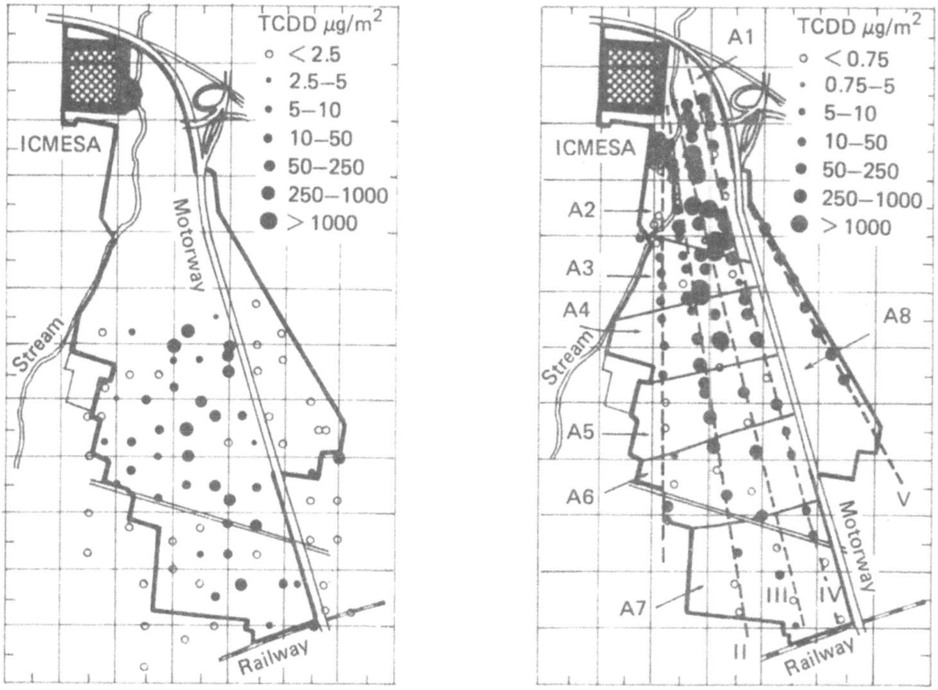

(*a*) (*b*)

Figure 3.4. Evolution in time of zone A TCDD levels. (*a*) The upper part (clear area) of the August 1976 map was determined on the basis of TCDD findings of various environmental samples collected before 26 July 1976. The boundaries of the lower part of zone A (dotted area) were established on the basis of numerous topsoil specimens sampled on 26 July and analyzed immediately afterwards. The north–south square grid (each square is 200 × 200 m) shown is intended to facilitate the comparison of TCDD levels in this map with those reported in (*b*), (*c*), and (*d*), but does not identify with original coordinates. Mapped site locations are accurate to ± 50 m. Early analytical findings were expressed as μg/100 g of sample. They are here converted into μg/m^2 to make data values consistent with those of other maps. Data are accurate to approximately ± 2.5 times the μg/m^2 numerical value obtained (not reported) which must be allowed for such conversion. (*b*) September 1976 map. Original polar coordinates are labeled with Roman numerals; the square grid has been superimposed as discussed above in (*a*). Soil specimens were collected on 11–13 August 1976. (*Source*: Pocchiari *et al.*, 1983.)

zones A and B were enclosed by a larger territory, zone R (1430 ha), exhibiting just-detectable threshold contamination levels. The borderline between zones B and R was where the average TCDD concentration in soil was found to be 5 μg/m^2 whereas zone R boundaries were set at nondetectable TCDD levels (formally, below 0.75 μg/m^2). The borderline

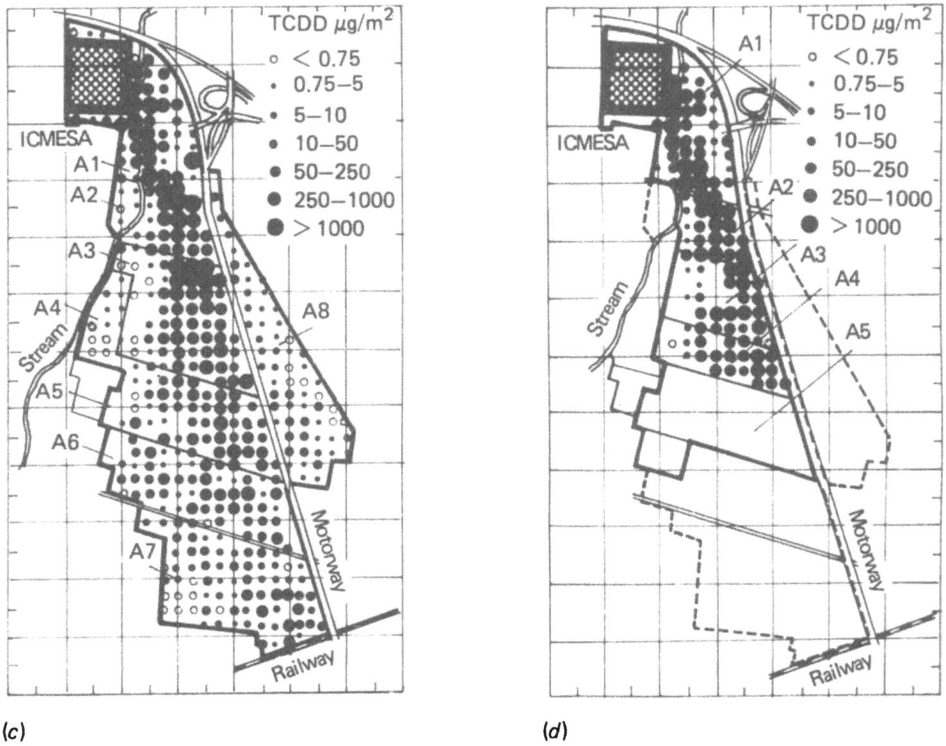

Figure 3.4. (*cont.*). (*c*) January 1977 map. Sampling sites were established on a 50 m square grid. The sampling campaign started in late September and lasted through December 1976. (*d*) March 1978 map. Sampling sites are as in case (*c*). Most soil specimens were sampled in the period from 15 December 1977 to 5 January 1978. For (*b*), (*c*), and (*d*) mapped site locations are accurate to ± 25 m. (*Source*: Pocchiari *et al.*, 1983.)

between zones A and B ran along the 50 $\mu g/m^2$ line. The inhabitants of zones B and R (*Figure 3.1*), were subjected to a number of hygiene regulations, including the prohibition to farm and consume local agricultural products and keep poultry and other animals.

3.4. Rehabilitation Activities

A very extensive monitoring program was carried out on:

(1) Soil and building surface distribution of TCDD.
(2) TCDD vertical distribution in the soil.
(3) TCDD levels in atmospheric particles.
(4) TCDD in ground and surface waters.

(5) TCDD in animals and cows' milk.
(6) TCDD in vegetation and grains.

The aim of this monitoring program was twofold, i.e. to understand the extent of contamination and to verify the effectiveness of the environmental rehabilitation works.

The ICMESA accident has, probably, caused the world's largest contamination of soil and buildings by TCDD. A number of contaminated areas, including some sections of zones A, B, and R, have been reclaimed using various detoxification techniques. Reclamation was relatively easier and particularly urgent for subzones A6 and A7, areas that had lodged approximately 490 people (about 60% of the total evacuated population) before the ICMESA accident. So far it has been achieved for the whole of zone A. In 1978, after reclamation, inhabitants of zones A6 and A7 were allowed to go back to their houses, whereas those of the remaining part of zone A have definitively settled somewhere else.

Farming land and yards were normally scarified to a depth of at least 20 cm and often more in order to reduce the topsoil level of TCDD to at least 5 $\mu g/m^2$. Contaminated topsoil was disposed of by transferring it to an underground waste deposit and was replaced by an equivalent amount of fresh earth. The extensive soil scarification carried out produced about 200 000 m^3 of contaminated soil. A number of alternatives were considered for the disposal of contaminated soil. The only practical solution was to dispose of the polluted soil in basins, the peripheral parts of which were damp-proofed with a layer of bentonite (a mixture of gravel and sand), overlaid with a sheet of plastic material (high-density polyethylene), and finally capped with clean soil. The procedure is similar to the one used for radioactive waste, for which several natural and artificial barriers are used to ensure that no radioactive elements may come into contact with the environment.

In the case of Seveso soil a variety of barriers were also available:

(1) The dioxin is bound to the clay, which is rather abundant in the soil of the polluted zones.
(2) In the disposal of the polluted soil resulting from the reclamation work, the most polluted is deposited in the central part of the basin while the less polluted is deposited around the core along the protective sheeting.
(3) The plastic sheet (high-density polyethylene, 2.5 mm thick) is welded so as to constitute a unique blanket.
(4) A 15 cm thick foundation built with sand and bentonite (reinforced concrete) has the appropriate characteristics of impermeability and plasticity.

In addition to all these barriers, the deposit cap is protected by a layer of "Gunite" (concrete reinforced with an iron net) to prevent possible damage from the outside. The entire deposit is covered with a 1 m thick layer of soil. A well with a drainage pump allows the extraction of water which gathers at the bottom of the deposit during the filling phase. The water is extracted with a pump and then analyzed. The barriers are controlled by inspecting the interior part of the basin through the well. Periodic checks of the deposit integrity have also been established (Noè, 1983).

In some cases where soil removal was not possible or desirable, soil plowing was used to dilute the contaminant and reduce the level of TCDD in the superficial soil layer that is the most important for direct exposure of human beings.

Smooth nonabsorbing surfaces of buildings were washed with surfactants and common solvents, while wall plaster finish and wooden floors were often subjected to various degrees of scraping. Lino floors, wallpaper, furniture, and loose objects which could not be cleansed were eliminated. Many interior surfaces were subsequently coated with paint or synthetic varnish. Agricultural lands with residual TCDD levels in the range 5–15 $\mu g/m^2$ were fenced off to prevent people from entering them and normally cultivated to speed up degradation of TCDD.

Detoxification of zone B implied scarifying town yards, plowing topsoil layers in order to dilute TCDD in the surface and then covering with fresh soil; in agricultural lands, cereals and other food crops were grown. Rehabilitation in zone R basically consisted of agricultural works.

Maximum permissible TCDD levels after rehabilitation were established by the Lombardy regional authorities as follows: in the 7 cm topsoil layer, below 5 $\mu g/m^2$; on exterior building surfaces, below 0.75 $\mu g/m^2$; on interior building surfaces, below 0.01 $\mu g/m^2$.

At present; the reclamation is complete in zone R and in zone A, but is still under development in zone B.

The company owning the ICMESA plant was ordered to begin, at its own expense, a decontamination project for the plant unit from which the toxic cloud originated. A number of possible options existed (Noè, 1983):

(1) Construction of a giant monolith (a concrete casting) to enclose both the equipment and the building containing department B.

(2) Dismantling and (i) construction of a small monolith to enclose the equipment only or (ii) chemical reclamation of the dismantled equipment.

(3) Dismantling of the equipment using the so-called "nuclear method", and successive removal of the highly contaminated material properly contained and packaged.

It was decided to dismantle the equipment with the "nuclear method". To dispose of the resulting low-contamination wastes, the underground deposit above-described for soil was utilized. The high-contamination wastes (e.g. the content of the exploded reactor) were sent to Switzerland where they were incinerated.

As a consequence of the rehabilitation work carried out so far, it has been decided that the hygiene regulations issued in zone R immediately after the accident (e.g. prohibition of farming and consumption of local agricultural products and the keeping of poultry and other animals, and prohibition to undertake construction works) no longer apply in view of the extensive reduction of TCDD levels that has occurred since 1976. In zone A a natural park is being established.

3.5. Impact on Health

A progress report on the health effects of the ICMESA accident was published by Pocchiari *et al.* in 1979. A number of epidemiological studies have been carried out or are still in progress to assess fully the impact on public health of the Seveso accident. The interpretation of these studies, however, is not always easy owing to several factors, the most important of which is the difficulty of identifying the exposed people and quantifying exposure.

Initial exposure of the Seveso population to the toxic cloud was acute and occurred mainly through inhalation and dermal routes. Such exposure was not only to TCDD but also to other chemicals including caustic soda and TCP. During the first ten days after the accident, before the discovery of TCDD in the contaminated environment and before the population was alarmed, there was an additional opportunity that people were exposed through ingestion of contaminated food grown locally and through contact with contaminated surfaces. After evacuation of inhabitants in zone A and enacting the hygiene restrictions in zones B and R, the likelihood of exposure was greatly reduced. In the meantime the soluble toxic substances released with the cloud were washed away by rainfall, while TCDD slightly penetrated the topsoil layer, where it was bound very strongly to soil particles.

Quantitative assessment of exposure is difficult not only because of the changing pattern with time, but also because of its dependence on the compliance of exposed persons with hygiene regulations, which is difficult to assess.

In order to improve the sensitivity of epidemiological analysis, the affected area has been broken down not only by subzones A, B, and R and an outside zone according to soil concentration of TCDD (*Figure 3.1*), but also in several other ways:.

(1) By global soil content of dioxin.
(2) By density of chloracne cases.
(3) By density of cases of acute skin lesions (313 cases observed in July 1976).

Mortality data in the Seveso area, available from 1975 to 1981 (Apricena *et al.*, 1983), reflect the mortality pattern of industrialized countries (cardiovascular diseases and cancer being the leading causes of death). The data do not indicate that the ICMESA accident appreciably altered the specific mortality rates by sex, age, and cause in the monitored area. No significant clusters were observed within the area, either in space or time, that could be attributed to the accident.

The ICMESA accident caused many cases of chloracne (Pocchiari *et al.*, 1979; Fara *et al.*, 1982). Of the 187 cases recorded in total, 50 were detected between September and December 1976 ("early" chloracne) and 137 in a subsequent screening between February and April 1977 ("late" chloracne). The large majority of cases of early and late chloracne occurred in zone A (6.3% and 2.1% respectively of the population originally present in this zone). About 88% of the cases were children under 15 years old. A third screening undertaken after January 1978 involved 32000 subjects and indicated a gradual decrease of chloracne cases (no new cases were detected after 1978). A significant association between the severity of chloracne lesions and the soil concentration of TCDD (measured close to the home of affected subjects) was assessed (Fara *et al.*, 1982). The clinical follow-up of chloracneic children was started in 1976 and was completed in June 1985. Results of the complete study are not yet known, but interim reports indicated a significant difference between a group of 146 chloracneic children and a 183-children age-matched control group with respect to symptoms of the gastrointestinal tract (lack of appetite, nausea, vomiting, abdominal pain, gastritis). Headache and eye irritation were also more frequent in chloracneic children (Fara *et al.*, 1982). Very extensive clinical laboratory investigations were carried out by Mocarelli *et al.* (1982, 1983a,b). The most consistent differences between the exposed children and those from the control areas concerned serum alanine aminotransferase and gamma-glutamyltransferase activity and cholesterol level. In all instances, these parameters were higher in exposed children. However, the differences were restricted to values inside the "normal range" and lessened with time after July 1976. The serum complement activity (CH50) of a rather small group of exposed children (about 50% chloracneic) was significantly higher than that of the controls; the exposed children also showed higher values of lymphocyte response to lectins (PHA, PWR) than controls and a tendency towards a higher number of lymphocytes in peripheral blood. It appears that available data are not suggestive of an immunological depression in the exposed children (Sirchia, 1982). Lastly, no major pathology that might be attributed to TCDD exposure has been reported by physicians in postnatal pediatric screening of babies born so far in zones A, B, and R after the ICMESA accident. Three surveys on the excretion of D-glucaric acid (an indicator of hepatic microsomal induction) were carried out in children in 1976, 1979, and 1981 (Ideo, 1983). Children with chloracne showed a significantly higher excretion of glucaric acid in 1976, and those from zone B in 1979. In 1981 no statistically significant differences were observed.

The Pregnancy Register began to operate systematically in 1979. Before this date information was gathered at the welfare clinics and hospitals in the zone. The population covered by the study included the about 15000 babies born after 1976 who at birth were resident in the 11 municipalities affected by the accident. The monitored unfavorable outcome of pregnancy included: abortion, stillbirth, death in the first week, death in the first year, severe malformation, underweight for gestational age, and slight malformation. A significant reduction of conceptions was detected in the third quarter of 1976 (immediately after the ICMESA accident). Moreover, higher spontaneous abortion rates were recorded in zone A + B than in zone non-A + B + R, with a peak of 33% in the first quarter of 1977. Lastly, in the 11 municipalities under surveillance, between July 1976 and June 1978, the abortion rate was higher in zone A + B + R than in the rest of the territory during the two years following the accident, whereas in 1979 and 1980, the abortion rates in zone A + B + R were lower than those in zone non-A + B + R (Bianco *et al.*, 1983). The malformation rates (both total and severe ones) vary widely according to the parameter used to define the risk areas, and probably also because of the small size of the population under study. A statistically significant ($p < 0.05$) excess was detected only for angiomas in the areas at risk defined on the basis of chloracne. The frequency of angiomas smaller than 4 cm^2 was significantly higher ($p < 0.001$) in the area with a higher frequency of chloracne, whereas that of angiomas larger than 4 cm^2 was not (Mastroiacovo *et al.*, 1984).

The internal medicine monitoring plan provided for half-yearly or yearly medical examination. As far as the correlation with risk areas is concerned, prevalence of hepatomegaly (modest degree, i.e. 1 cm on 1 fingerbreadth beyond the costal arch) was higher in some exposed groups. Transaminases rose in the latter half of 1976 in zone A and normalized later. Several hundred people in the 11 municipalities showed one or more abnormal liver function tests (e.g. γ-GT, ALT, and AP) but the significance of these data is difficult to assess.

An ad hoc cancer registry has been established for the 11 municipalities under surveillance (this registry is expected to be in operation at least until 1997). The rates obtained so far have been compared with those of the Varese Cancer Registry. The data available so far indicate that there was no significant increase for any type of tumor, except for "malignant tumors of the connective tissue of the soft tissues" and lymphosarcoma, reticulosarcoma, Hodgkin's disease and other lymphoid tumors" (Fini *et al.*, 1982). However, the number of cases reported for these types of tumor is so small that the observed variations could be due to chance.

A high association of prevalence of peripheral nerve impairment (PNI) with chloracne was detected, suggesting an association of PNI with other possible TCDD effects (Filippini *et al.*, 1980, 1981). A difference was also assessed between the residents in the most polluted area of Seveso (37.5% of the evacuated population) and an unpolluted town (Cannero) with respect to prevalence of symptoms of PNI, which were more frequently reported in the Seveso area.

Several other special studies have been carried out on adults. D-glucaric acid excretion in the urine of adults was determined in 1978 and found to be higher in the exposed subjects than in nonexposed controls. Immunological and cytogenetic screenings were also carried out on selected groups, particularly from zone A, zone B, and control groups. In general, the differences were small and not statistically significant (Mottura *et al.*, 1980; Sirchia *et al.*, 1980).

3.6. Conclusion

The Seveso accident has had profound effects on public policy relating to chemical safety in Europe and possibly worldwide. It had a considerable impact on both government and public opinion in a number of countries and convinced them of the need for stricter control of hazardous chemicals throughout the whole life cycle. As a consequence, the development of several legislative systems concerning chemical safety was considerably accelerated. Good examples of this are the three following EEC Directives that have had a major impact in improving chemical safety in Europe:

(1) The EEC Council Directive to prevent major accidents which might result from certain industrial activities and to limit their consequences for man and the environment (EEC Directive 82/501).
(2) The sixth amendment of the EEC Council Directive 67/548 on approximation of laws, regulations, and administrative provisions

relating to the classification, packaging, and labeling of dangerous substances (EEC Directive 79/831).

(3) The EEC Directive on supervision and control of transfrontier shipment of hazardous waste.

References

Abate, L., Basso, P., Belloni, A., Bisanti, L., Borgna, C., Bruzzi, P., Dorigotti, G., Falliva, L., Fanuzzi, A., Formigaro, M., Maggiore, G., Marni, E., Meazza, L., Merlo, F., Puntoni, R., Rosa, A., Stagnaro, E., Vercelli, M., and Santi, L. (1982), Mortality and birth defects from 1976 to 1979 in the population living in the TCDD polluted area of Seveso, in O. Hutzinger, R.W. Frei, E. Merian, and F. Pocchiari (eds), *Chlorinated Dioxins and Related Compounds*, 571–587 (Pergamon Press, Oxford).

Apricena, M., Ghioldi, R., Pentoni, R., Stagnaro, E., Vercelli, M., Santi, L., Dorigotti, G., and Meazza, L. (1983), *Mortality Study in the Seveso Area, 1975–1981*, Report to the Lombardy Region Authority (Special Office for Seveso, Lombardy Region, Milan).

Bianco, V., Meazza, L., and Remotti, G. (1983), *Obstetric Monitoring in Brianza of Seveso from 1975 to 1981*, Report to the Lombardy Region Authority (Special Office for Seveso, Lombardy Region, Milan).

Fara, G.M., Del Corno, G., Bonetti, F., Caramaschi, L., Dardanoni, L., Favaretti, C., Giambelluca, S.E., Marni, E., Mocarelli, P., Montesarchio, E., Puccinelli, V., and Volpato, C. (1982), Chloracne after release of TCDD at Seveso, Italy, in O. Hutzinger, R.W. Frei, E. Merian, and F. Pocchiari (eds), *Chlorinated Dioxins and Related Compounds*, 545–560 (Pergamon Press, Oxford).

Filippini, G., Bordo, B., Crenna, P., Massetto, N., Musicco, M., Boeri, R., and Zecchini, A. (1980), *Cross-sectional Neurological Survey of a Population living in a TCDD Polluted Area compared with an Unexposed Population*, Report to the Lombardy Region Authority (Special Office for Seveso, Lombardy Region, Milan).

Filippini, G., Bordo, B., Crenna, P., Massetto, N., Musicco, M., and Boeri, R. (1981), Relationship between clinical and electro-physiological findings and indicators of heavy exposure to 2,3,7,8-TCDD, *Scand. J. Work Environ. Health*, **7**, 257–262.

Fini, A., Gennar, Y., Puntoni, R., Rosa, A.M., and Belloni, A. (1982), *Study of Cancer Incidence at Seveso*, Report to the Lombardy Region Authority (Special Office for Seveso, Lombardy Region, Milan).

Ideo, G. (1983), *Final Report on the Results of D-glucaric Acid Determinations done at Seveso in Adults and Children in the Period 1978–1982. Urinary D-glucaric Acid Excretion in the Seveso Area: 5 Years of Experience*, Report to the Lombardy Region Authority (Special Office for Seveso, Lombardy Region, Milan).

Mastroiacovo, P., Spagnolo, A., Segni, G., and Meazza, L. (1984), *Rapporto sulle malformazioni congenite registrate nell'area di Seveso nel periodo 1977–1982*, Report to the Lombardy Region Authority (Special Office for Seveso, Lombardy Region, Milan).

Mocarelli, P., Marocchi, A., Brambilla, P., Gerthoux, P.M., and Young, D.S. (1983a), *Clinical Laboratory Manifestations of Exposure to Dioxins in Children: A Five Year Study of the Effects of an Environmental Disaster at Seveso, Italy,* Definitive Report No. 1, Report to the Lombardy Region Authority (Special Office for Seveso, Lombardy Region, Milan).

Mocarelli, P., *et al.* (1983b), *Study of Laboratory Tests of Pregnant Women from TCDD Polluted Areas, 1976–1980,* Progress Report No. 3, Report to the Lombardy Region Authority (Special Office for Seveso, Lombardy Region, Milan).

Mocarelli, P., Marocchi, A., Brambilla, P., and Gerthoux, P.M. (1982), *Laboratory Data on Chloracneic Children 2–12 Years Old from Zone A, 1976–1980. Evidence for TCDD Receptor*?, Progress Report No. 2, Report to the Lombardy Region Authority (Special Office for Seveso, Lombardy Region, Milan).

Mottura, A., Zei, G., Nuzzo, F., Crimaudo, C., Giorgi, P., Veneroni, P., Paggini, L., Mocarelli, P., Fraccaro, M., Nicoletti, B., and De Carli, L. (1980), *Evaluation of Results of Chromosome Analyses on Lymphocytes of TCDD Exposed Subjects After the Seveso Accident,* Report to the Lombardy Region Authority (Special Office for Seveso, Lombardy Region, Milan).

Noè, L. (1983), Reclamation of the TCDD-contaminated area at Seveso, in F. Coulston and F. Pocchiari (eds), *Accidental Exposure to Dioxins. Human Health Aspects*, 69–74 (Academic Press, New York).

Pocchiari, F., Silano, V., and Zampieri, A. (1979), Human health effects from accidental release of tetrachlorodibenzo-*p*-dioxin (TCDD) at Seveso, Italy, *Ann. N.Y. Acad. Sci.*, **320**, 311–320.

Pocchiari, F., di Domenico, A., Silano, V., and Zapponi, G. (1983), Environmental impact of the accidental release of TCDD at Seveso (Italy), in F. Coulston, and F. Pocchiari (eds), *Accidental Exposure to Dioxins. Human Health Aspects*, 5–35 (Academic Press, New York).

Pocchiari, F., Silano, V., and Zapponi, G.A. (1986), Chemical risk management process in Italy. A case study: the Seveso accident, *Science of the Total Environment*, **51**, 227–235.

Sirchia, G. (1982), *Immunological Monitoring of Persons Exposed to TCDD in Seveso*, Progress Report, Report to the Lombardy Region Authority (Special Office for Seveso, Lombardy Region, Milan).

Sirchia, G., *et al.* (1980), *Immunological monitoring of Subjects Exposed to TCDD at Seveso,* International Workshop of the National Academy of Sciences (NAS, Washington, DC).

CHAPTER 4

Compensation for Exceptional Environmental Damage caused by Industrial Activities

H. Smets [1]

4.1. Introduction

Incidents causing severe environmental damage, sometimes referred to as "ecological disasters", only receive attention from time to time because, so some people think, they are isolated occurrences, although this is not confirmed by the statistics. While some disasters are remembered, most are quickly forgotten or were never widely reported. And, as Deschamps (1972) points out, "the best way to ward off misfortune is never to discuss it. To study it might be tempting fate. *The scientific investigation of disasters is a sign that man has mastered his ancestral fears and taken charge of his own destiny* [emphasis added]". Recent interest in the risks of technology [2] is an indication of changing attitudes but has shed little light on compensation for potential damage in the vicinity of fixed or mobile industrial plant.

In this chapter, damage means off-site bodily injury and material damage, including the cost of measures to limit or make good such damage. Damage and injury to the plant itself and to workers, as well as to users, is therefore excluded, as are environmental damage caused by natural disasters and by pollution from widely dispersed sources, and payments made before a polluting installation is set up to "compensate" for permanent deterioration of the environment [3]. The study does not cover damage linked to military or defence activities. It is therefore confined to environmental damage associated with industrial emissions of pollutants or energy.

This damage can affect man (death, injury, immediate and later medical costs, evacuation costs, loss of earnings, disamenity), fauna and flora, land, and property. It may be costly to clean up or dispose of waste and to restore the environment to its original state.

This study will be confined to the most serious cases of damage to the environment, ignoring incidents involving under FF 1 million [4] and those whose effects on man and environment do not appear very significant. It will be mainly concerned with "disasters" involving deaths, large-scale evacuation, or expensive clean-up.

In Section 4.2 we examine oil pollution at sea, environmental damage due to accumulations of toxic substances or energy, water and air pollution, noise, and radioactive pollution, in order to see how much damage can be caused in exceptional conditions. Section 4.3 will deal with the special arrangements introduced to provide compensation for exceptional damage.

4.2. Exceptional Damage to the Environment

4.2.1. Damage to the marine environment by oil spills

The marine environment suffers considerable damage in the event of large spills of crude or other oils, especially accidents to tankers, offshore wells, coastal reservoirs, or seabed pipelines.

Damage Caused by Oil Transported by Sea

Several tanker accidents have led to heavy expenditure to counteract oil pollution. The cost varies according to how much oil is spilled, weather conditions, where the pollution takes place, and what measures are taken. Furthermore, such accidents can involve serious economic losses for neighboring populations.

In each year from 1974 to 1983, between five and 37 tanker accidents involved spills of over 675 t (average: 19 per year). Between 1974 and 1982 accidents costing over $250000 (at current prices) in pollution damage averaged 17 per year. Since 1974, seven accidents have cost over current $10 million. The two most expensive accidents have been the *Amoco Cadiz* (220000 t) and the *Tanio* (17000 t) off the Brittany coast [5].

Compensation in respect of the *Amoco Cadiz* is to be determined by a Chicago court. French sources [6] suggest it will reach at least 1978 FF 800 million. French and United States (US) economists have produced a study of the economic consequences of that accident [7]. The main heads of damage related to:

(1) Clean-up costs, FF 445–490 M (excluding tax).
(2) Marine resources, FF 140 M.
(3) Loss of amenity (residents and tourists), FF 53–342 M.
(4) Brittany tourist industry, FF 16–251 M.
(5) Secondary regional effects, FF 25–26 M.

On top of these, account must be taken of costs borne by the local authorities (over FF 100 million) and losses incurred by them. For the *Tanio*, compensation claimed amounted to FF 516 million including FF 490 million for government clean-up costs. Reimbursable costs will probably be between FF 360 million and FF 516 million. Government and private claimants have already received FF 245 million and are continuing proceedings.

Table 4.1 breaks down tanker spill costs between 1974 and 1980. The five most expensive spills are estimated to have cost $120, $80, $56, $51, and $36 million (1983) (*Table 4.2*). *Table 4.3* shows that most oil pollution damage took place in Europe.

Table 4.1. Oil spill costs, 1974–1980.

Cost per slick (1983 $ M)	*Slicks*	*Total cost* (1983 $ M)
0.94–3.75	43	84
3.75–15	24	191[a]
15–60	7	214[a]
60–240	2	200[a]
Total	76	689

[a]The three highest cost categories had approximately the same total cost.
Source: OCIMF figures, July 1982, amended by estimating the overall cost for the two most expensive accidents at $200 million instead of $325 million (1983).

Damage Caused by Offshore Oil Installations

Although there have been relatively few accidents involving offshore oil installations as compared to maritime transport, highly spectacular accidents resulting in considerable expenditure have nevertheless occurred. Unfortunately, information is incomplete (see *Table 4.4*) especially with regard to the Persian Gulf. The most serious cases of pollution would seem to be those at Santa Barbara ($45–85 million in 1969) and Funiwa in Nigeria [8]. The biggest spill (500 000 t at Ixtoc One) polluted the American coastline over 1000 km away but does not

Table 4.2. Main tanker accidents since 1967 (compensation of $20 M and over for pollution).

Date	*Country affected*	*Ship (Flag)*	*Quantity spilled* (10^3t)	*Estimated costs*[a] (1983 $ M)
1978	France	*Amoco Cadiz* (Liberia)	220	120
1980	France	*Tanio* (Madagascar)	6	80
1979	Sweden, Finland, USSR	*A. Gramsci* (USSR)	6	56
1980	Cuba	*Princess Anne Marie* (Greece)	6	51
1979	Ireland	*Bételgeuse* (France)	27	36
1967	France, UK	*Torrey Canyon* (Liberia)	121	21
1976	France	*Boehlen* (GDR)	11	20
1976	Spain	*Urquiola* (Spain)	101	20
1973	Puerto Rico	*Zoe Colocotroni* (Greece)	8	20

[a]Cost estimates based on compensation awarded (or likely to be paid when outstanding claims are settled). Real costs are generally higher than compensation paid.

Table 4.3. Compensation for oil spills, 1970–1981.

Country	*Compensation* (1982 $ M)	
Norway, Sweden, Finland	32	
France	267	
UK	34	402 (56%)
Ireland	48	
Spain	21	
Greece	18	
Turkey	6	
USSR	21	
USA	123	140 (19%)
Canada	17	
Japan	53	(7%)
Rest of world	80	
Total OECD countries	619	(86%)
Total other countries	101	(14%)
Total	720	

Source: Interpretation of Temple *et al.*'s data on expected or paid claims (1982 $) (Temple *et al.*, 1983).

seem to have caused an ecological disaster even though so much oil was involved.

Table 4.4. Major offshore oil spills.

Date	*Place (Company)*	*Volume (t)*	*Costs*[a] *(clean-up + 3rd party)*	*Remarks*
1969	Santa Barbara, California (Union Oil)	4000	$45 to 85 M	Damage to coastline (7 km) Compensation paid: $9.5 M
1970	Main Pass, Louisiana (Chevron)	9000	0	No coastal pollution (20 km) Well closure: $15 M
1970	Baie Marchand, Louisiana (Shell)	8000	0	Slight coastal pollution Well closure: $35 M
1971	Laban, Persian Gulf	14000	0	–
1975	Ekofisk, North Sea, Norway (Phillips)	21300	NKr 26 M[b]	No coastal pollution but payment to the authorities of some NKr 5 M
1979	Ixtoc One, Mexico (Pemex)	500000	USA: $12.5 M Mexico:	Settled out of court for $4.14 M with the USA (public and private sectors) Well closure: $50 M
1979	El Tigre, Venezuela	13000	?	–
1980	Funiwa 5, Nigeria (Texaco)	35000	At least $8 M	Serious coastal pollution
1980	Hashah 6, Saudi Arabia (Aramco)	13200	?	Serious coastal pollution
1980	Ras Tanura, Bahrain (Aramco)	3000	$1 M	–
1980	Ron Tapmeyer, Persian Gulf	14000	$7 M	–
1983	Nowruz/Ardeshir, Iran, Persian Gulf	100000	?	War

[a]Clean-up costs and damage to third parties do not include the cost of stopping the oil flow, removing the platform, replacing equipment destroyed and compensation for lost oil. The costs of cleaning up in the vicinity of the installation are normally paid by the operator and are often unavailable.
[b]NKr, Norwegian kroner.

Other Damage

Considerable damage can also be caused by the bursting of oil storage tanks situated near the coast. Some 8500 t of oil was spilled on a coastal area in Japan (Mitzushima), following the bursting of a storage tank. Clean-up costs amounted to $43.3 million, and in 1975 Mitsubishi paid $56.7 million as damages to third parties.

4.2.2. Damage to the environment due to accumulations of dangerous waste

An aspect that has recently grown to considerable proportions is environmental damage resulting from the build-up of dangerous waste at waste disposal sites and the progressive contamination of the soil around industrial installations.

Older Hazardous Waste Dumps

In many European countries and in the USA, numerous industrial waste dumps have been found to constitute serious health hazards and have therefore had to be cleared. Given the quantities of toxic wastes that have accumulated, the cost involved can reach several tens or even hundreds of millions of dollars for a single site. In the USA, legal action has been taken against a large firm to cover the costs of clearing a Colorado dump, for which the Defense Department has already earmarked $500 million. Present estimates suggest that the USA will be paying out more than $16 billion over the next decade or so to clear older dumps [9]. One of the most famous examples is Love Canal, where 2500 people had to be evacuated in 1978 and more than $53 million (including $30 million for rehousing) paid out because of cancer risks.

Europe has also been affected, and at Lekkerkerk in the Netherlands, it cost $70 million to evacuate 870 people from the vicinity of a dump [10]. In the FRG, a large dump in the Georgsweder suburb of Hamburg is causing concern and may also involve significant expenditure. In France, costs have been relatively low so far (ten million francs per site at most) [11].

In Japan, residues from numerous disused mines have been left above ground. Between 1920 and 1962, 3000 t of arsenic waste poisoned more than 140 people. Under a recent court decision [12], 22 casualties were awarded Yen (Y) 507 million ($2.3 million).

Tables 4.5 and *4.6* give an indication of actual or estimated costs in a number of serious cases in Europe and the USA. Although it is too early to assess the exact extent of the costs associated with old waste dumps, their financial impact would appear to be such as to make this

Table 4.5. Decontamination costs for hazardous waste dumps: Serious cases (over $4 M/site).

Case	*Cost ($ M)*	
Colorado, USA (law suit claim: $1950 *M*)	over 500	
Times Beach, USA (estimate)	235[a]	($35 M spent)
Lekkerkerk, Netherlands	70	(claim for $17 M)
Bridgeport, New Jersey, USA	55	
Love Canal, USA	53	
Louisiana, 2 sites, USA	50	
Dordrecht, Netherlands (G 150 M)	over 50	
Seymour, Indiana, USA	45	($7 M spent)
Velsicol/Gratiot County, St Louis, Michigan, USA	38.5	
Gouderak, Netherlands (G 110 M)	36	
Niagara Falls, S. Area, New York, USA	30	
Wilsonville, Illinois, USA	25	
Waukegan, Illinois, USA	21	
Iron Mountain, Redding, California, USA	16.8	
Swartz Creek, Michigan, USA	14	
Southwest Philadelphia, USA	12.8	($8 M spent)
Vickery, Ohio, USA	10	
Kent, Washington, USA	10	
BT Kemi site, Sweden	7	
Picillo Farm, Coventry, Rhode Island, USA	6.5	
Green-up, Illinois, USA	5.6	
FRG, one site	4.8	
Sylvester, Nashua, New Hampshire, USA	4.3	
Ventrol/Velsicol, Woodbridge, New Jersey, USA	4	
Tokyo, Japan (hexavalent chromium)	4	
Grand Prairie, Texas, USA	4	

[a]The figure of $235 M represents pollution by spreading oil containing dioxin in 1971, not by dumping.

one of the major environmental problems in the 1980s. According to the Netherlands Minister of the Environment [13], clean-up costs could amount to $700 million a year for 13 years in Europe. Just clearing the old dumps could absorb 0.25% of gross national product (GNP). In the FRG, $2.2 billion would be needed to clean up about 1400 dumps. For the USA, some 2000 older dumps would cost approximately $16 billion to clean up at an annual rate of $2 billion. There is some prospect that such a program might be implemented only partially, in view of the very high cost.

Gradual Contamination of the Soil

A problem similar to that of disposal sites is the build-up of pollutants near industrial installations. In Morocco, 31 children are said to have

Table 4.6. Committed and actual expenditures[a] for clean-up of toxic waste in the USA.

Expense per site ($ M)	*No. of sites*	*Total* ($ M)
Committed federal expenditures (31 March 1985)		
1.25–2.5	23	40.1
2.5–5	17	56.7
5–10	8	54.3
10–20	3	51.9
20–40	1	34.4
Subtotal	52	237.4
Total	416	407.6
Actual federal expenditures (31 March 1985)		
0.8–1.6	22	23.8
1.6–3.2	10	23.6
3.2–6.4	3	12.7
6.4–12.8	3	24.0
Subtotal	38	84.1
Total	416	133.3

[a]These expenses represent only a fraction of total clean-up cost. Many decided-upon clean-up activities have not yet materialized in financial commitments. The average cost per site is now estimated to be $8.5 M.

died of lead poisoning owing to waste from a leadworks being dumped in the middle of a village (Mekouar, 1984).

In Belgium, in a suburb of Antwerp, it cost Belgian francs (BF) 100 million (about $22 million) to eliminate lead-contaminated dust which had settled on the ground and in houses in the vicinity of a leadworks.

In the areas surrounding old aluminum works in France, the FRG, Switzerland, and Greece, soil contamination by fluorine effluent was severe enough to destroy much livestock, obliging the operators to compensate farmers. However, compensation paid came to under FF 100000 per works, per annum: at Lannemezan, in 1974, fluorine killed 200 cows; compensation in Greece totaled $30000 for the loss of 720 animals in six years.

Gradual Contamination of Marine Sediments

Continuous pollution of coastal waters can cause an excessive build-up of pollutants in sediments which may be very expensive to dredge. In the bay of Tokuyama, 450000 m^3 of mercury-contaminated sediments were removed at the expense of the two firms responsible (Toyo Soda Kogyo and Tokuyama Soda) costing a total of Y 10 billion in 1975. Similar works were undertaken for Minamata Bay (75% borne by the firm and 25 % by the authorities).

4.2.3. Damage to the environment caused by release of stored energy

Serious damage to surrounding areas has been caused by the storage of energy mechanically (water) or chemically (explosives, gas). *Table 4.7* shows that the number of lives lost as a result of dams bursting can exceed 1000. Worldwide, the probability of a dam bursting is 2×10^{-4} per year, i.e. between 1 and 2 cases a year in the world involving deaths and devastation over wide areas. In the USA five dams burst between 1918 and 1958, killing 1680 people (Okrent, 1979). For Western Europe and Japan, dam burst risks are lower (0.2×10^{-4}) [14].

Environmental damage can also be caused by loose earth movements. In Aberfan (Wales) a slag heap wiped out part of the village (144 dead, one school and 18 houses destroyed).

The chemical and oil industries have also caused several disasters involving the explosion or combustion of dangerous substances [15]. *Table 4.8* gives details of disasters involving over 50 deaths (employees and third parties). *Table 4.9* shows that over the last 50 years there have been at least 24 such accidents worldwide of which 14 involved fixed land-based plant; and of these, seven involved chemicals. Over the same period, of eight accidents causing over 200 deaths, five were in OECD countries. It is also noteworthy that most recent serious accidents have occurred in the Third World (six out of seven for land-based installations) and involved oil or gas.

Furthermore, explosions have been responsible for severe damage destroying buildings or industrial plant. In Buenos Aires, Argentina, in 1967 a propane fire destroyed 400 houses; in Escombreras, New York, in 1969 an oil explosion meant evacuating 5000 residents; in Flixborough, the UK, in 1974 a cyclohexane explosion involved the evacuation of 3000 residents and destroyed 100 houses. Carriage of explosive substances can also involve heavy damage [16]. In 1974 in Los Angeles, FF 250 million of damage was caused by an exploding tank of organic peroxide. Serious damage has also been caused by munitions and explosives (carriage, storage, manufacturing), by fireworks, and by gas explosions in buildings and urban supply networks.

These figures suggest that it is release of stored energy which has involved the severest damage to the environment.

Table 4.7. Accidents involving dams, 1959–1983.

Date	*Place*	*Country*	*Deaths*	*Damage* ($ M)
1959	Vega de Tera	Spain	144–400	–
	Malpasset	France	421	68
1960	Oros	Brazil	1000	–
1961	Kiev-Babu Yar	USSR	145	–
1962	Sunchon-Hyokiri	South Korea	250	–
1963	Vajont[a]	Italy	2118	30
	Quebrada La Chapa[b]	Colombia	250	–
	Balwin Hills (California)	USA	3	50
1967	Nanaksagar	India	100	–
	Sempor	Indonesia	200	–
1969	Pardo	Argentina	100	20
1972	Toledon	Colombia	60	–
	Canyon Lake (South Dakota)[a]	USA	231	115
	Buffalo Creek (West Virginia)	USA	129	26
1976	Tetons (Idaho)	USA	11	407
	Del Monte	Columbia	80	–
	Santos Tomas[a]	Philippines	80	–
1977	Limpopo	Mozambique	300	–
	Toccoa (Georgia)[a]	USA	39	3
	Euclides de Cunha[a]	Brazil	–	60
	Armando Salles de Oliverra[a]	Brazil	–	60
1978	Bakhera	Nepal	500+	–
1979	Morvi-Macchu 2[a]	India	15000 (?)	–
1980	Oressa	India	1000	–
1981	Kernataka	India	120	–
1982	–	Liberia	200	–
1983	Cundinamarca	Colombia	150	–

[a]These accidents were caused by heavy rainfall, rivers in spate, or earthslides (Vajont).
[b]Natural dam.
Sources: *Quid*, 1982, p 1149; *Sigma*, 1976–1984; Goubet, 1979.

4.2.4. Environmental damage caused by air and water pollution

Pollution near industrial plant caused by emissions of harmful pollutants during the production process, from chimneys, and as effluent discharged into rivers is a well-known side effect of industrialization. Although this often occurred in the past, the situation is nowadays distinctly better; however, as the Bhopal disaster reminds us, the risks are still present.

Table 4.8. Industrial accidents causing more than 50 deaths.[a]

Year	*Place*	*Enterprise*	*Cause*	*Deaths*[b]	*Other damages*
1907	Pittsburg, USA		Explosion in a steelworks	59	Several persons unaccounted for
1917	Petrograd, USSR		Factory explosion	100	
1921	Oppau, Germany	BASF	Fertilizer factory explosion (ammonium nitrate)	561	1900 casualties; damage to the town
1933	Neuenkirchen, Germany		Gas explosion in steelworks	63	Several hundred casualties; 70 houses destroyed
1939	Zarnesti, Rumania		Leak of 25 t of chlorine in a factory	60	300 casualties
1942	Tessenderloo, Belgium		Explosion in a chemical plant (ammonium nitrate)	200	1000 casualties
1943	Ludwigshafen, Germany		Factory explosion of 16.5 t of butadiene	57	439 casualties
1944	Cleveland, USA		Explosion of 4300 m^3 of confined LNG[c] fire ball	136	350 casualties; streets swept by burning gas; windows broken; 79 houses, 2 factories, and 79 cars destroyed; \$6.8 M
1947	Texas City, USA	Grandchamp	Explosion of a ship with a cargo of ammonium nitrate (1 750 t)	532	200 unaccounted for; 300 casualties; serious damage to city (DM 100 M)

(*cont.*)

[a]Not including accidents in the USSR, accidents involving explosives and munitions, mining accidents, gas distribution accidents, transportation accidents (passengers).
[b]Deaths of workers and third parties.
[c]LNG, liquefied natural gas; LPG, liquefied petroleum gas.
Sources: Lagadec (1981a,b); Andurand (1979); UBA (1983).

Table 4.8. (*cont.*) Industrial accidents causing more than 50 deaths.[a]

Year	*Place*	*Enterprise*	*Cause*	*Deaths*[b]	*Other damages*
1948	Ludwigshafen, FRG		Explosion of unconfined dimethyl ether	245	3800 casualties; FF 80 M; explosion and fire in a factory caused by a wagon; damage at 8 km distance
1948	GDR		Steam generating station explosion when pulverizing coal	50	76 casualties
1956	Minamata Japan	Chisso	Mercury discharge into river and bay	250	Over 100 000 alleged mercury poisoning casualties
1957	Bahrain	Seistan	Cotton-wool explosion	57	
1970	Osaka, Japan		Explosion of confined gas in an underground railway construction site	92	
1978	Los Alfaques, Spain		Explosion of liquefied propylene in transport by lorry	216	200 casualties at a camping site; FF 144 M compensation
1978	Xilatopec, Mexico		Explosion of 10 000 l of LPG[c] following multiple pile-up involving lorry and 12 vehicles	100	150 casualties
1978	Huimanguilla, Mexico		Gas pipeline fracture	58	
1979	Istanbul, Turkey	Independenta	Ship–tanker collision	75	95 000 t of oil on fire

(*cont.*)

[a]Not including accidents in the USSR, accidents involving explosives and munitions, mining accidents, gas distribution accidents, transportation accidents (passengers).
[b]Deaths of workers and third parties.
[c]LNG, liquefied natural gas; LPG, liquefied petroleum gas.
Sources: Lagadec (1981a,b); Andurand (1979); UBA (1983).

Table 4.8. (*cont.*) Industrial accidents causing more than 50 deaths.[a]

Year	*Place*	*Enterprise*	*Cause*	*Deaths*[b]	*Other damages*
1979	Bantry Bay, Ireland	*Bételgeuse*	Explosion of tanker at berth	50	
1979	China	Bohai	Offshore rig collapse	72	
1980	Norway	*A. Kielland*	Offshore rig collapse	123	
1980	Alaska		Fire at oil-rig construction	51	
1980	Canada	*Ocean Ranger*	Offshore rig collapse	84	
1982	Tacon, Venezuela		Oil explosion and fire at power station	145	Fire in neighborhood; 1 000 casualties
1984	Cubatao, Sao Paulo Brazil	Petrobras	Petrol explosion following pipeline fracture	508	Fire in a shanty town (3 000 inhabitants) built illegally on Petrobras lands (Petrobras claims no more than 90 deaths)
1984	San Juan, Ixhuatepec, Mexico	Pemex	Explosion of LPG[c] reservoirs (90 000 bbl)	452	Fire in shanty town (4 248 casualties, 31 000 homeless, 300 000 evacuated, flames 300 m high, 300 houses destroyed)
1984	Bhopal, India	Union Carbide	Release of 50 t of methyl isocyanate	2 500	About 34 000 eye casualties; 200 000 people left the area voluntarily when the plant was recommissioned
1984	Gahri, Ohoda, Pakistan		Explosion of a natural gas pipeline	60	

[a]Not including accidents in the USSR, accidents involving explosives and munitions, mining accidents, gas distribution accidents, transportation accidents (passengers).
[b]Deaths of workers and third parties.
[c]LNG, liquefied natural gas; LPG, liquefied petroleum gas.
Sources: Lagadec (1981a,b); Andurand (1979); UBA (1983).

Table 4.9. Distribution of industrial disasters (over 50 dead, see *Table 4.8*).

	Area	*Land-based plant*	*Surface transport*	*Offshore oil*	*Maritime transport*	*Total*
Oil and gas						
1934–1969	OECD	1	0	–	0	1
	Others	0	0	–	0	0
1970–1984	OECD	1	0	3	2	6
	Others	5	1	1	0	7
Subtotal		7	1	4	2	14
Chemicals						
1934–1969	OECD	4	0	–	1	5
	Others	2	0	–	1	3
1970–1984	OECD	0	1	–	0	1
	Others	1	0	–	0	1
Subtotal		7	1	–	2	10
TOTAL		14	2	4	4	24
of which OECD		6	1	3	3	13

Air Pollution

Some industrial plants have been the cause of continual or accidental air pollution serious enough to give ground for complaint by people in the locality and, in some cases, to require special protection measures (closed windows, evacuation, ban on agricultural produce grown close to the factory, etc.). The pollutants most often involved are fluorine, lead, asbestos, mercaptan, hydrogen sulfide, certain pesticides, sulfur oxide, chromium, chlorine, and ammonia [17, 18]. Owing to the localized nature of damage, the cost seldom exceeds FF 2 million per incident. To take one example, a sulfur oxide cloud damaged 700 vehicles (at a cost of FF 1.2 million) in the car park of a French firm.

Nevertheless, there have been some very costly incidents. Compensation for pollution caused in the USA by Trail Smelter in Canada finally came to $428 000 (1940 prices).

In 1950, at Poza Rica (Mexico) a hydrogen sulfide leak put 320 people in hospital and caused 22 deaths. The evacuation of 240,000 people at Mississauga near Toronto in 1979 because of a pollution hazard caused by a derailed chemicals shipment cost several tens of millions of US dollars and full compensation has yet to be paid [19].

In 1974, accidental discharge of 460 kg of chlorine in a Japanese factory in Yokkaichi caused eye and mucous irritation to 10 000 local residents [20].

The Seveso dioxin leak in 1976 caused 736 people to be moved out and affected 1800 ha. The burial of 250000 m^3 of contaminated soil cost about lire (L) 30 billion. Off-site expenditure amounted to more than Swiss francs (SF) 300 million [21].

In 1984, over 2500 people were killed in the Bhopal (India) accident by an accidental escape of 30 t of methyl isocyanate [22]. Compensation payable varies, depending on the assumptions on which estimates are made, from many tens of millions of dollars to several billion dollars. This accident is the largest pollution disaster ever recorded. Figures in *Table 4.10* give an idea of the cost of a similar disaster in Europe. The Indian Government refused a $350 million settlement out of court.

Table 4.10. Estimated cost of an ecological disaster (toxic gas).

Type of expenses	*Estimated cost* (FF M)	
Compensation:[a]		
250 dead	50	
125 80% handicapped	125	80%
500 50% handicapped	225	
625 20% handicapped	80	
Emergency hospital and medical expenditures for 10 000 patients	10	
Special medical expenditures (medium term)	20	
Subsequent medical attention (long term)	20	
Evacuation of 30 000 residents (5 days)		
Loss of earnings	20	
Evacuee accommodation	20	
Miscellaneous	30	
Total	600	

[a]The compensation figures are based on average payments for road accidents in France (1982). No account is made for economic losses to trade.

Published information on air pollution incidents has been fuller as from December 1984. In the space of two months, nine serious incidents were recorded [23], three of them in countries of the Organization for Economic Cooperation and Development (OECD). In France in 1984, a disaster was narrowly averted [24].

Air pollution can also be caused by the combined activities of several enterprises. In 1971, in the Lacq region of France, market gardeners in Gave de Pau received FF 3.3 million for air pollution damage caused by four firms.

At Yokkaichi, Japan, six petrochemical companies and one electricity company were held jointly liable for poor air quality (sulfur oxides, etc.) and ordered to pay damages to bronchial casualties in the vicinity (130 casualties, Y 569 million in 1973). In 1975, a Japanese steel company paid Y 2 billion ($6.6 Million) to air pollution victims under the Compensation Act (see [25] and Section 4.3.6).

Pollutant emissions from industrialized regions have also led to serious pollution in exceptional weather conditions. In 1930 there were 63 deaths in the Meuse Valley and in 1947 in Donora (Pennsylvania) 43% of the population (14 000 people) were affected and 20 deaths recorded. In 1962 in the Ruhr, high pollution is thought to have caused an increase in the number of deaths (over 100 people).

Water Pollution

There are many instances of rivers being badly polluted by industrial waste or by sludge when reservoirs behind dams are emptied, but in general the damage involved is not very great [26]. Tens of thousands of fish can be lost and aquatic life destroyed over several kilometers without the cost of damage exceeding FF 1 million. In France, freshwater fishing associations seem to obtain about FF 5000 per kilometer polluted [27]. Among the highest figures recorded are those for the pollution of the Sandrine (Garonne basin) by a metal processing factory, which cost FF 803 090 in 1977, and for the accidental spilling of 20 t of acrolein into the Rhone in 1976 (300 t of fish killed), which seems to have been settled out of court for under FF 3 million.

In 1982 at Saint Fons, the accidental discharge of 1.7 t of hydroquinone caused the accidental death of 60 t of fish [28]. In 1983, the bursting of an earth dyke for a potassium waste dump at Lvov in the USSR polluted the Dniester for 500 km, causing considerable damage.

River pollution damage can be considerable when drinking water installations are affected or when water is used for irrigation. In 1984, Pec Engineering were ordered to pay FF 3 million on account for polluting the Upper Rhine with toluene, with the result that wells in four villages (5000 residents) had to be shut off.

Leaks from storage tanks and pipelines can be very costly if not detected quickly, and can seriously damage the water table, as can waste from which toxic liquids seep or drain away, particularly after a storm. In France such incidents have cost several million francs, especially where they have affected drinking water supply (FF 4 million in one case).

In the Rhine salt case, the Alsace potash mines were held liable for salt damage in the Netherlands. Damages have not yet been assessed but the claim in 1975 was for gulden (G) 45 million ($15 million).

Industries have also caused serious damage by discharging waste in coastal areas. For instance, fishing was seriously affected at the mouth of the Seine. In Corsica damages in respect of red sludge discharged by Montedison into the Mediterranean in 1972 has been assessed in 1985 at FF 680 000, while 20 Italian fishermen received FF 174 000 as compensation.

In 1974, 250 t of tetraethyl lead went down with the *Cavtat* 15 km off the coast of Italy. In 1977, the Italian authorities spent FF 57 million to retrieve the drums. In 1981, the salvage of a dangerous cargo from the *Klearchos* which went down near Sardinia cost $9 million. In 1984 the accidental loss of 80 drums of Dinoseb in the North Sea cost over G 2 million in search and retrieval expenses. In the Netherlands, disposal of 51 drums of chlorinated products lost 25 km offshore in 1979 by an Iraqi vessel cost over G 1.75 million.

In Turkey the city of Istanbul received $7 million in 1983 from the airport authority and $3 million from a banking group as compensation for coastal water pollution for which they had been responsible.

In the USA in 1975, release of pesticides (kepone) into the James River and Chesapeake Bay in Virginia caused serious damage. More than $12 million was paid to fishermen in compensation and the Allied Chemicals Company was fined $13.2 million. This fine was reduced on appeal to $5 million, but the company donated $8 million to an environment protection fund. In another case, the Dupont Company offered to pay the State of Virginia $2 million in settlement of litigation arising from the presence in a river of mercury discharged more than 30 years previously. A pollution of underground water at Jackson Township (New Jersey) by a municipal waste dump caused damage estimated at $15.8 million by a court (see [29]).

In Canada, rivers flowing through Indian reserves were polluted by mercury from a pulp and paper plant. The Federal Government paid $1.6 million in 1982 to Whitedog Indian Reserve and $4.4 million in 1984 to Grassy Narrows (River Wabigoon). In the latter case, negotiations are in hand to obtain additional compensation from the company [30].

The most serious water pollution damage seems to have occurred in Japan following the build up of toxic waste over the period 1950–1970. A group of petrochemical companies paid compensation of Y 200 million to the Tokuyama Bay fishermen for fishing losses between 1957 and 1961. Dumping of heavy metal waste into a river by the Ashio Company polluted rice fields [30]. An out-of-court settlement of Y 1.55 billion was made to cover damages for the period 1952–1971. Two outbreaks of chronic arsenic pollution from mining wastes caused 155 casualties (*Table 4.11*).

"Itai Itai", a disease that affected some 100 people, was caused by cadmium pollution of a river. A first payment of Y 148 million was made in 1972 and was followed by other payments.

Mercury pollution of rivers and bays caused what was known as the Minamata disease. The worst case was due to discharges by the

Table 4.11. Water pollution in Japan.

Type of poisoning	*Firm date of first observation*	*Casualties compensated in 1983*	*Fatalities to 1983*	*Claims still pending in 1983*
Minamata (mercury)	Chisso (1956)	1370	595	5605
Minamata Niigata (mercury)	Showa Denko (1965)	553	135	48
Itai-Itai (cadmium)	Mitsui (1955)	37	80	–
Toroku Miyazaki (arsenic)	Sumitomo (1972)	107	32	–
Shimano (arsenic)	Sasagadani (1970)	9	12	–

Source: Environment Agency (1984).
Note: Not all fatalities of compensated victims were caused by pollution.

Chisso Company into Minamata Bay. Under the first judgment in 1973, 138 victims were awarded Y 937 million ($3.4 million). By 1977, Chisso had paid a total of Y 36 billion to 1348 victims. In 1983, 1909 victims had received compensation and nearly 5368 claims were pending. In another mercury pollution case, in Niigata, the Showa Denko Company paid Y 270 million ($0.92 million) to 76 victims in 1971. By 1983, compensation had been paid to 685 victims. These figures would suggest that total compensation paid to victims of water pollution in Japan up to 1983 could well exceed Y 60 billion, i.e. Y 6 billion a year for ten years. It should be noted that this represents the cumulative effect of discharges over the period 1950–1970.

Environmental Noise Nuisance

Noise can cause serious damage, especially near airports or railways. Aircraft, airport, and rail operators have sometimes been compelled to reduce disamenity for buildings or make good falling property values. In the FRG nearly $300 million had been spent by 1981 to soundproof houses near airports. *Table 4.12* shows that outlays have been considerable in some cases. After 14 years litigation, 3828 victims of Osaka airport noise were awarded Y 1.3 billion as settlement in 1984. In France, one airport's noise victims were awarded FF 2.7 million [31].

4.2.6. Damage caused by radioactivity

Serious damage may be caused by "nuclear" plants and radioactive materials for medical and industrial use. We shall not deal here with damage caused by nuclear weapons just as we have excluded damage associated with explosives and munitions under Section 4.2.3.

Table 4.12. Compensation paid for noise nuisance (airports, railways).

	Compensation
Soundproofing (airports)	
Heathrow and Gatwick, UK (1966–1978)	4.25 M[b]
(1978–)	19.0 M[b]
40 FRG airports (1974–1978): civil	DM 12.7 M
military	DM 7.7 M
Japanese airports (1979)	
Housing	Y 50 600 M
Public buildings	Y 9 900 M
Paris airports (Roissy and Orly) (1973–1983)	FF 67.2 M
Copenhagen, Denmark 1976–1980	DKr 198 M[a]
1981–1986 (500 houses)	DKr 125 M
Fall in property values (airports)	
Japan (1979): Rehousing	Y 12 700 M
Los Angeles, USA (1972–1976): Purchase	$300 M
Paris, France (1973–1983): Purchase	FF 114 M
Fall in property values (railways)	
Shinkansen train Japan (railways)	
Rehousing	Y 757 M
Purchase	Y 488 M

[a]Dkr, Danish kroner.
[b]Pounds sterling.
Sources: *Noise Abatement Policies*, OECD, 1980; Aeroport de Paris; Danish Authorities.

With the exception of the Chernobyl disaster, nuclear installations have rarely caused serious damage in their immediate vicinity, owing to precautions taken and the size of sites [32]. The most serious reactor accidents have been at Windscale (1947), where enough radioactivity was released to lead the authorities to destroy milk produced in the area, and at Three Mile Island (1979), where evacuation was advised as a precaution against any significant radioactivity leak. While the Three Mile Island accident involved high compensation ($33 million) [33], the damage to third parties in the Chernobyl case is considerably larger.

Radioactive effluent from reprocessing plant leaking into the sea at Sellafield (1983) caused enough contamination for the beach to be closed off to the public. The cost of decontaminating the beach is not known but was probably not very high.

A strange recent case was the contamination of an American submarine hull by a drum of radioactive waste dumped at sea. In the USA in New Jersey (Essex County), an $8 million program was launched in

1985 to counteract radium wastes affecting 80 houses (radon contamination).

Transport of radioactive substances (e.g. uranium, plutonium, irradiated fuels, and waste) has caused a few instances of road and rail contamination but at comparatively low cost (less than $1 million). The loss of 450 t of uranium hexafluoride (the *Mont-Louis*) may have caused some slight radioactive pollution but the drums were soon recovered at a cost of some tens of millions of French francs.

Radioelements for medical and industrial applications (e.g. radioactive cobalt sources, needles, etc.) have caused serious accidents to members of the public when found after being lost or stolen. Deaths and serious burns have occurred. The most expensive case seems to have been in 1983 and 1984 in Mexico, where 450 Ci of radioactive cobalt contaminated over 6000 t of steel and castings [34]. Apart from injuries (some severe, affecting about 100 people) and radiation-induced leukemia, extensive material damage must be anticipated if hundreds of houses built with the contaminated material have to be demolished.

The fall of the Soviet satellite, *Cosmos 954*, on Canada caused radioactive contamination of the soil in an uninhabited area. Costs of retrieval and decontamination amounted to Canadian $16 million. The Canadian authorities claimed $6 million and finally accepted $3 million.

Lastly, mention should be made of damage caused by the natural radioactivity of uranium. In the USA, uranium mining byproducts used as building sand in the 1950s were discovered in the 1980s to be dangerous. Work costing several hundred million dollars is in hand to remedy the situation. In 1984, work was done on 61 properties (for a total cost of $1 million) [35].

4.2.7. Worst-case environmental damage

Tables 4.13 and *4.14* show the heaviest damage to the environment mentioned in this report for the period 1959–1984. The largest forced evacuation was 240000 people at Mississauga and 13 other forced evacuations involved over 4000 people (*Table 4.15*).

In terms of deaths, dam accidents have been the most disastrous. Apart from Minamata disease, the Bhopal accident, and the Chernobyl disaster, *no instance of accidental pollution has caused more than 50 deaths*. In terms of costs, the most expensive so far has amounted to $200 million (excluding dam bursts of up to $740 million). With one or two exceptions only, pollution accidents have never caused third party damage valued above that seen in air crashes or industrial fire or explosion [36]. It should therefore be possible to provide the same kind of third party liability cover as that available for wide-bodied airliners.

Table 4.13. Third party damage: number of incidents recorded, 1959–1984.

Type of incident	*Number of incidents*
Dam	4 incidents with over 1000 deaths
Explosion: factory	3 incidents with over 100 deaths
transport	2 incidents with over 100 deaths
Oil: tankers	5 incidents exceeding $10 M (1983)
offshore	4 incidents exceeding $10 M (1983)
Waste disposal	13 incidents exceeding $20 M (1983)
Air pollution: plant	3 incidents exceeding $10 M (1983)
Water pollution	7 incidents exceeding $10 M (1983)
Radioactive pollution	1 incident exceeding $20 M (1983)

Theoretically there is no limit to the amount of environmental damage that a company could cause, because some more serious accident can always be imagined. Bhopal confirms that an accident can always be on a larger scale than any previously experienced. However, it is observed that serious damage only rarely exceeds certain amounts, which can be put at a few million dollars for accidental water and air pollution, some tens of millions for oil tanker pollution, old waste dumps, and toxic pollutant leaks, and several hundred million dollars for major disasters (dam bursts, explosions, large-scale toxic gas leaks). But at the same time it *has to be recognized that environmental damage caused by certain industrial activities cost tens if not hundreds of millions of dollars* per incident. These figures will probably increase over time, partly because of inflation and because firms are getting bigger, and also because more types of damage are being compensated, awards are becoming larger, and it is increasingly easy to identify victims.

The high cost of this damage could seriously embarrass most firms, though the problem should not be exaggerated. Losses in the form of destruction of plant, compensation payable to injured employees, and other direct and indirect financial losses may well far exceed the civil liability risk in respect of pollution [37]. When these items are covered by insurance, the premium for third party damage will be only a modest proportion of the total premium, especially that for damage to plant [38].

4.2.8. Disaster costs as a proportion of the total cost of environmental damage

The high cost of damage caused by certain industrial firms to the environment sometimes gives the impression that it would be financially

Table 4.14. Third party damages: worst-case costs, 1959–1984.

Type of Incident	*Costs*[a] (1983 $ M)	*Third party deaths*	*Evacuations*
Dam	600 (Tetons)	15 000? (Morvi)	–
		2 118 (Vajont)	–
Earth slip	–	144 (Aberfan)	–
Explosion:			
plant	12.2 (Signal Hill)	452 (Mexico)	3 000 (Flixborough)
pipeline	–	508 (Cubatao)	3 000 (Callao)
transport	103 (Eagle Pass)	216 (Alfaques)	2 500 (Hagerstown)
Oil:			
tankers	120 (Amoco Cadiz)	–	–
offshore oil	120 (Santa Barbara)	–	–
coastal reservoir	200 (Mitzushima)	–	–
Waste disposal	84 (Lekkerkerk)	31 (Morocco)	2 500 (Love Canal)
Soil pollution	30 (Times Beach)	–	–
Sedimentary pollution	64 (Tokuyama)	–	–
Air pollution:			
plant	138 (Seveso)	2 500 (Bhopal)	17 000 (Taft)
transport	57(US Railroad)	–	240 000 (Mississauga)
Water pollution	300 (Chisso)	250 (Chisso)	–
Radioactive pollution	33 (Three Mile Island)	–	–

[a]The cost figures are compensation for work already carried out. The costs were corrected for inflation in the incident country and then converted into 1983 US dollars.

difficult to insure and provide compensation for the consequences of such events. In fact, available statistics show that the total financial burden of major disasters is not all that large a proportion of total accidental damage to the environment.

As an example, for tanker spills (*Table 4.1* and *Figure 4.1*), those costing more than $60 million represent 28% of total oil spill costs. For natural disasters, those involving over $400 million represent 27% of incidents costing more than $50 million [39].

For other kinds of disaster, e.g. serious fires [39], those costing over $100 million represent 35% of the cost of all disasters costing over $25 million. The proportions are similar for deaths caused by fires and explosions, and for accidents in mining, at sea, and on the railways [38].

Table 4.15. Some large-scale[a] evacuations, 1967–1984.

Date	*Place*	*Number evacuated*	*Enterprise*	*Product*[b]	*Comments*
1984	Mexico	300 000	Factory	LPG	452 deaths; panic
1979	Mississauga (Canada)	220 000	Railway	70 t chlorine	
1984	Bhopal (India)	200 000	Chemical plant	MIC	Panic after 2500 deaths
1979	Three Mile Island (USA)	5000	Nuclear power station	Radioactivity	Panic and official advice
1969	Glendora (USA)	30 000	Railway	Vinyl chloride	
1980	Somerville (USA)	23 000	Railway (collision)	49 m^3 phosphorous trichloride	3000 poison victims
1982	Taft (USA)	17 000	Plant	Acrolein	
1975	Heimstetten (FRG)	10 000	Warehouse	Nitrogen oxide	
1976	Baton Rouge (USA)	10 000	Plant (natural gas explosion)	90 t chlorine	Mississippi closed for 80 km
1978	Manfredonia (Italy)	10 000	Plant	Ammonia	
1979	Crystal City (USA)	6 000	Warehouse (fire)	Herbicide, pesticide	
1969	Escombreras (USA)	5 000	Refinery explosion	Petrol	4 deaths, 3 injured
1973	Fort Wayne (USA)	4 500	Railway (fire, collision)	Vinyl chloride	

(*cont.*)

[a]Over 2 000 evacuated.
[b]LPG, liquefied petroleum gas; MIC, methyl isocyanate.
Source: UBA (1983).

Table 4.15. (*cont.*) Some large-scale[a] evacuations, 1967–1984.

Date	*Place*	*Number evacuated*	*Enterprise*	*Product*[b]	*Comments*
1979	Crestview (USA)	4500	Railway (derailment)	Ammonia, chlorine	14 poison victims
1980	Barking (UK)	3500	Plant (fire, explosion)	Cyanide	12 poison victims
1978	Youngstown (USA)	3500	Railway (sabotage)	30 t chlorine	8 deaths, 114 poison victims
1974	Flixborough (UK)	3000	Plant (explosion)	60 t cyclohexane	28 deaths, 104 houses destroyed, $70 M damages
1984	Callao (Peru)	3000	Pipeline	Tethraethyl lead	
1984	Matamoras (Mexico)	3000	Fertilizer factory	Ammonia	200 hospitalizations
1967	Newton (USA)	2800	Railway (collision)	50 t chlorine	
1968	Hagerstown (USA)	2500	Truck	Propane	
1973	Market Tres (USA)	2500	Railway (collision)	LPG	
1978	Love Canal (USA)	2500	Waste dump	Various	
1984	North Little Rock (USA)	2500	Railway	Ethylene oxide	
1973	Greensburg (USA)	2000	Railway (leak)	Chlorine	8 poison victims
1978	Regensburg (FRG)	2000	Plant (fire)	Nitrous oxide	40 poison victims

[a]Over 2 000 evacuated.
[b]LPG, liquefied petroleum gas; MIC, methyl isocyanate.
Source: UBA (1983).

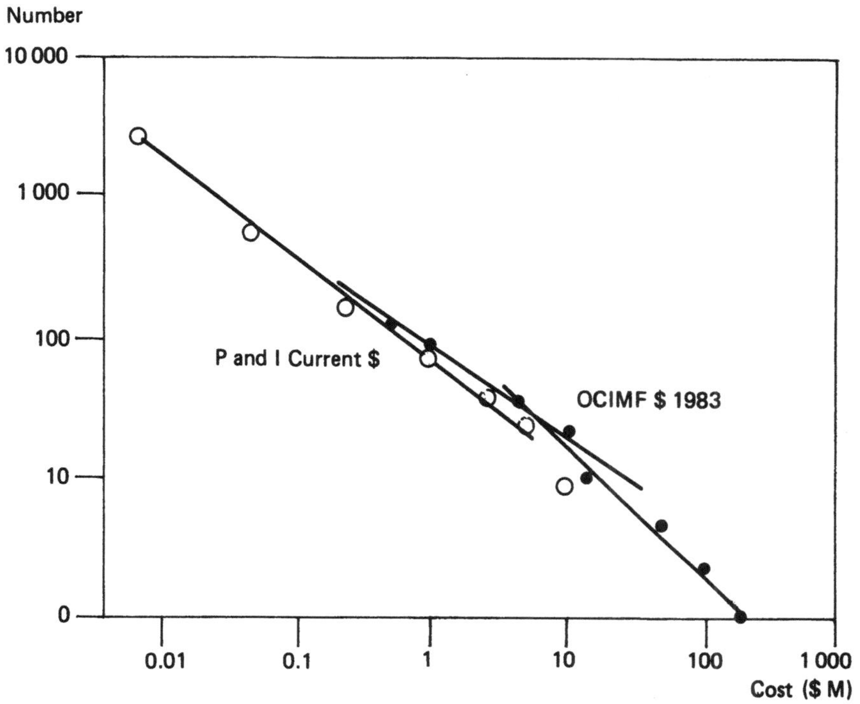

Figure 4.1. Number of oil slicks costing in excess of $*x* million.

Statistics available for the frequency and scale of the various kinds of serious accident [39, 40] show that frequency diminishes in proportion to severity (*Figures 4.1* and *4.2*) and that the financial weight of the combination of disasters belonging to the different severity brackets stays roughly constant. This statistical "law" [41] seems to prevail for serious accidents caused by human activity, but does not apply beyond certain severity thresholds where no incident has occurred [42]. And there are other kinds of disaster to which the "law" does not apply – in particular, natural disasters for which the loss of life associated with the most serious and rarest cases far exceed the losses for all other cases.

For accidents obeying the statistical "law", the financial weight of the maximum severity class (i.e. accidents ranking as disasters) equals the weight of each of the lesser severity classes. If there are three severity classes, the disaster-scale accidents represent 33% of the combined costs of accidents in all three classes. Since accidents can generally be classified into more than three severity classes, it follows that ***disaster-scale accidents cost less than 25% of all accidents***. So it follows that compensation for disasters introduces no important economic constraint into compensation systems and that the cost of

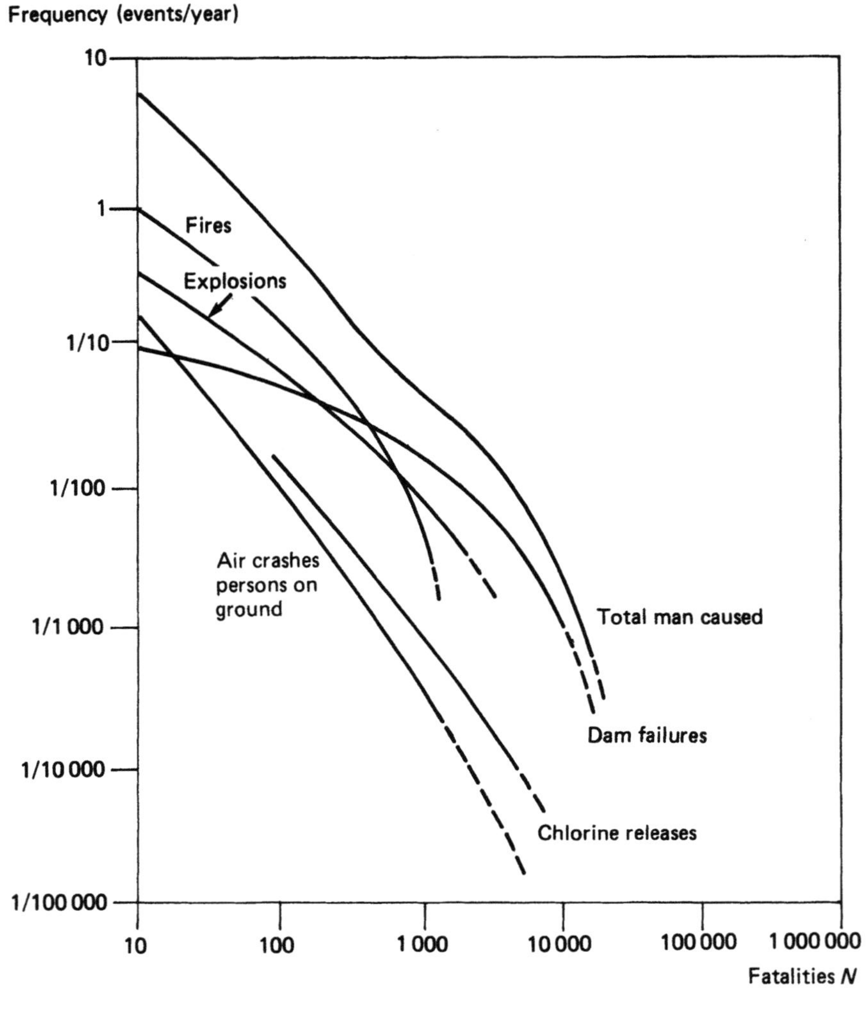

Figure 4.2. Frequency of fatalities due to man-caused events (accidents per year killing over N people). (*Source*: Rasmussen, 1974.)

compensating all accidents is not much higher than the cost of compensating most accidents. Hence compensation ceilings for third party damage have no intrinsic economic justification although they facilitate management of the risk insured and make it possible to satisfy a maximum of claims of less cost. Nevertheless, they are necessary to prevent an exceptional or unexpectedly costly event from ruining an insurance system based on a certain distribution of accidents.

This justification of ceilings in actuarial terms applies in the first place to the insured who can choose higher or lower liability coverage but should not be used to reduce compensation for victims in the most dramatic cases of a disaster-scale accident. Nevertheless, in certain disasters the state may have to stand in for the person liable, pay compensation to victims, and prevent the disaster from ruining the firm. Here some mechanism for state aid to firms (industrial disaster funds) would be better than setting a ceiling on their liability, thereby prompting victims to look to the state for compensation not paid by the party liable, or even to claim against financially powerful firms whose liability may not be obvious.

4.2.9. Economic impact of the cost of pollution accidents

All the available indications show that the total cost of *compensation for environmental damage* caused by industry is no more than a *small fraction* of the cost of measures to prevent environmental damage [43].

For instance, compensation for tanker spills is about $140 million per year worldwide, whereas prevention measures cost over $1 billion per year. In other words, the cost of damage is about 15% of the costs of prevention.

The very high cost of cleaning-up older waste dumps in the United States (over $1 billion per year) is about 10% of industrial waste disposal expenditure in the United States, but no more than a few percent of US industry pollution prevention costs. In France, the annual cost of cleaning up older dumps is apparently well under FF 25 million, while French industry spends FF 2.9 billion on waste disposal (including some FF 600 million for toxic wastes). The economic impact of older dumps in France is therefore ten times less than the impact in the USA.

The total cost of compensation for accidental air and water pollution in France is thought to be some tens of millions of francs per year [44] as against the FF 7.2 billion per year (1983) French industry spends on preventing air and water pollution. Damage accordingly costs less than 1 % of the cost of preventing pollution [45]. The cost of accidental water pollution damage in France represents only a few percent of the pollution charges collected by the French Water Basin Agencies [46].

In Japan in 1982, the total cost of compensation for air and water pollution is thought to have been about Y 106 billion, i.e. 16% of the Y 648 billion Japanese industry spends on pollution prevention. This compensation cost is inflated by the very high amount paid to air pollution victims which would be borne in part by social security in France (health effects).

It seems fair to conclude that *the cost of pollution damage to industry is low compared with the costs of pollution prevention.* In practice, many polluters seem to take compliance with new pollution control standards much more seriously than possible liability for failure to comply with such standards. The risk of stricter controls and more stringent operating conditions affects the decision-maker more than possible compensation liability.

OECD studies of the impact of pollution control costs upon various industries show that for the most polluting industries they only represent between 5% and 15% of production costs and that for other industries the figures are still lower. Taking the highest figures for compensation and pollution control costs, the compensation cost represents 2% at most of the production cost. A more realistic figure would be between 0.1% and 1% of production costs. Thus, in maritime oil transport, total compensation is 1% of the shipping cost and 0.1% of the oil cost. This means that doubling compensation would have very little economic influence on the price level.

4.2.10. Pollution disasters and natural disasters in industrialized countries

The total cost of pollution disasters seems very much smaller than the total cost of natural disasters in the industrialized countries. For the USA in 1983, compensation and insurance paid out for natural disasters amounted to $2.55 billion while total compensation for environmental damage to third parties due to industrial activities in 1983 in the USA was probably well below $400 million (the costs of cleaning up older dumps are not included in that figure because the damage had been caused previously). France has in ten years had only two very costly pollution incidents (oil spill), at a total cost of some FF 1.5 billion. On an annual basis, pollution disasters therefore cost FF 150 million as against natural disasters which cost insurers some FF 3 billion.

Effective compensation for pollution disasters would therefore not appear to be an unreasonable aim, but it would require special government action. Victims might receive compensation from insurance companies, which would pass on the cost to industry (firms' civil liability, industry charge, etc.). Such a system might cost up to FF 300 million per year in France if it were to include all accidental industrial environmental damage. The financial impact of such a system would be the same as a 5% increase in the cost of preventing industrial pollution (0.01% of GNP). Compensation for such damage would appear to be financially supportable to the extent that appropriate machinery were introduced [47]. Initially, guaranteed compensation could perhaps be limited to FF 200 million per incident, including FF 100 million for private victims. In the USA insurance cover was available for third party pollution liability up to a ceiling of $30 million per event (*Table 4.16*). In Europe, insurance pools offer smaller coverage for

Table 4.16. Maximum cover on leading insurance markets for nonsudden environmental damage in the USA and Canada.[a]

	Maximum cover (1983 $ M)	
Insurer	*per incident*	*per year*
Shand, Morahan, & Co., Inc.	30	60
London (through Alexander Howden)	30	60
Stewart Smith, Inc.	25	50
Hartford Steam Boiler and Inspection Co.	20	40
American International Group	20	20
Swett and Crawford Management Co., Ltd.	20	20
Dryden & Co., Inc.	15	30
Hartford Insurance Group	10	10
Pollution Liability Insurance Association	–	6

[a]1982 premium volume was $28 M; while 1983 premium volume was $40 M. The 1984 situation is distinctly less favorable: less cover is available (about $10 M) and premiums have soared.

Source: Risk Science International, *Financial Responsibility under the Comprehensive Environmental Response Compensation and Liability Act of 1980, Summary of Issues,* June 1983.

environmental damage (in France FF 20–30 million, in the Netherlands G 7.5 million).

4.2.11. Conclusions

Compensation for serious environmental damage does not substantially increase the total cost of compensation, since the latter represents only a small proportion of the cost of pollution control, and the cost of pollution control has a minimal impact on prices. Accordingly, compensation has no appreciable economic impact and there is *no economic argument for depriving victims of compensation for damage caused by industrial activities*. A substantial improvement to compensation systems is therefore an economically feasible target.

However, the high cost of certain accidents and pollution situations can pose difficulties, because firms could not easily pay compensation of millions of dollars. Costs of severe damage would therefore have to be *shared* among potential polluters, consumers, and taxpayers to avoid their being borne partly by the victims or ruining the firm responsible.

4.3. Compensation for Exceptional Environmental Damage

4.3.1. Compensation for serious pollution

Over the last 30 years, compensation arrangements have improved for certain specific kinds of pollution and for environmental problems generally. The main improvements have been in the rules governing liability, in authorizing collective legal action, and in extending the range of damage eligible for compensation.

In most countries, liability for pollution varies according to the type of pollution. For instance, liability for pollution of fresh water in France is strict and unlimited, while for maritime pollution by tankers it is strict but limited. Air pollution liability is based on a fault, but if the pollutant is highly toxic, liability is based on risk. In some countries no-fault liability for environmental damage has become the norm. Other countries have created an obligation to clean up and dispose of pollutants upon administrative order (with no financial limit) but have not changed the civil liability system. Lastly, it has been recognized that the state might be required to compensate victims when the damage results from its own default.

For the most serious kinds of pollution, the main improvement has been to ensure that victims are fully compensated notwithstanding the high cost of the damage. For this purpose, insurance has been made compulsory, compensation funds have been established (industry-wide or at budget level), and parent companies have been made liable for the obligations of their subsidiaries.

Compensation Ceiling Under Sharing Arrangements

Compensation payable by firms is necessarily limited by their financial capacity, taking account of the insurance cover available to them from an industry pool or via the insurance and reinsurance market. These risk-sharing systems can function in a balanced way provided total claims do not fluctuate unduly, i.e. provided that a very costly claim does not overshadow total claims arising over a period of time. This means that no pool system can cover the very rarest and most costly events [48]. In other words, there will always be a compensation ceiling [49] and the higher the total cost of damage covered (spreading the risk among a great number of insured) the higher the ceiling will be.

For tanker spills, maritime insurers in 1983 felt able to provide cover against civil pollution liability up to $50–100 million, i.e. an amount equal to the total oil pollution damage claims paid out by insurers over about two years. The maximum insurable compensation might be attained about once every five or ten years [50].

Applying the same principle to other types of accidental pollution, compensation ceilings could be determined for pools. Beyond sums between $100 million and $1 billion for damage caused by industrial activities to the environment, the state would probably be prompted to intervene financially if it wished to guarantee compensation for victims (disaster funds, etc.) because industry would not succeed in obtaining insurance cover or organizing to share the costlier risk [51].

Compensation ceilings are linked to economic magnitudes, which are themselves determined by the cost of living. Accordingly, a ceiling can vary from year to year. Although it is fixed at a particular moment in accordance with a particular insurance market, it should not be allowed to assume the status of a legal norm, nor to become an instrument for gradually lightening the liability of polluters to the detriment of victims or the state. This unfortunately occurs when the law limits liability (e.g. nuclear and oil liability) and even more so in an international convention which can only be amended with the agreement of all parties.

In the following sections we examine measures to provide better compensation for victims in the event of a major incident occurring. We start with nuclear energy and oil, where problems first arose.

4.3.2. Compensation for damage caused by radioactivity

In 1960 the first international system specially designed for large-scale environmental damages was adopted, under OECD auspices [52]. The system applies to accidents caused by nuclear plant. It established a strict liability regime for Europe, concentrating on the operator, with limitation of liability to $15 million. For damages between $15 and $70 million the state steps in, and for damages between $70 and $120 million states party to the Supplementary Brussels Convention (1963) would provide compensation.

The Paris and Brussels Conventions were revised in 1982 but the Protocols will not come into force for several years. When they do, maximum compensation payable by the operator and the state will increase from Special Drawing Rights (SDRs) 70 million to 170 million and compensation payable by states will be within the band of SDRs 175–300 million instead of SDRs 70–120 million.

In the USA, the operator's liability is currently limited to $595 million of which $160 million is covered by insurance and $435 million by the reactor operators [53]. In the FRG, operators are liable up to DM 500 million (of which DM 300 million is covered by insurance) and the government covers compensation between DM 500 million and DM 1 billion. In Switzerland, the operator's liability in the event of accident is up to SF 300 million and for damage in excess of SF 1000 million, while the state covers the SF 300 million to 1 billion band. *Table 4.17* shows the wide diversity in compensation and liability ceilings for OECD countries.

Switzerland abolished the limited liability system and the FRG recently did the same. In the USA the compensation ceiling will shortly be revised upwards to at least $1 billion, i.e. the value of the plant (for a group of some 80 nuclear power stations). In Japan, the operator's liability has never been limited.

Table 4.17. Maximum amounts of compensation for large nuclear installations, by country.

Country	*Nuclear Operator*: *Maximum liability and required financial guarantee* (national currency unit, M)	*Nuclear Operator*: *Approximate equivalent equivalent in million US dollars*[a]	*Compensation ceiling with additional state intervention* (national currency unit and US$ M)
Austria	500 (1964)	23	
Belgium	1 000 (1980)	16	
Canada	75 (1970)	57	C$150 (US$113) unless government decides otherwise
Denmark	75 (1974)	6.7	SDRs 120
Finland	42 (1972)	6.5	SDRs 120
France	50 (1968)	6.3	FF 600 ($63)
FRG	Unlimited liability, 500	160	DM 1000 ($323)
Italy	7 500 (1975)	3.9	L 43 750–75 000 ($22.8–39.3)
Japan	Unlimited liability, 10,000	41	To be decided by Parliament
Netherlands	200 (1984)	57	G 1000 (1979) ($286)
Norway	70 (1972)	77	SDRs 120
Spain	350 (1961)	2	SDRs 120; appropriate measures to be taken by government
Sweden	500	56.5	SKr 3 000 ($102)
Switzerland	Unlimited liability, 300	118	SF 1 000 ($393) plus additional sum if necessary
UK	20 (1983)	24	SDRs 300 (to be decided by Parliament)
USA	160	160	$595

[a]Value on 30 November 1984.

Severe damage due to radioactivity has occurred in circumstances not covered by the special nuclear energy regime and in particular as a result of the dumping, loss, or theft of radioactive substances. In these cases the ordinary liability rules apply with all the uncertainties they may entail, especially if the persons liable are unknown or insolvent.

4.3.3. Compensation for damage due to the transport of oil

In 1969 and 1971 special liability and compensation rules were introduced for the sea transport of oil [54] based on the earlier arrangements for nuclear energy. *Table 4.18* compares the two systems and shows that, for the same annual turnover, maximum liability and average risk are also similar. In the long run, maximum compensation payments will be similar too. More surprising is the fact that the number of serious accidents per year of operation is the same for tankers and nuclear reactors.

Features of the 1969 system are the shipowner's strict liability limited to SDRs 14 million for large tankers and a smaller amount for ships of under 105 000 t registered tonnage. Not all liability is channeled exclusively to the shipowner, and victims may therefore seek compensation from the actual operator, the salvager, the shipbuilder, or the state. This system was supplemented in 1971 by a compensation fund financed by oil importers in the contracting countries. Since 1979, the compensation ceiling has been SDRs 45 million irrespective of ship size. Shipowners from countries that have ratified the Convention setting up the fund have had their maximum liability reduced from SDRs 14 million to 8.3 million. It may be noted that this system is more in the nature of an insurance for the damage to the oil importing countries themselves than supplementary liability insurance paid for by importing countries for the benefit of victim countries.

The liability and compensation system set up in 1969 and 1971 entered into force in 1976–1978 and has proved satisfactory in every case except for the two oil spills where a compensation ceiling was greatly exceeded, this being a consequence of the steady erosion of the value of the amounts set in 1969 and 1971 for liability and compensation. In 1984 these were worth less than a quarter of their 1969 value.

Given the inadequacy of the compensation normally available for the *Amoco Cadiz* oil spill in 1978, victims sought compensation outside the terms of the 1969 Convention and took their case to the Court of Chicago which held that liability for the accident lay not only with the operator, Amoco Transport, but also with the parent company, Standard Oil of Indiana. What happened in this case greatly influenced the

Table 4.18. Comparative liability and compensation for oil tankers and nuclear reactors (1984 SDRs M).

	350 000 t oil tanker	*1200 MW nuclear reactor*[a]	*Reactor/tanker ratio*[a]
Total loss value[b]	100	1100	11
Annual turnover	150	150	1
Maximum operator's liability[c]:			
in force 1984	14	5–15	1
decided (subject to ratification)	59.7	15	0.25
Maximum compensation[c]:			
in force 1984	45	120	2.7
decided (subject to ratification)	200	300	1.5
Total damage caused by 100 units over the last ten years[d]	25[e]	(18)[f,h]	(0.7)
Number of cases involving more than SDRs 15 M for 100 units over ten years	0.36	(0.58)[h]	(1.6)
Maximum damage over ten years	120[g]	(30)[f,h]	(0.25)

[a]Figures in parentheses are not statistically reliable.
[b]Value of ship when new and of cargo or of nuclear reactor in 1984.
[c]On the basis of the international conventions and amendments.
[d]Damage calculated regardless of any liability or compensation limitations.
[e]Annual average: about $88 M for 3500 oil tankers.
[f]Presumed damage cost for Three Mile Island accident (SDRs 30 M) and 1725 reactor-years over ten years.
[g]Presumed damage cost for *Amoco Cadiz*.
[h]These figures will be much higher when Chernobyl data are included.

revision of the 1969 and 1971 Conventions. Another significant incident was the *Antonio Gramsci* oil spill (5000 t near Riga in the USSR) for which a Soviet court assessed "ecological" damage at 33 million pounds sterling, a figure bearing no relation to assessments that might have been made by courts in Western Europe.

The International Maritime Organization (IMO) Conference adopted two Protocols in 1984 to revise the 1969 and 1971 Conventions. The liability and compensation figures were steeply increased after difficult negotiations during which shipowners (and certain countries) sought to limit the increase in their liability to a figure simply reflecting inflation between 1971 and 1984.

A special procedure for revising liability and compensation amounts was also adopted. It is not an indexing formula and will not allow the amounts to be modified frequently. *Tables 4.19* and *4.20* show that the shipowners' maximum liability will decrease from SDRs 14 million in 1969 to about SDRs 3.2 million by 1991 and maximum compensation from SDRs 30 million to 14.6 million over the same period. Thereafter the position of victims will improve until, at best, amounts equivalent to those adopted in 1969 and 1971 are reached.

The channeling of legal action for compensation has been tightened. In future it will no longer be possible to sue the actual

Table 4.19. Changes over time in shipowners' maximum liability.[a]

Year	*Decision*	*Amounts in force*	*Notes*
Actual situation			
1969	14	0	Signing of the 1969 Convention ($14 M)
1976	9	9	Entry into force of the 1969 Convention
1983	5.2	5.2	
1984	16.9	4.9	Adoption of the 1984 Protocol (increase from $14 M to $48.8 M, current prices)
Possible future situation			
1991	11.3	3.2	
1992	10.6	10.6	Implementation of the 1984 Protocol Start of the revision of the 1984 Protocol
1993	10.0	10.0	
1994	16.9	9.5	Revision of Protocol amount from $48.8 M to $87.4 M at current prices (maximum permitted rate of 6% over ten years)
1996	15.1	8.4	
1997	14.2	14.2	Implementation of the 1994 revision of the 1984 Protocol.

[a]Figures for a 105 000 t tanker in millions of 1969 US dollars, assuming a 6% inflation rate as from 1984.

Table 4.20. Changes over time in maximum compensation (tankers).[a]

	Decision		*Amounts in force*		
Date	*Current prices*	*1969 prices*	*Current prices*	*1969 prices*	*Notes*
Actual situation					
1971	30	27.1	–	–	Signing of the 1971 Convention
1976	34.4	22.2	–	–	Entry into force of the 1969 Convention
1978	37.2	20.9	37.2	20.9	Entry into force of the 1971 Convention
1979	58.1	29.3	58.1	29.3	Revision of the 1971 Convention (× 1.5)
1983	50	18.4	50	18.4	
1984	140–207	48.5–71.8	47	16.3	Revision Protocol
Possible future situation					
1987	140–207	40.7–60.2	63	18.3	Second revision of the 1971 Convention ceiling
1991	140–207	32.3–47.8	63	14.6	
1992	140–207	30.5–45.1	140	30.5	Partial entry into force of the 1984 Protocol Start of the revision of the 1984 Protocol
1993	140–207	28.7–42.4	140	28.7	
1994	250–350	48.5–67.9	207	40.2	Full entry into force of the 1984 Protocol Revision of the 1984 Protocol
1996	350	60.5	207	35.8	
1997	350	57.0	250	41.1	Partial entry into force of the 1994 revision
1999	350	50.7	350	50.7	Full entry into force of the 1994 revision

[a]In millions of US dollars, assuming a 6% inflation rate as from 1984.

operator, the salvager, or the state except in cases of intentional fault. Surprisingly, however, it will be possible to proceed against the shipbuilder. The shipowner will benefit from limitation of liability except in cases of intentional fault, whereas previously the limitation did not apply where there was "actual fault or privity of the owner." To prevent excessive claims, damage qualifying for compensation has been defined more clearly so that courts will no longer be able to proceed exclusively on the basis of the national law applicable. In accordance with the new law of the sea, damage occurring within the economic zone has been made subject to compensation, as has the cost of preventive measures prior to an oil spill.

The Protocols adopted in 1984 will increase victims' compensation when they enter into force in 1990–1992 or later. In the interim, governments will have recourse to the 1969–1971 system and, in some cases, equivalent systems introduced voluntarily by shipowners (TOVALOP) and oil companies (CRISTAL) [55].

National compensation systems have also been introduced, especially in the USA. They guarantee compensation up to as much as $200 million in some instances [56].

4.3.4. Compensation for damage due to offshore oil installations

Offshore installations are governed by the liability rules laid down by the authorities of the country on whose shelf they operate. In the UK liability is strict but limited. Many countries make liability both strict and unlimited, on the grounds that offshore operations are ultrahazardous. It is noteworthy that Norway has dealt with the possibility of an operator's proving insolvent after an accident by insisting that parent companies underwrite the operator (in this connection see the *Amoco Cadiz* ruling).

The purpose of the Convention on Civil Liability for Oil Pollution Damage resulting from Exploration for and Exploitation of Seabed Mineral Resources, which concerns countries bordering the North Sea, is to set up a no-fault liability system with a SDRs 40 million limit for pollution damage and measures to prevent or minimize pollution (with the exception of well-control measures and measures taken to protect, repair, or replace an installation). Not all the countries concerned have signed this Convention, and none has ratified it. This is because, for many of them, ratification would mean that a lower amount of compensation would be available.

Oil companies operating offshore in the North Sea have formed a pool within the framework of the Offshore Pollution Liability Agreement (OPOL). In the event of an accident, OPOL guarantees a maximum of $60 million compensation for pollution damage and preventive measures.

It should be noted that neither of these two systems covers the cost of what has to be done to stop an oil blowout, which is often higher than the costs that are covered. The economic losses suffered by offshore operators through platform accidents have so far been much higher than the cost of damage to the environment.

4.3.5. Clean-up costs for accumulations of dangerous waste

When an accumulation of waste (e.g. in an abandoned dump) is found to be a source of danger to the environment and it is decided to remove the waste for disposal in ways deemed acceptable today, the covering of the considerable removal and disposal costs raises very serious problems. The companies who generated the waste, and those operating the dump who accepted it, may have acted in ignorance of the dangers involved, and sometimes even in compliance with the administrative provisions in force at the time. Those liable, when identifiable, were not always insured for civil liability and their ability to pay is often very limited. Moreover, they could be held jointly liable for waste dumped by others. Given that removal and disposal costs may easily exceed several million dollars for each enterprise involved, governments understandably find themselves obliged to contribute financially to the solution of some of these environmental problems.

In the USA, "Superfund" totaling $1600 million was set up for five years in 1980 under the Comprehensive Environmental Response, Compensation and Liability Act (CERCLA). Figures quoted in the current debate about the renewal of this Act suggest that some $7.5 billion will be paid into Superfund for the second five-year period. Of these resources, 86% is from taxes on chemicals and oil products, and the balance from the federal budget. Its main purpose is to finance the removal and disposal of waste from abandoned sites.

Under CERCLA, any person dumping waste is held to be liable for it, without proof of fault, and must meet any costs incurred by the government in connection with the waste and indemnify the government up to a maximum of $50 million for any damage caused to publicly owned natural resources. Any such person deliberately causing damage or infringing mandatory waste disposal provisions loses the benefit of the $50 million maximum. Should he fail to remove waste when ordered to do so, he can be sentenced to punitive damages equal to three times the cost of removal. Surprisingly, the current CERCLA does not amend the liability rules regarding compensation for damage caused by waste to private persons.

Over the last few years, several large US enterprises have been made to pay for the clearing of old disposal sites at a cost of several million dollars per site. However, it seems that only a quarter of the cost will be met by the enterprises liable, the balance coming from the Superfund.

In other OECD countries, the setting up of a fund to cover the costs of cleaning up old waste dumps has not so far been deemed necessary [57]. When the waste producer can be identified and has the necessary resources, he is often ordered to meet the cost. As for the future, the trend is to prevent dangerous wastes building up and to increase the extent to which the authorities are responsible for dumping and other disposal facilities. In the USA, moreover, operators of waste disposal facilities are obliged to meet their liability (including all clean-up costs outside their property) up to a ceiling of $1 million for each sudden incident, and $3 million for each nonsudden incident.

One unanswered question is whether a waste producer effectively transfers all liability attaching to the waste when he hands it over to a government-approved disposal agent. Another is how to organize viable funding systems to guarantee compensation, after site closure, for environmental damage caused by dumps operated in conformity with the legislation and regulations. One such fund has been set up in the USA and financed to the sum of $200 million by a tax on waste delivered to dumps.

4.3.6. Compensation for damage caused by air pollution

Damage to health by air pollution is covered in many countries by sickness insurance (social security). In Japan, where such systems are less developed, a serious problem has arisen in highly polluted industrial areas. Legislation has been enacted to provide compensation for victims in certain areas by means of a tax on industrial sulfur emissions (80%) and on car emissions (20%). The yield rose from Y 35 billion in 1975 (20 755 compensated victims) to nearly Y 100 billion in 1982 (85 541 compensated victims).

As it can sometimes be difficult to establish a causal link between air pollution damage and the person responsible, the Netherlands has established a compensation fund to cover air pollution damage which would not otherwise be compensated. Up to now, this fund has intervened only very rarely and for modest amounts [58].

4.3.7. Compensation for damage due to noise

Compensation for people living near airports has been awarded on the basis of no-fault liability on the part of the airlines or because of abnormal neighborhood disturbance of the airport (e.g. Orly in France, Osaka in Japan).

A more effective method is to pay for soundproofing and provide compensation for erosion of property values out of public funds or under special legislation (Land Compensation Act in the UK). Compensation funds financed by noise levies have been introduced in France, in the UK (Manchester), the Netherlands, Switzerland, and Japan [59].

4.3.8. Conclusions

Because of the scale of environmental damage which could be and has been caused by radioactivity, oil slicks, discharges of dangerous waste, air pollution, and noise, some governments have established *special*

compensation systems to protect both the industries concerned, and public and private victims. These systems, which rely on insurance and contributions by government or the industries concerned, should make it possible to collect larger sums and to compensate victims more rapidly without having first to prove that any particular party has been at fault.

These arrangements could be extended to other countries and other problems, so that, in the event of serious environmental damage, victims should receive compensation without delay and without having to rely on government help (except in certain special circumstances). In other words, one could try to implement the "polluter pays" principle more fully, so that the cost of the damage is borne by the firm responsible for it or, rather, by all firms in the industry concerned, by sharing out the risks among them all. The risk can be shared out in this way via insurance and reinsurance [60] or via a fund constituted from levies on the products of the industrial sector concerned (*Table 4.21*).

To facilitate compensation, it would be necessary for government to step in to introduce the industrial levy designed to feed a compensation fund, and to insist that polluting plants have full insurance cover for civil liability. This is because experience has shown that some firms have been unable to pay up in the event of costly pollution, while some have gone out of business. Experience has also shown that victims may in some cases not obtain satisfactory compensation, through inability to identify the polluter or to establish the chain of causality, and may have great difficulty in obtaining the funds necessary to pursue their claim.

The problem with major disasters is that they impose a heavy if not impossible financial burden on the firm liable, and on victims and perhaps even insurers who may not have sufficient finance. Consequently, government may be called upon to intervene to save the firm or help victims. Such intervention may be justified if the government has tolerated or authorized a dangerous activity without providing for compensation. It may also be justified if the disaster is so great as to exceed existing compensation ceilings [61]. Lastly, it may result from the fact that a government may sometimes implement measures which are not necessarily justified but correspond to a "political" need; or the government may not have been able to evaluate the dangers of a particular form of pollution or land use which only proved to be dangerous several years later.

To reduce the frequency of government involvement and to prevent the taxpayer from having to bear costs which should be met by those benefiting from the technology concerned, the government might encourage the introduction of systems based on insurance and reinsurance to guarantee compensation for environmental damage of up to

Table 4.21. Civil liability in certain high-risk activities (up to 1984).

Type	*Number worldwide*	*Maximum third party liability*	*Premiums*
Nuclear reactors	USA 82 Other countries 182	$595 M $16–383 M Max. damages paid: $33 M	Annual premium for $10 M liability coverage: *ca.* $0.1 M/reactor
Oil tankers	3127 over 10000 DWt	SDRs 45 M Max. damages paid[a]: $50 M of 1980 (*Tanio*)	Annual premiums: shipowners, *ca.* $50 M; oil importers, *ca.* $50 M
Oil rigs	700	Pollution: $60 M or unlimited Max. damages paid: $30 M (Santa Barbara)	
Civil aviation	8950 aircraft	$100000/passenger or unlimited (USA) Max. damages paid: $100 M/crash	Premiums: $185 M/year for civil liability and $440 M/year for hull insurance
Satellites	Several dozen insured per year	Unlimited Max. damages paid: $3 M (*Cosmos*)	1975–1982 accidents: $192 M 1983 accidents: $85 M 1984 accidents: 3 ($300 M) Premiums: 10–20%
Dangerous plant (pollution)	6000 to 12000 according to definition (according to Seveso Directive 1500 in the EEC)	Unlimited Max. damages paid: $300 M (Chisso)	EIL[b] Insurance in the USA Premiums: $40 M in 1983, $65 M in 1984 Max. cover: $30 M per accident in 1983, $20 M in 1984
Dams	10000	Unlimited Max. damages paid: $600 M (Tetons)	Accident frequency: 2×10^{-4} to 2×10^{-5} per year

[a]Compensation to be paid for the *Amoco Cadiz* could be higher.
[b]EIL, environmental impairment liability.

several tens or hundreds of millions of dollars per incident. No doubt, this kind of initiative will only be taken when public opinion realizes that the present situation is rather unsatisfactory and that serious damage can be caused by firms with limited financial resources.

In the meantime, widely varying compensation systems are to be found. Some may require insurance cover for unlimited no-fault liability while others give victims of serious pollution little chance of obtaining compensation.

A first step forward would be to authorize "dangerous" activities only on condition that satisfactory liability cover is obtained (financial security or insurance) especially for firms with only moderate financial resources [29]. A second move would be to harmonize liability systems by the general adoption of no-fault liability which already applies for certain types of pollution in all countries [62].

Since such developments would impose extra costs on those potentially liable and would run counter to certain "traditional" legal principles, they will come about only very gradually and perhaps only at international level so as to share out the high risks involved. Yet there is no argument based on economics or equity to refute the principle that those who create a hazard should take responsibility for it, given that the small extra cost involved will in any case ultimately be borne by the consumer.

Notes

[1] The author is solely responsible for the opinions expressed here.

[2] On technological risks, see Lagadec (1979; 1981a,b).

[3] Many accidents causing environmental damage seriously impair the health of workers and users, and facilities themselves, but those consequences are not examined here. Aspects not discussed include decontamination and decommissioning of polluting installations, and compensation for workers and members of the armed forces affected (e.g. asbestosis, which has caused 50000 casualties in the USA with possible compensation of between $5 B and $30 B; agent orange; $180 M for US civilian and armed forces victims of nuclear tests in Nevada, USA); compensation for victims of thalidomide, mercury wheat in Iran, toxic oils and adulterated foods in Spain and Japan, and Bendectin in the USA ($120 M); intrauterine device (IUD) users (7700 cases, $259 M); gas distribution explosions; damage by ships to port facilities; etc. Similarly, we do not consider damage caused by volcanic eruptions, floods, hurricanes, road noise, acid rain, or diffuse pollution due to phosphates, nitrates, or pesticides, excess mortality in polluted cities, the purchase of the right to pollute in fishing areas, expropriation prior to the construction of roads or airports, etc.

[4] FF, French francs.

[5] Very full statistics on tanker accidents have been drawn up by the P and I Clubs (Revision of 1969 Civil Liability Convention and 1971 Fund Convention, IMO–LEG, February 1984). Studies of these statistics have been produced by the OCIMF (position papers, IMO–LEG documents, 1982, 1983, 1984) and by the author (Smets, 1984).

[6] The figure of 1983 $120 M for the *Amoco Cadiz* spill compensation reflects the following estimates:

FF 800 M, Minister for the Sea, France (*Le Monde*, October 1982).
FF 1 B, Environment Ministry Official (*Int. Env. Rep.*, 1984, p 188).

The original claims were much higher but do not seem to fit in the framework of this study. French claims in the *Amoco Cadiz* case (*Le Monde*, 13 March 1985) are given in *Table 4.22*. According to Standard

Oil of Indiana (*Wall Street Journal*, 6 March 1985) the total claim is only for $665 M.

Table 4.22. French claims in the *Amoco Cadiz* case.

Government	1978 FF 478 M	$263 M
Various private claims	–	$218 M
"Syndicat mixte"	1978 FF 980 M	$287.8 M
Department of Côtes du Nord	1978 FF 13.7 M	
Department of Finistère	15.9	
43 Côtes du Nord communes	339.5	
47 Finistère communes	454.8	
Sea fishermen	57.2	
Shellfish producers	75.5	
Nature associations	6.3	
Tourism industry	16.3	
TOTAL		$769 M (1985)

[7] See National Oceanic and Atmospheric Administration (1983) and OECD (1982).

[8] Pollution of the Niger delta in 1980 (270 000 bbl of oil after an explosion) involved an oil slick covering 100 km of coast and extending 30 km inland, seriously affecting a population of 250 000 dependent upon fishing. The Nigerian Government provided about $6 M in aid and Texaco voluntarily gave $1 M [*Earthscan*, **3** (8), 1980].

[9] In the USA, the cost of cleaning up old dumps is thought to be between $8.4 B and $26 B (see *Pesticide and Toxic Chemical News*, 1984, p 5). In December 1984 the US Environmental Protection Agency estimated that the approximately 1800 priority sites would by themselves cost the government $12 B to clean up and anticipated that those responsible would pay something of the same order. The clean-up cost is expected to rise from $290 M per year to $1 B per year. In 1985 the US Government proposed a new $5.3 B Superfund for five years, two-thirds of which would be financed by a levy on wastes.

[10] According to the Dutch Minister, Mr. Winsemius, clean-up costs for hazardous dumps in the Netherlands will amount to several hundred million dollars and the Netherlands Government decided to spend $33.3 M in 1983 (*Int. Env. Rep.*, 1983, p 390). At Lekkerkerk, 270 houses were evacuated gulden (G 200 M); at Dordrecht 106 houses will have to be demolished and the total cost of the operation may reach G 750 M. At Gouderak, 99 houses will have to be demolished (333 people, G 20 M) and soils will have to be shifted (G 100 M) (*Ambio*, **13**, 1984, p 71. The Netherlands is expected to spend G 200 M per year over 15 years to eliminate old dumps. In 1984, the Netherlands Government had already spent G 600 M to clean up older dumps and had commenced legal proceedings against Duphar, Shell, Phillips, and Argrunal amounting to a total of G 125 M.

[11] According to the French Environment Ministry (*Les anciens dépôts industriels, May 1984*), the cost of eliminating older dumps (of which there were 62 in 1978) varies from a few tens of thousands of francs to

several hundred thousand francs. Two dumps (Carling and Mulhouse) will cost several million francs. The same report cites the following clean-up costs: Lillebonne (FF 5 M), Saint Sébastien (FF 4.8 M), Mulhouse (FF 3.7 M), Dannemarie (FF 3 M), Abscon (FF 1.5 M), Mailleraye (FF 1.2 M), Traitt (FF 1.2 M), Montjalin (FF 1.2 M), Gravenchon (FF 1 M), Roumazière (FF 3 M). The total of cases exceeding FF 1 M amounts to some FF 29 M for a period of about five years (i.e. an annual cost of over FF 6 M). Overall, the 92 older listed dumps may represent costs of the order of FF 50 M (at most FF 10 M per year). Such costs are very low in comparison with the costs of eliminating toxic wastes in France (FF 600 M a year). However, the situation is not very satisfactory because, in France in 1984, the official services were aware of the destination of only 1.1 M out of the 2 M tons of dangerous waste produced (*Presse Envir.*, 28 September 1984).

[12] See *Int. Env. Rep.*, 1979, p 851 and 1984, p 137, The Sumitomo Mining Company case.

[13] Opening Speech at the OECD Conference on Environment and Economics, 18 June 1984.

[14] According to Goubet (1979), the total risk of dam bursts in Western Europe and Japan is calculated at 0.2×10^{-4} (1 for 50000 dam-years) mainly due to the risk of submersion. According to Goubet, the burst costing the most lives before 1979 was at South Fork River, Pennsylvania, which is thought to have killed between 2000 and 4000 in 1889. Other older disasters have been at San Francis (400–2000 killed, 1929), Iruka (1200 killed, 1868), Möhne (1200 killed, 1942). Between 1966 and 1979, 17 of the 21 dams to have burst in the world were of earth and only one was in Europe (Spain).

[15] For accidents caused by chemicals in factories and transport, see Andurand (1979). In 1958, a refinery accident at Signal Hill (California) caused damage to neighboring properties estimated at $3.5 M. In 1974, an organic peroxide fire in a Los Angeles warehouse caused $5 M in damage to the neighborhood. In 1973, a liquefied petroleum gas fire at a Staten Island plant caused 40 deaths and $5 M worth of damage.

[16] Of the accidents involving the road transport of hazardous materials, those at Los Alfaques and Xilatopec were far more severe than the accidents at St Amand Les Eaux, France, 1973 (4 deaths, 2 missing, and 37 casualties), Liévin, France, 1976 (6 killed, 20 hospital casualties), and Martelange, Belgium, 1973. Los Alfaques and Xilatopec must be regarded as exceptional in comparison with all road transport accidents in which the actual cargo caused high damage to third parties. But they appear less exceptional if all transport accidents are considered, some of which do cause heavy loss of life (e.g. buses falling into ravines, vehicles out of control in a crowd). The most costly road transport accident in the USA seems to have been at Eagle Pass, Texas, where a liquefied petroleum gas tanker exploded (17 deaths, 34 casualties) and caused $50 M worth of damage.

[17] The Indian tribe St Régis claimed $50 M compensation for fluorine pollution by the Reynold Metal Co. and Alcoa plant on the Cornwall Isle of the St Lawrence (*Int. Env. Rep.*, 1980, p 101).

[18] In 1928, in Hamburg, 10 t of phosgene escaped from storage in a chemical plant. Damage to the vegetation extended up to 14 km from the

plant owing to a large 2 km cloud, causing 11 deaths and 200 poisoning casualties – a forerunner of the Bhopal accident.

[19] At Mississauga, economic losses are thought to have amounted to $25 M per day of evacuation. The accident caused the compulsory evacuation of 75000 families of which 71200 stayed with relatives or friends and 3800 in reception centers [see the study by P. Lagadec (1984), based on the final report of Burton *et al.*, 1983]. Canadian Rail Pacific is thought to have paid $10 M to the evacuees and the accident to have caused $70 M worth of damage (mainly economic losses due to the shutting down of activities). The costs of cleaning up an acrylonitrile spill from a train in Indiana in 1977 amounted to $1 M. A derailment in Texas causing a propane fire entailed damage of $4 M to properties in 1977. (See Swaigen, 1981.) When ammonia poisoning killed two people in their own homes as the result of a derailment on the Louisville and Nashville Railways in the USA the compensation paid was $52 M (Pfennigstorf, 1982).

[20] See *Int. Env. Rep,*, 11 April 1979 and 14 March 1984.

[21] The Seveso accident (1.6 kg of dioxin) contaminated an area of 1800 ha. The population affected was 37235 (including 735 in the most heavily polluted 110 ha and 4699 in the vicinity). Immediately after the accident 3300 chickens and 12 livestock animals died. Total animal losses were 80000 chickens and 650 livestock.

After the accident the government enacted Laws Nos. 615 (19 August 1975) and 688 (8 October 1976) which provided fiscal facilities for pollution victims, an advance of L 40 B to the Lombardy region, grants to workers, pensioners, farmers, craftsmen, and shopkeepers, and exemption from social security contributions. These were advances on compensation payable with a right of subrogation. The expenditure anticipated in 1977 (*Gaz. ufficiale della Regione Lombardia*, Suppl. No. 28, 14 July 1977) amounted to L 121.6 B (SF 328.6 M), listed in *Table 4.23*. In addition to these were the costs of decommissioning the plant and eliminating the pollution in its vicinity. In

Table 4.23. Breakdown of expenditure on compensation for Seveso.

	(1977 L B)	*(1977 SF M)*
Welfare assistance to displaced persons, accommodation, schools, etc.	15.1	40.8
Contributions to craftsmen, farmers, shopkeepers, and industrialists	13.7	37.0
Health inspection and other medical outlays	18.1	48.9
Improvement of contaminated zone (removal of contaminated earth, provision of fresh earth, elimination of pollution, planting, etc.)	66.4	179.5
Civil engineering	7.8	21.1
General	0.5	1.3
TOTAL	121.6	328.6

Table 4.24. Breakdown of costs to Hoffman-Laroche of third party damages.

	(1984 SF M)
Italian Government	15
Lombardy region (clean-up, health programme)	147
Compensation to residents and firms	58
Compensation to local authorities	25
Miscellaneous	58
Total	303

December 1980, Givaudan (Switzerland), the Italian government and the region of Lombardy reached a settlement amounting to L 103 B. The accident cost Hoffman-Laroche over SF 300 million in third party damages, broken down in *Table 4.24*. Additional costs were for cleaning up of the plant itself and for financial losses (loss of capital, compensation to workers, lost income, etc.). Soil decontamination costs in zone A (the most heavily polluted) amounted to some SF 46 M (60 ha). Restrictions on the agricultural use of 1800 ha lasted six years. In 1984, only a few cases of chloracne persisted and a 40 ha park had been established in the contaminated area. The cost of eliminating plant wastes (41 drums) is thought to be SF 3 M, paid out by Mannesmann Italia which received SF 500 000 for its attempts to recover the wastes after their disappearance (*Int. Env. Rep.*, 11 July 1984, p 220).

[22] The gas that leaked at Bhopal was methyl isocyanate, which is five times more toxic than phosgene and 50 times more toxic than chlorine which were used as gas weapons during World War I. Methyl isocyanate was well known to be very dangerous, having asphyxiated a Bhopal employee in 1981. Apart from the 2500 people killed by asphyxia and other acute effects, the consequences of the Bhopal accident have not yet been completely determined. According to *European Chemical News* (December 1984) and *UNEP News* (January 1985), 200 000 persons are thought to have inhaled the gas, 120 000 to 150 000 have received treatment from 1000 doctors, 20 000 are thought to have been seriously poisoned, and thousands may expect minor eye or lung disorders. Nervous system, liver, and kidney disorders may occur later. The long-term effects are as yet uncertain, but several miscarriages, still-births, and nonviable births have already occurred. Ascertaining what caused the observed lung disorders will be complicated by the fact that the exposed population was not in an excellent state of health to begin with. Large numbers of animals were also killed (over 1500 buffalo, cows, and goats). People in Bhopal panicked both after the disaster and again when the plant was recommissioned. Over 150 000 people are thought to have left Bhopal on two occasions. The toxic gas is thought to have spread over 50 km^2. The damage caused by this disaster cannot be assessed at this stage as it depends to a large extent on the number of handicapped and the extent of their disability, and also on whether Indian or US criteria are used. If compensation is calculated on the basis of current legal practice and salary levels in India, it may not amount to more than a few hundred million dollars. In that case the

Bhopal disaster will not cost the parties liable, or their insurers, much more than compensation paid in the USA by insurance companies after two wide-bodied airliner accidents.

Union Carbide is understood to have up to $200 M civil liability insurance cover, its Indian subsidiary up to rupees (Rs) 25 million. The 51% owned subsidiary operates 13 factories in India, whose turnover is Rs 2.12 B with profits of Rs 150 M, making it one of India's 20 largest firms.

[23] In September 1984 a bromine leak at Fawley (Southampton, UK) affected between 60 and 70 people and will be the subject of compensation. In the USA in 1984, a methyl isocyanate leak at Middleport, New York, caused eye trouble to nine children, and an accidental pesticide leak at Linden, New Jersey, put 160 people in hospital. In Switzerland, between 2 and 3 kg of bromine leaked in Geneva in November 1984. During the same month 29 people (local residents and employees) in Liverpool (UK) were poisoned by a chlorine leak when two chemicals were mixed. Two months later, 16 were still receiving treatment. The accident was due to a labeling error. Civil proceedings are in hand and a fine of 3600 pounds sterling has been imposed. In the USA in December 1984, a defective wagon carrying 80000 l of ethylene oxide at Little Rock, Arkansas, obliged the authorities to evacuate 2500 people for a night. In December 1984, 3000 people were evacuated for 24 hours from Callao, Peru, following a tetraethyl lead pipeline break. At Matamoras, Mexico, in December 1984, 3000 people had to be evacuated and 200 received hospital treatment when 35000 l of ammonia leaked at a fertilizer factory. In January 1985, 42 people had to be hospitalized when cholorine leaked in a textile plant at Trichur, India, and four Gujarat villages were troubled by a gas leak from a chemicals plant, while at Javalpur 100 people had eye and throat irritations after inhaling gas leaking from sodium hydrosulfate drums contaminated by water. In January 1985 at Karlskoga, Sweden, 30 t of sulfuric acid vapor formed a 3 km^2 cloud above the town. The accident, caused by cold weather in the Bofors–Nobel plant, involved evacuating 300 people and hospitalizing 20. The 35000 residents were advised not to leave their homes and to keep doors and windows closed. In January 1985, 15 t of ammonia leaked at a national fertilizer plant at Cubatao, Brazil, resulting in the evacuation of 5000 people and hospitalization for 300. At Westmalle, Belgium, in February 1985, a cloud of chlorine and hydrochloric acid put 25 people in hospital. The town center was closed off to traffic and residents were advised to stay at home with doors and windows closed. In February 1985 an ammonia leak in a prawn processing plant in North Sumatra injured 130 workers, some requiring hospital treatment. In March 1985 a mesitylene cloud leak from a Union Carbide facility at Charleston made ten people ill, putting four in hospital.

[24] In France in September 1984, a truck lost drums containing 22 t of sodium cyanide on a motorway at Lyons. About 15 t of cyanide spilled onto the road surface and had it come into contact with rain would have threatened the city with large-scale poisoning. Fortunately the dangerous substances were recovered 10 minutes before rain began to fall.

[25] Total compensation paid out for air pollution in Japan between 1973 and 1982 is thought to be of the order of Y 650 B, i.e. the equivalent of over $3 B, to some 85 000 victims (nearly $5000 per year per person).

[26] Water pollution accidents in the USA between 1976 and 1980 caused on average 687 substantial fish losses per year (averaging 4000 fish per accident). Of those accidents, 21.2% involved farming, 17.2% industry, 24.2% municipal undertakings (e.g. electricity) and transport, 12% had other causes and 25.3% were for unknown cases [see *Environmental Quality*, 1983, Table A29 (Fourteenth Annual Report of the Council on Environmental Quality, Washington, DC)]. That distribution shows that industry in the broad sense is not responsible for most fish destruction and that in at least one-quarter of the cases no compensation will probably be paid (diffuse agricultural pollution). In the FRG, the average number of losses of dangerous chemicals (other than from tankers) polluting water during storage and transport was 208 per year between 1979 and 1981. In 1981, 1504 accidents (oil and other dangerous substances) involved 7259 m^3 of which 5633 m^3 were recovered. Surface water pollution occurred in 650 cases. In France there were 19 main pollution accidents in industry in 1983.

[27] Deprimoz (1984). According to Deprimoz, there are three potential damage levels:

(1) Pleasure fishing: between FF 100 000 and FF 1 M.
(2) Drinking water pollution: FF 5–10 M.
(3) Toxic air pollution: FF 100 M.

The French market can offer cover of up to FF 30 M or even FF 50 M. In 1983, some 400 French firms, mainly small and medium sized, took out a contract called GARPOL with a maximum cover of FF 20 M. Most French firms insure against accidental pollution liability under their existing civil liability policy.

[28] The death of 60 t of fish following a spill of 1700 kg of hydroquinone into the Rhone in 1982 cost Rhone-Poulenc FF 120 000 in compensation to fishing associations as "good neighbor" payments (2 F/kg). The firm also paid a FF 12 000 fine and FF 16 000 damages to nature protection bodies. According to "Données économiques de l'environnement" (*Doc. Franc.*, 1984), the cost of the damage was FF 1 M.

[29] One consequence of very costly pollution cases could be for gradual and even accidental pollution to be excluded from insurance policies covering the liability of firms (as already for radioactive pollution). The pollution risk would then have to be covered by an additional policy which might seem costly and might not be taken out by all hazardous undertakings that already have liability insurance. As a result, cover of pollution risk would be even worse than at present.

At the end of 1984 the US Insurance Services Office developed a model operator's liability policy which excludes all pollution damage. The result is that pollution risk cover will only be available under special pollution policies. That decision was partly related to the Jackson Township (New Jersey) case ($15.8 M) in which gradual pollution of groundwater was found to be an accidental situation.

In this case, the jury awarded, in November 1983, $8.2 M for medical surveillance costs, $5.4 M for impairment of quality of life, $2 M for emotional distress, $104 000 for expenses in securing new water supplies, and $92 000 for nuisance damages. After appeal, the awards for medical surveillance and emotional distress were nullified. Further appeals are being introduced. In the Jackson Township case, contamination of the water supply system (well water) was caused by the operation of a municipal waste disposal facility. No illness or injuries to health were established. The Jackson Township was insured for sudden and accidental releases. The court found that unintended pollution damage from gradual contamination constituted an accident and qualified for coverage (see Baran, 1985).

In Minnesota, as a result of the MERLA Act making firms liable for wastes produced since 1960, insurers have decided no longer to cover gradual pollution (environment impairment liability insurance) and 27 firms are no longer insured (see *Int. Env. Rep.*, 13 June 1984, p 199). According to the *Wall Street Journal* (20 March 1985), the pollution insurance market has collapsed in the USA.

[30] A study ($300 000) of the decontamination of the River Wabigoon shows that it will cost $20 M to extract the mercury-contaminated sediments. The contamination, discovered in 1970, could involve the liability of a paper mill (see *Int. Env. Rep.*, 11 July 1984, p 221, and 12 September 1984, p 291). When Reed sold the mill to Dryden it was agreed that Reed would not be responsible for damages beyond $15 M (mercury pollution of the Wabigoon) (see *Int. Env. Rep.*, 1980, p 101).

[31] In France, three airlines were ordered to pay FF 2.73 M in 1979 to a commune in the neighborhood of Orly airport. Two other cases are before the courts for damage in the neighborhood of Roissy airport. It has been decided in principle that the air companies are liable, but four years later the assessment of damage has not been completed. In another case, 87 applicants were invited in 1982 to contribute FF 7000 each towards costs of expert witnesses (medical, FF 1000; acoustic, FF 3000; property valuer, FF 3000). An appeal has been lodged against this decision, which will be very influential in regard to collective damage. Payments have been made from 1981 to finance soundproofing for 88 private buildings near Paris airports (FF 1.2 M). Other payments have been made for the soundproofing of 79 schools (FF 58 M) and eight medical/welfare institutions (FF 8 M). FF 114 M were spent to purchase 377 buildings for demolition.

[32] Exhaustive statistics for compensation associated with civil nuclear accidents in the USA (Marrone, 1984) highlight the Three Mile Island accident as a unique case. Other accidents cost less than $700 000 (18 accidents between $7500 and $75 000, nine accidents between $75 000 and $750 000 from 1962 to 1982). The Three Mile Island accident was unique because it is the only insured reactor to have polluted its neighborhood with radioactivity. Furthermore, the cost of compensation reflects the panic caused by the accident rather than the physical effect of the radiation emitted.

[33] The damages paid to third party victims of the Three Mile Island accident were:

(1) $1.3 M: evacuation costs of 3170 families (including $92 400 for loss of earnings by 636 claimants).

(2) $20 M: financial losses of residents within a 25-mile radius.
(3) $5 M: costs of medical surveillance of residents within the 25-mile radius (paid to a "public health fund").
(4) Payments to local authorities for disruption of their activities.

There are still many unsettled proceedings for personal injury (stress, cardiac disorders, genetic defects, etc.) associated more with the panic caused than with radioactivity as such (1.4 mrem on 2 M people). (Marrone, 1985). In February 1985, a further settlement was signed with 280 claimants for an unknown amount in excess of $4 M. It is noteworthy that evacuation costs were paid only in respect of a small number of families out of all those deciding to leave their homes (nearly 50000). For the utility concerned, the accident had a much greater financial impact. The cost of the Three Mile Island accident for the undertaking is estimated at $4 B (*Nuclear News*, January 1981): i.e. $1 B for decontamination; $430 M for repairs; $1.5 B for substitute electricity supplies; two reactors' down time, $950 M; financial costs, $40 M. The cost of substitute electricity is thought to have been $14 M per month.

[34] A 450 Ci cobalt-60 source dumped near Ciuadad Juarez (Mexico) seriously irradiated some 140 persons and in 1983 contaminated 6000 t of steel and the production of a Mexican cast-iron factory. The Aceros de Chihuaha company might be fined Pesos 7 M ($435000) for air pollution and physical injury. Over 1000 t of steel were exported to the USA. Several hundred houses in Mexico may have to be demolished. The costs of storing the contaminated steel are in excess of $1 million (see *Int. Env. Rep.*, 14 March 1984, p 73; 8 August 1984; and *Science*, **223**, 1984, p 1152).

[35] In the USA wastes from uranium mining were used as a building material in the 1950s but were found to be slightly radioactive. The US Government has undertaken a substantial decontamination program applying in 1984 to over 61 properties in Colorado ($17500 per property) (*New York Times*, 14 May 1984).

[36] The collision between two Boeing 747s at Tenerife in 1977 cost $161 M in insurance and the Chicago DC-10 crash in 1979 cost $122 M. Between 1978 and 1983, accidents to wide-bodied aircraft cost up to $52 M for the aircraft and $100 M for civil liability. For the four costliest accidents in the USA, average compensation was $470000 per death. Thus the loss of a wide-bodied aircraft in the USA carrying 300 passengers could cost $141 M for civil liability. In Europe, the cost would be lower because the compensation limit per passenger is tending to be established at Special Drawing Rights (SDRs) 100000 (Montreal Protocol No. 3). In France, the limit has been FF 500000 since May 1982. Since 1983 losses in aviation insurance amounted to $700 M, a 40% increase in premiums has been announced.

[37] This obviously does not apply to major accidents, which can cause damage well in excess of the value of the plant affected. The Bhopal plant had cost some $30 M, much less than the compensation to be anticipated. The same applies to the ICMESA plant at Seveso. However, the pollution risk is still low compared to the risk of losing the plant. Insurance for the pollution risk may cost a polluting firm 0.1% of its turnover.

The highest risks can be insured against, since Union Carbide whose liability was insured up to $200 M declined an extension to $300 M which would have cost $30 000 (0.03% of additional cover). Certain leading chemical corporations in the USA have insured for civil liability up to $800 M. In France, Rhone Poulenc is thought to be insured up to FF 500 M against third party liability (with the first FF 100 M exempt).

[38] For example, a nuclear power station operator in Belgium or Japan faces a nuclear liability premium that is about ten times lower than the cost of insuring the power station itself. The liability oil pollution premiums paid to the P and I clubs (mutual insurance companies) represent under 17% of the P and I premiums for tanker owners' civil liability. Of all the insurance premiums paid by an owner, pollution liability accounts for under 5%, the cargo being insured by the shipper. According to a study by M and M Protection Consultants, *Major Oil Industry Losses Exceeding $1 Million 1964–80*, the average (insured) loss for offshore facilities between 1969 and 1980 was $97.5 M per year. This figure is well over ten times the civil liability for pollution. In aviation, material losses are twice as high as civil liability. Total premiums are 0.6% of fleet values ($102 879 M in 1983). Total pollution civil liability premiums in the USA for 1983 were $40 M, while property and casualty insurance premium volume amounted to $110 B.

[39] Among the costliest accidents for insurers and reinsurers, *Sigma* (October 1982) cites:

(1) The San Francisco earthquake (1906): $350 M (i.e. $41.6 B nowadays).
(2) Hurricane Betsy (1965): $715 M (i.e. $2.2 B nowadays).
(3) Hurricane Frederick (1979): $752 M (i.e. $1 B nowadays).

Natural catastrophes between 1970 and 1981 caused insurers to pay out compensation as in *Table 4.25*. According to Bertz (1984), for

Table 4.25. Compensation for natural catastrophes, 1970–1981.

Compensation range	*No. of incidents (Total compensation)*
$50–100 M:	13 ($964 M)
$100–200 M:	9 ($1156 M)
$200–400 M:	4 ($1175 M)
$400–800 M:	23 ($1232 M)

Source: *Sigma* (current US dollars).

natural disasters between 1960 and 1983 the cost distribution is as in *Table 4.26*.

According to *Sigma*, 21 fire accidents exceeded $25 M between 1970 and 1981 (current US dollars):

$25–50 M:	12 ($476 M)
$50–100 M:	6 ($359 M)
$100–200 M:	3 ($463 M)

Table 4.26. Cost distribution for natural disasters, 1960–1983.

Loss range	*Assessed losses*	*Insured losses*
\$50–100 M	–	9 (\$585 M)
\$100–200 M	–	7 (\$838 M)
\$200–400 M	9 (\$2424 M)	7 (\$1896 M)
\$400–800 M	18 (\$10043 M)	4 (\$2395 M)
\$800 M–1.6 B	11 (\$12420 M)	1 (\$825 M)
\$1.6–3.2 B	9 (\$21250 M)	–
Over \$3.2 B	2 (\$13300 M)	–
TOTAL	49 (\$59437 M)	28 (\$6539 M)

Source: Bertz, 1984 (current US dollars).

An enquiry by M and M Protection Consultants into material damages between 1951 and 1980 in the oil and chemicals industry (explosions, fires) shows that the 65 costliest accidents break down as follows (1981 US dollars):

\$15–30 M: 34 accidents (\$665 M)
\$30–60 M: 21 accidents (\$839 M)
\$60–120 M: 10 accidents (\$813 M)

According to the *Recueil de données statistiques sur l'assurance automobile en France* (*French Car Insurance Statistics*, 1982, Table 81, Assemblée Générale des Sociétés d'Assurance contra les Accidents), sums paid in respect of civil liability for material damage and personal injury in car accidents in France are given in *Table 4.27*.

Table 4.27. Sums paid for civil liability for material damage and personal injury in France.

Amount (FF)	*Number* (%)	*Costs* (%)
1000–10000	68.4	29.4
10000–100000	8.5	28.1
100000–1 M	0.9	27.9
over 1 M	*ca.*0	11.7

Fires in France for the period 1982–1984 expressed in 1984 francs approximately obey the law of equal weight for successive classes of severity between FF 5 M and 40 M (see *Table 4.28*). Beyond FF 40 M, the distribution law is different because losses are limited and there are so few insured installations liable to cause high losses. The sum of the direct risks of the two upper brackets (FF 1.4 B) is close to the amount of the average direct risks (FF 1.2 B) of the four preceding brackets. Generally speaking, when the number of accidents in successive brackets beyond FF X M declines according to the square of the ratio between the maxima and minima of successive brackets (A), the

Table 4.28. Number and costs of fires in France, 1982–1984.

	Direct risks		*Operating losses*	
Costs bracket (FF M)	*No.*	*Cost* (FF M)	*No.*	*Cost* (FF M)
5–10	153	1 072	42	260
10–20	109	1 547	20	265
20–40	34	954	6	162
40–80	25	1 302	6	337
80–160	9	984	2	252
160–320	2	431	0	0
TOTAL	332	6 290	76	1 276

Source: Assemblée Plenière des Sociétés d'Assurances Incendie et Risques Divers, Paris, 1985.

total weight of all accidents in excess of FF X equals $1/(A-1)$ times the total weight of each bracket below X if the equal-weight law applies to brackets below X. For $A = 2$, the weight of accidents in excess of X is equal to the weight of accidents between $X/2$ and X. For $A = 10$, the weight of accidents in excess of X equals 1/9 of the weight of accidents between $0.1X$ and X.

[40] On the basis of the report by Fryer and Griffiths (1979), the distribution of serious man-made accidents can be shown to comply approximately with the equal weight law for successive classes of gravity (in terms of deaths) for four types of human activity (*Table 4.29.*).

The Fryer and Griffiths report also shows that natural disasters (storms, floods, earthquakes, avalanches) do not obey the equal weight law when classes of successive gravity are measured in deaths, because the number of deaths in the maximum gravity class is nearly the same as the total number of deaths for all classes. In this case, whether a compensation system remains in balance depends very much on whether or not a grave event occurs.

On the basis of *Table 4.7*, dam burst accidents are broken down in *Table 4.30*.

The Rasmussen report (1974) gives details of the distribution of disasters as a function of severity. Distributions for meteorites (deaths), explosions, dam bursts (deaths), fires (deaths and damage), and accidental chlorine emissions during shipment (deaths) are characteristic of the law of equal weight for all classes of gravity. Conversely, distributions for hurricanes (death and damages) and earthquakes (death and damages) give relatively greater weight to the severer events. Rasmussen also shows that the distribution for combined man-made disasters (in dollars) corresponds to the equal-weight law for damages between \$10 M and \$1 B (see also *Figure 4.2*).

[41] If damages are divided into three classes of gravity ($1-A$, $A-A^2$, and A^2-A^3) and if the number of events in those classes is $B/A^{0.5}$, $B/AA^{0.5}$, $B/A^2A^{0.5}$, each class contains A times fewer events than its predecessor but those events are A times costlier, so that the combined cost of events in each class is the same. The relative weight of the higher class is 33%. If compensation is paid for all events with a

Table 4.29. Deaths due to serious man-made accidents.

Severity (number of deaths)	*Number of accidents*	*Total deaths*
(1) *Fires and explosions*		
100–200	15	2010
200–400	4	1122
400–800	2	952
800–1600	1	1200
1600+	1	1700
TOTAL	153	12084
(2) *Maritime accidents*		
100–200	37	4774
200–400	17	4480
400–800	5	2591
800–1600	1	1500
1600–3200	1	2750
3200+	1	6000
TOTAL	281	31277
(3) *Mining accidents*		
50–100	21	1460
100–200	11	1515
200–400	4	1179
400–800	3	1296
800+	1	3700
TOTAL	97	11000
(4) *Railway accidents*		
50–100	33	2350
100–200	20	2670
200–400	5	1204
TOTAL	188	10277

Source: Fry and Griffiths (1979).

Table 4.30. Deaths due to dam burst.

Severity (number of deaths)	*Number of accidents*	*Total deaths*
100–200	7	959
200–400	4	1000
400–800	3	1383
800–1600	2	2000
1600–3200	2	3755
3200+	1	15000 (?)

limit of A^2, the uncompensated fraction of total damages equals $(A^{0.5} - 1)/3A^{0.5}$. For $A = 2$, that fraction is 9.8% and for $A = 10$, 23%. The fraction of events not fully compensated is $1/(A^2 + A + 1)$, i.e. 14% for $A = 2$ and 0.9% for $A = 10$. If $A = 10$, 99.1% of accidents are fully compensated but only 77% of damages are compensated. Given that the assumption that there will never be an event costing more than A^3 does not seem demonstrable (at least as regards third party liability), it may be useful to provide a compensation ceiling such as A^3 to forestall the possibility of any one very rare event profoundly modifying the financial balance of an insurance system.

[42] It has been asked whether the Bhopal disaster was statistically predictable. Since there have in 50 years been 16 "terrestrial" accidents in the chemicals and oil industry causing from 50 to 500 deaths (including eight causing over 200), it is in accordance with the statistical law mentioned that an accident should also occur involving between 500 and 5000 deaths. However, referring only to serious air pollution, there have been no accidents even ten times less serious than at Bhopal and accordingly the Bhopal disaster constitutes a genuinely exceptional event. Among a combination of pollution accidents costing up to several million dollars, coverage for a single $100 M accident can cause difficulties for the insurers.

[43] The fact that prevention costs are very much higher than damage costs in no way means that it would be preferable to reduce prevention and pay compensation to victims. The reason is that compensation payments correspond to only part of the damage caused, and courts probably assess them very much lower than economists would. Furthermore, to arrive at an appropriate level of prevention expenditure, we calculate the damage caused by reduction in the level of prevention and the benefit in terms of lower prevention outlays. A comparison between the cost of imperfect prevention and the cost of damage caused by the absence of perfect prevention is no help as to how appropriate prevention outlay may or may not be.

[44] Statement by Mr J. Graveleau (1982), who reports 300 accidents in 18 months costing several tens of thousands of francs and states, "a million francs is commonplace." (See discussion of Siskind, 1982.)

[45] For 1979, accidental pollution of French rivers is thought to have cost FF 2.5 M while the costs of preventing such pollution were FF 15.5 B (1981) of which FF 2.27 B were shouldered by industry (Siskind, 1982). In the FRG, the chemicals industry spent DM 27 B ($8.44 B) over ten years on environmental protection, i.e. 10% of total investments and 2% of sales, (*Wall Street Journal*, 6 February 1985. (*Note*: FRG industry spent DM 7.23 B on environmental protection in 1978.) There is every reason to suppose that the FRG chemicals industry has not been paying DM 100 M every year (4% of the cost of controlling pollution, or 1% of sales) as compensation for the pollution it has caused.

[46] In 1981 the French Agences de Bassin received FF 1914 B of which FF 532 M were for industrial pollution. If the costs of accidental water pollution damages were covered by an additional fee, the maximum extra cost for industrial polluters would be well under 5 % of the pollution fee they are already paying.

[47] According to the French Environment Ministry (*L'industrie et l'environnement en 1984: L'état des gros rejets*, August 1984), 50000

factories in France "are sources of serious environmental pollution and hazards". In fact most of the hazards and pollution are concentrated in only a few of them. Half the organic pollution, for instance, comes from 377 sources, half the toxic water pollution from 55, and half the sulfur dioxide discharges from 50. Two thousand of the largest sources are participating in a self-monitoring exercise. Assuming that those 2000 sources annually pay FF 50 M compensation (higher than in fact), the average compensation would be FF 25 000 per factory in respect of accidental industrial pollution. Such pollution on average causes no more than a few third party deaths per year in France. In 1982, 1360 people died as a result of industrial accidents. Mining accidents cause a few dozen deaths per year while silicosis causes 700 per year. Full and prompt compensation for physical injury to third parties through accidental pollution would not constitute an additional burden in comparison with the compensation already paid to workers for industrial accidents.

[48] It is noteworthy that nuclear civil liability does not satisfy this criterion because it is based on a very low number of costly accidents while there is no high number of inexpensive accidents. Similarly space insurance (losses associated with satellite launch failures) fluctuates considerably because the market is narrow and the loss in the event of failure can be so high. Between 1975 and 1981, premiums paid amounted to $107 M and in 1982 two accidents together cost $90 M. The present capacity of the market is thought to be about $100 M per launch. Aviation civil liability also fluctuates widely because accidents in the USA have now become very expensive (four accidents costing over $50 M between 1978 and 1983: $51, 70, 85, and 100 M). In systems with ex post facto premium payments (e.g. Cristal and FIPOL for oil pollution), the premium varies widely from year to year because the systems cover the most serious and unusual forms of pollution.

[49] In France, road carriers (hazardous goods) have unlimited civil liability insurance cover and can cause costly accidents. The very few such accidents do little to disturb the motor insurance system (FF 26 536 M in civil liability premium for 1982 and FF 17 788 M in damages). The cost of an accident like the one at Los Alfaques (Pts 2.4 B or FF 144 M) can be compared with carriers' civil liability premiums in France (FF 1.41 B including FF 470 M for lorries over 20 t). In the USA, the 1980 Motor Carrier Act requires minimum cover of $5 M for hazardous goods transport.

[50] While maritime insurers announced that they could cover tanker owners' civil liability up $100 M, the P and I Clubs (mutual insurance companies) offered their members insurance of up to $300 M in cases where the limitation of liability under the 1969 Convention did not apply.

[51] The ceiling depends on the country, the industry concerned, the risks covered, and demand. Where insurance for environmental risks is compulsory, there should be enough of a market to attain high ceilings.

[52] The Paris Convention was intended, according to its preamble, both to ensure "adequate and equitable compensation" and "to ensure that the development of the production and uses of nuclear energy ... is not ... hindered." The maximum liability (SDRs 15 M) has probably made it possible to reduce the cost of insurance for nuclear operators

(although no incident has reached that figure). The figure was set in the light of insurance market conditions in the early 1960s and remains in force in 1985 whereas the market could very easily offer cover of $75 M.

[53] It seems probable that the $595 M limit will be doubled in 1987 so that the compensation ceiling is of the same order of magnitude as the sum assured in respect of damage to the facility.

[54] After the *Torrey Canyon* accident in 1967 (117 000 t of oil, $15 M clean-up costs), governments signed the following conventions:

(1) International Convention on Civil Liability for Oil Pollution Damage (1969).
(2) Convention on the Establishment of an International Compensation Fund for Oil Pollution Damage (1971).

[55] The CRISTAL plan offers maximum compensation of $36 M for victims in states that have not ratified the 1971 Convention.

[56] In the USA, several enactments provide for special compensation machinery for oil damage (Trans-Alaska Pipeline, Outer Continental Shelf Liability Act, Deep Water Port Act). There are also state and national funds to compensate for oil damage [Florida, Maryland, Maine, Texas, New Jersey, North Carolina, Washington; also South Africa, Finland (Mk 100 M), Canada, New Zealand, Australia, Israel, Japan, Ontario]. Turkey has introduced unlimited liability for shipowners.

[57] In Belgium, a law creating a national fund was adopted in 1974 to finance the disposal of wastes whose producer would not do so owing to bankruptcy. The costs involved were to have been borne by all waste producers. However, the fund has not yet been established. In the Netherlands, a Soil Clean-up Interim Act, which entered into force on 15 January 1983 for a term of six years, provides government funding for the disposal of older wastes by provinces and local authorities. The government may also proceed against the parties responsible. The financial security requirement applies to disposers in the FRG, Austria, Belgium, Canada, and the USA. In Belgium, Canada, and the USA, the disposer has to take out insurance whereas in Germany and Austria a condition for granting a disposal permit may be that an insurance has been obtained. For waste carriage, special insurance is required in the FRG, Austria, Belgium, Canada, the USA and Sweden. In Italy, a financial guarantee is required for a permit to store hazardous wastes temporarily or permanently (see Article 5.3.5 of the implementing provisions of Article 4 of Decree No. 915 of 10 September 1982, *Gazz. Uff.*, Number 253, 13 September 1984).

[58] For a description of the Netherlands fund, see Thiem (1981). The fund was established in 1972, prompted by a 1965 smog causing serious agricultural damage in the Rijnmond area of Rotterdam. It is financed by industry (35%), domestic heating (15%) and motorists (50%).

[59] In France, the parafiscal passenger tax system in force from 1973 to 1983 has been replaced by a noise levy as from 1 January 1984. Under a French Bill, the Paris airport levy system is to be extended to all other airports. The Bill may also cover issues of liability and compensation. Noise levies apply in the Netherlands (G 7 M per year) and Switzerland ($2.9 M in 1982). In the FRG, quieter aircraft qualify for a rebate on the landing tax.

[60] The capacities of the reinsurance market should not be underestimated. For air transport, their capacity is thought to be $800 M. Certain chemical and oil companies are insured for up to $1 B per event. When there is a long time lapse between the generating event (leak) and the damage, public fund systems may be preferred to insurance systems.

[61] Leaving aside payments which may be covered by the general budget, the highest compensation ceilings in force are found in the USA, i.e. $595 M for nuclear energy and $200 M for oil. In Europe, the highest are SF 300 M (Switzerland, nuclear power) and DM 500 M (the FRG, nuclear power). These figures suggest that the damage level beyond which the government should intervene is at present between $100 M and $200 M per accident. These levels also correspond to the maximum amounts ever paid out for third party damage by insurers (and reinsurers) after an accident affecting the vicinity.

[62] In December 1984 a Bill laid before Parliament as part of the Netherlands Soil Protection Act provided that environment and agricultural ministers could "require that certain activities potentially hazardous to the soil be insured to cover the risk of soil pollution." It was stated that "the Government wanted to be sure that the community would not pay for contamination caused by individuals" (see *Int. Env. Rep.*, 13 February 1985). Similar provisions might be made for all hazardous industries (guarantee or insurance).

M.J. Florio, Chairman of the Sub-Committee on Energy and Commerce of the US House of Representatives, has announced that a bill is being considered with the aim of prescribing no-fault liability for physical damage caused by toxic chemical products (*Int. Env. Rep.*, 13 February 1985, p 34).

In 1984, the German Federal Court issued a decree requiring polluters to prove that they had taken the necessary precautions to forestall damage (*World Environment Report*, 23 January 1985).

The US Environmental Protection Agency is contemplating compulsory insurance cover for hazardous waste incinerators, of $50 M to $500 M (*Int. Env. Rep.*, 1985, p 92). The insurers have argued that $500 M is too high, there being so far few incinerator ships (*Int. Env. Rep.*, 1984, p 393).

References

Andurand, R. (1979), Le Rapport de sûreté et son application à l'industrie, *Annales des Mines*, July, 115.

Baran, M.S. (1986), *Chemical Industry Hazards* (International Institute for Applied Systems Analysis, Laxenburg).

Bertz, G. (1984), Research and statistics on natural disasters in insurance, *Geneva Papers on Risk and Insurance*, **9**, No. 31.

Burton, S., Victor, P., and White, A. (1983), *The Mississauga Evacuation*, Final Report to the Ontario Ministry of the Solicitor General (The Institute for Environmental Studies, University of Toronto).

Deprimoz, J. (1984), Consideration of the extent of insurable risks and how they are to be covered, Seminar on the Legal and Institutional Aspects of Transfrontier Movements of Hazardous Wastes, OECD, Paris.

Deschamps, M.A. (1972), Catastrophe et responsabilité, *Revue Francaise de Sociologie*, **13**, 376.
Environment Agency (1984), *Quality of the Environment in Japan* (EA, Government of Japan, Tokyo).
Fryer, L.S., and Griffiths, R.F. (1979), *Worldwide Data on the Incidence of Multiple Fatality Accidents*, SRD-R-149 (HMSO, London).
Goubet, A. (1979), Risques associés aux barrages, *La Houille Blanche*, No. 8, 475–490.
Graveleau, J. (1982), in *Les enterprises et les atteintes à l'environnement* (Centre d'Etudes et de Perfectionnement de l'Assurance, Paris), 80.
Lagadec, P. (1979), *Futurible 2000*, Special Issue, November.
Lagadec, P. (1981a), *Le Risque technologique majeur – Politique, risque et processus de développement* (Peragamon Press/*Futuribles*, Paris). In English: *Major Technological Risk – An Assessment of Industrial Disasters* (Pergamon Press, Oxford, 1982).
Lagadec, P. (1981b), La civilisation du risque – Catastrophes technologiques et responsabilité sociale (Editions du Seuil/*Science ouverte*, Paris). In Spanish: *La civilizacion del riesgo – Catastrofes technologicas y responsábilitad social* (Editorial Mapfre, Madrid, 1984).
Lagadec, P. (1984), *L'accident de Mississauga, Toronto, 10–16 Nov., 1979* [Laboratoire Econométrie, Ecole Polytechnique, Min. Envir. (SEI)].
Marrone, J. (1984), *Bull. Droit Nucléaire*, No. 33, June.
Marrone, J. (1985), The liability claims experience of the American nuclear pools and their response to the Three Mile Island accident, *Nuclear Third Party Liability and Insurance* (OECD, Paris).
Mekouar, A. (1984), Droit et non-droit des déchets industriels au Maroc, *Les Déchets Industriels et l'Environment*, Proceedings of International Seminar on Industrial Wastes and the Environment in Comparative and International Law, Limoges, May, 1984 (PUF, Paris).
National Oceanic and Atmospheric Administration (1983), *Assessing the Economic Costs of Oil Spills: The* Amoco Cadiz *Case Study* (NOAA, Washington, DC).
OCIMF (1982, 1983, 1984), Position Papers, IMO–LEG documents (Oil Companies International Maritime Forum, London).
OECD (1982), *The Cost of Oil Spills* (Organization for Economic Cooperation and Development, Paris).
Okrent, D. (1979), Risk benefit evaluation for large technological systems, *Nuclear Safety*, **20** (2), March, 148.
P and I Clubs (1984), Revision of 1969 Civil Liability Convention and 1971 Fund Convention, IMO–LEG, February.
Pfennigstorf, W. (1982), *Insurance of Environmental Risks*, American Bar Foundation, No. 1 (American Bar Association).
Rasmussen (1974), *Reactor Safety Study*, WASH 1400 (US Atomic Energy Commission, Washington, DC).
Siskind, B. (1982), *Les enterprises et les atteintes à l'environnement* (Centre d'Etudes et de Perfectionnement de l'Assurance, Paris).
Smets, H. (1984), Dommages causés par les marées noires, *Environment Policy and Law*, **12**, 13–36.
Swaigen, J.Z. (1981), *L'indemnisation des victimes de la pollution au Canada* (Centre d'edition du Jour du Canada, Ottawa).
Temple, Barker, and Sloane, (1983), *Cost Benefit Analysis of Possible US Adherence to Two international Conventions on Liability and Compensation for Oil Pollution Damage* (Lexington, MA).

Thiem, V. (1981), Environmental damage funds, in *Compensation for Pollution Damage* (Organization for Economic Cooperation and Development, Paris).
UBA (1983), *Handbuch Störfälle* (Schmidt, Berlin).

4.D. Discussion

J. Deprimoz [1]

A great merit of the chapter by Smets is to set forth the large variety of situations likely to be termed "disasters" involving exceptional pollution damage. Among such disasters, the distinction must be made not only in terms of causes (explosions, dam or dyke burst, pipe leakages or shipwrecks, etc.) but also according to *their consequences* which chiefly depend on the transmitting vector (soil, water, air). Some of them will mainly cause property damage and loss of use of property up to complete clean-up (black tides, ground pollution arising from dumping sites); others will mainly cause bodily injury and will also necessitate measures for evacuating surrounding populations who are exposed to bodily injury (air pollution).

4.D.1. First remark

For an estimation of what possible part private insurance can reasonably play in the compensation of major environmental impairment losses caused by industrial activities it seems appropriate to distinguish the following types of losses:

(1) Losses after which "bodily injury" prevails.
(2) Losses after which material damage prevails.

Losses with Bodily Injury Prevailing

The distribution of an empirical aggregate amount of FF 600 million shown in *Table 4.10* of Smets' paper for an ecological disaster caused by poisonous gas seems reasonable enough. Figures are based on an assumed average compensation of FF 200000 for each death (costs + lump sums or annuities paid to the widow and two children), of FF 1 million for each permanent disability exceeding 80%, and of FF 128000 for each permanent disability between 20% and 50%. Other items, estimated at FF 120 million, correspond to the refund of medical expenses or, as the case may be, to the refund of evacuation costs.

These values would not be significantly modified by the fact that victims will benefit, in their respective countries, from a legal system for health protection and for the payment of disability annuities or death benefits as the bodies in charge of such systems (health service and/or social security) will probably make use of their rights of recourse against the polluter up to 100% of indemnities paid by them.

Starting from the above figures, it can be noted that *the number of injured victims to be directly compensated* by the private insurer covering "polluter's liability" *will significantly decrease* where this insurer reduces the per loss commitment limit down to an amount noticeably under FF 600 million. Accordingly, even if the aggregate value of health and evacuation costs refund is brought down from FF 120 to FF 50 million (the evacuation of 30 000 people living in the surroundings is in fact rather on the pessimistic side), the number of victims entitled to compensation by payment of capitals or annuities would be reduced as in *Table 4.D.1*.

Table 4.D.1.

Assumed insurance coverage coverage (FF M)	*No. of victims entitled to compensation*			
		Disabled		
	Deceased	>80%	>50%	>20%
600 – 120 = 480	250	125	500	625
300 – 50 = 250	130	65	260	325
100 – 50 = 50	26	13	52	65

At present, the different national markets for pollution insurance will only offer, for all combined types of damage, covers generally limited as shown in *Table 4.D.2*. Currently, these amounts are sufficient

Table 4.D.2. Limits to pollution insurance coverage.

USA	Maximum commitment limit of PLIA Pool:	*ca.* FF 90 M
Italy	Maximum commitment limit of ANIA Pool:	*ca.* FF 75 M
France	Maximum commitment limit of GARPOL Pool:	*ca.* FF 30 M
Netherlands	Maximum commitment limit of MAS Pool:	*ca.* FF 30 M

for the coverage of almost all losses involving bodily injury of low and average importance, but they will not be sufficient in the case of true catastrophes due to the accidental release of noxious gases in the atmosphere (the Bhopal accident, for instance).

Accordingly, the following question is raised in terms of economy: *How many tens or hundreds of bodily injured victims* after a pollution accident are necessary to have this accident considered as catastrophic and to have the currently available insurance cover *taken over* by another financial guarantor?

Losses with Material Damage Prevailing

It seems that the total cost of Love Canal exceeded $50 million and that of the Lekkerkerk case $70 million for the evacuation and rehousing of surrounding populations. Such figures are, of course, largely above the currently available covers offered by private insurance.

Conversely, and according to examples given by Smets, most river and groundwater pollution losses will not in general exceed an amount of FF 1–10 million. Consequently, insurance coverages limited to FF 20–30 million for each loss would be satisfactory in a great many cases.

4.D.2. Second remark

The financial impact of high-cost losses, which owing to their low frequency represent only 25–30% of the total weight of environmental losses (Note 39 of Smets' paper), can be considered as a possible charge only if it is *spread* over the whole community of environment-polluting *industries*.

As long as the "law of great numbers " is not applicable, i.e. as long as all concerned polluting industries are not bound to join together to form a very large mutual group to cover risks, the necessity for each industrial firm (or its own particular insurer) to "carry for itself" the cost of an eventual disastrous loss implies a direct threat of debt – not to mention insolvency – leading to bankruptcy. In other words, the corrective factor based on low frequency *can be included in a macroeconomic calculation only* and not in calculations applied to one or a couple of industrial firms.

Would it be an appropriate remedy to provide that the liability coverage of polluting, transporting and waste disposal firms *is made compulsory* in order to give private insurance all appropriate means for obtaining substantially increased commitment limits (e.g. several hundred million French francs)?

We may feel rather dubious of this result if we contemplate the annual amount of earned premiums which can be raised under this type of compulsory insurance. The current premium amounts collected on the French market from the liability insurance of waste-producing industries (chemicals, petrochemicals, leather and skins, metal works, industrial cattle breeding, etc.) will apparently never exceed FF 1 billion. An average 5% overpremium gathered for the coverage of pollution hazards "from cradle to grave" would only produce, for a one-year period including all the activities connected with waste handling, approximately FF 50 million. In round francs, these earned premiums would only allow the correct settlement of one loss of FF 100 million every two years, and of one loss of FF 500 million every ten years.

Combined with the need for insurers to make up for the "law of great numbers" at a given date, appears the need to benefit from a long-term insurance agreement, and this leads to our third and final remark.

4.D.3. Third remark

To satisfy requirements for a cover spread over several years in order *to meet late claims* brought long after the date of handling of noxious waste and their burial, the insurer must be in a position to rely on *premiums collected over a long-term period*. Therefore, the stoppage of premium payments after the producing or reprocessing firm has ceased its activities represents a major obstacle which must be overcome before any offer of long-term coverage is possible.

At present in France, a solution to this difficulty is being considered as regards hazards linked with the firms operating class I disposal sites licensed under article 9 of the Law dated 15 July 1979. Those firms are now contemplating the setting up of a "Professional Guaranty Fund" which should be made up of contributions paid by all firms working on French territory. The contributions would be used not only for the *settlement of the costs of controlling and putting back in order* disposal sites after shutdown (compare the "Post Closure Liability Trust Fund" in the USA) but also as security *for the payment of insurance premiums* applicable to pluriannual liability covers against third parties and concerning claims notified with five years' notice (or perhaps more) after shutdown of disposal sites, even if the firm or firms operating these sites have ceased their activities when claims are notified.

Thus, assuming that this fund acts as a *del credere agent* for the payment of premiums, subsequent coverage could be more easily underwritten from private insurers. In any case, however, beyond the insurers' maximum limit of commitment and as regards claims notified after the contractually admitted period, there will still remain a gap of cover. So a "financial relay" is to be found for taking over the total compensation of victims. The SERVANT Report handed over to the French Ministry of Environment in February 1984 has proposed a governmental relay.

Note

[1] These discussion comments were delivered and interpreted at the Conference by Jean Collart.

PART TWO

Problem Context

CHAPTER 5

Insurance and Compensation as Policy Instruments for Hazardous Waste Management

P.R. Kleindorfer and H. Kunreuther

5.1. Introduction

Public policy regarding the transport, sale, and processing of hazardous materials has become an area of acute concern for industry, government and the public. The magnitude of the problem is large. In the USA, for example, the Environmental Protection Agency estimates (see US Environmental Protection Agency, 1980) that more than 50000 chemical substances are currently in commercial circulation, and this number is growing at a rate of 500–1000 new substances per annum. Hazardous waste streams stemming as by-products from the industrial use of these substances are now estimated at upwards of 50×10^6 t per annum in the USA (see Baram, 1982, Ch 9) and 30×10^6 t per annum in the European Economic Community (see Brusset and Rochevolles, 1979).

Given the number and heterogeneity of these hazardous materials, an immense problem faces public policymakers attempting to determine an appropriate mix of regulatory and market forces to assure a viable and safe infrastructure for the transport and use of hazardous materials. The key in this effort is to balance the benefits associated with the use of hazardous materials in producing products and services consumed by society with the costs associated with understanding and controlling their potentially negative side effects.

This chapter will be concerned with a small part of this overall problem, namely the choice of waste and transport technology by firms generating wastes and the siting of hazardous waste treatment and disposal facilities. Our approach might best be called a "decision process perspective" (see *Figure 5.1*). The key in this perspective is the mediating effect of decision processes of actors on consequence evaluation as these actors respond to regulatory and policy options.

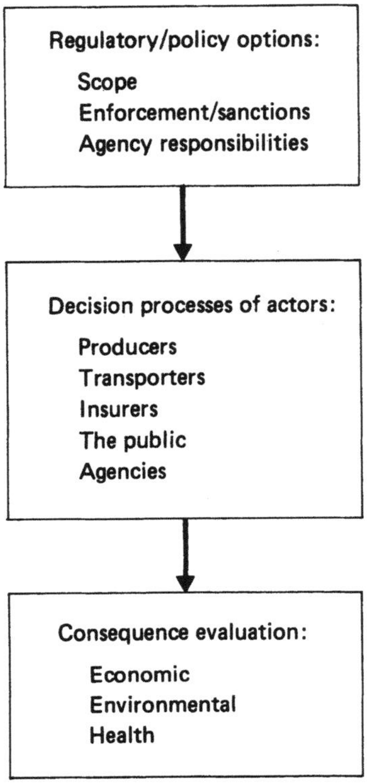

Figure 5.1. Research framework.

The following three areas of choice will be of special interest here:

(1) Choice by waste generators of how much waste to produce.
(2) Choice by waste generators and transporters of the means and location for disposal of waste.
(3) Choice by developers and communities of locations and control procedures for siting hazardous materials disposal facilities (HMDFs).

These three areas interact strongly with each other and with regulatory policy. The key policy instruments of interest in this chapter are compensation and insurance. The underlying motif is to understand how compensation and insurance can help balance the social incidence of risks, costs, and benefits of hazardous waste management. We proceed as follows.

In Section 5.2 we consider the problems of waste generators in more detail, concentrating on the effects of insurance and liability laws on the behavior of waste generators in their choice of waste reduction technology as well as means of transporting and disposing of waste. The basic theme of this section is that insurance and liability regulation concerning environmental and safety aspects of a firm's operations can strongly influence the firm's choices of how much waste to produce and what means it chooses to transport and dispose of the resulting waste. We also briefly discuss ongoing subsidies and taxes (i.e. compensation) to the firm or the public for waste-related activities.

In Section 5.3 we consider the problem of siting hazardous waste facilities. However, as indicated above, and as is abundantly clear to nearly everyone by now, the actual siting of such facilities is a very difficult task, fraught with public opposition and scientific uncertainty (see, for example, Morell and Magorian, 1982; O'Hare *et al.*, 1983; and Wynne, 1984). In view of this, we discuss how compensation and insurance may be used to facilitate communication among affected stakeholders in the siting process to achieve some measure of improved efficiency and equity. Several illustrative cases from practice are noted as encouraging signs that compensation and insurance can help solve the current stalemate between communities and beleaguered siting authorities in developing a viable treatment and disposal infrastructure for hazardous waste.

The concluding section couples the results of Sections 5.2 and 5.3 concerning the waste generator's decisions and the ultimate siting of treatment and disposal facilities. We reflect in our conclusions the difficulties of balancing environmental and public health concerns with the financial and property rights structures of industry and the affected public. We argue that compensation and insurance can play an important policy role in achieving this balance.

5.2. Compensation, Liability, and Insurance Issues for Waste Generators and Transporters

Transportation, treatment, and disposal of hazardous waste have three major consequences which policymakers must balance: economic effects, environmental effects, and health effects. In this section we examine how compensation and insurance might affect these consequences at the level of waste generators and transporters. The approach we will use is to model the decision process of a waste generator or transporter in a very general way, following Shavell (1984), concentrating on the incentives which anticipated liability and compensation may be expected to have on the total risk exposure a generator or transporter will accept.

Consider a profit-oriented firm which generates or transports hazardous waste. Suppose that this firm must choose between several alternatives which affect the probability of an accident involving this waste. The firm must decide how much it will spend on safety or waste reduction measures to avoid such an accident. We denote by x the amount the firm decides to spend on such measures and by $p(x)$ the probability of an accident. This probability depends on x of course: the more the firm expends on waste reduction/safety, the less is $p(x)$.

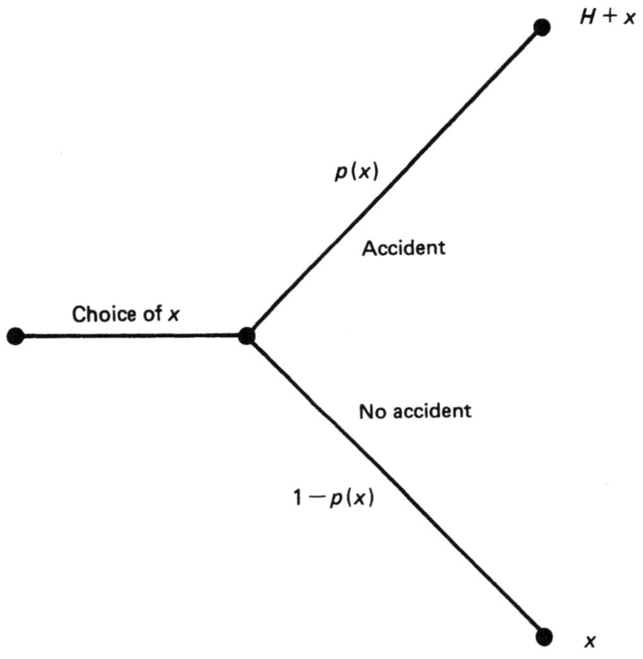

Figure 5.2. Decision tree for waste control decision.

Suppose that, if an accident occurs, the cost or harm resulting will be H, [1]. We can depict the outcomes of this process as in *Figure 5.2*. In the figure, the firm first chooses x; this results in an accident with probability $p(x)$ and costs $x + H$ and no accident with probability $1 - p(x)$ and costs x. The expected social cost (ESC) of the firm's decision on x will be $p(x)(x + H) + [1 - p(x)]x$, or simply

$$\text{ESC} = x + p(x)H \ , \tag{5.1}$$

the sum of the firm's outlays to reduce the probability of an accident plus the expected cost of an accident.

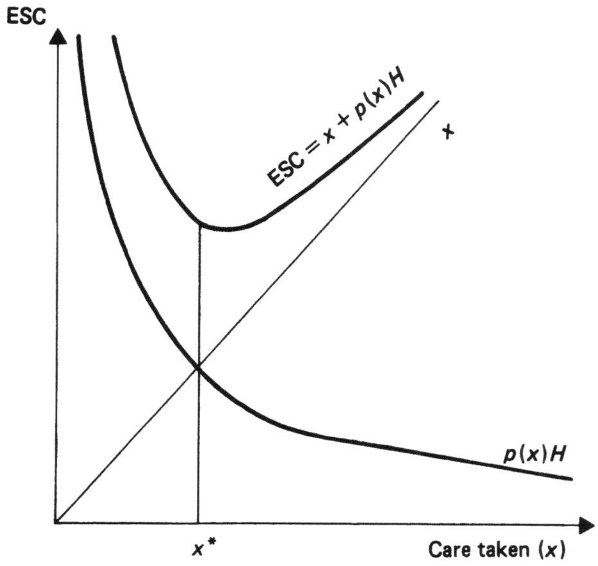

Figure 5.3. Expected social costs of waste control.

If an omniscient regulator were to choose x, then (s)he would do so by minimizing expected social costs (ESC) as shown in *Figure 5.3*, where we graph ESC = $x + p(x)H$ and show its minimum x^*. But the firm may not perceive its costs to be coincident with social costs. Indeed, a profit-oriented firm would only face that portion of the social costs which results from fines or other sanctions it could expect to ensue from an accident. We might represent this formally as a compound "lottery" as in *Figure 5.4*.

Figure 5.4 represents the following scenario. If there is no accident, which happens with probability $1 - p(x)$, then the firm pays nothing except the cost x of whatever measures it undertook. If there is an accident, then the firm pays the full amount of damages resulting from the accident with probability $q(x, y)$, and pays nothing with probability $1 - q(x, y)$. We can think of $q(x, y)$ as the probability of the accident "being noticed" and litigated with assessed damages of H. The variable y can be thought of as the level of effort expended by regulators to monitor and litigate damage suits. From *Figure 5.4*, the costs seen by the firm as it chooses x are

$$\text{COST} = x + p(x)q(x, y)H \ , \tag{5.2}$$

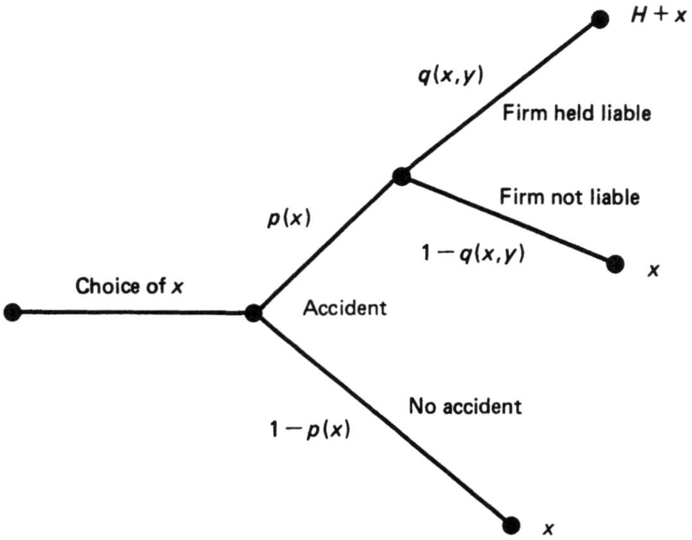

Figure 5.4. The waste generator's decision problem.

which are less than the true social costs $x + p(x)H$, unless the probability $q(x, y)$ of being held fully liable for damages is 1. Assuming that $q(x, y)$ is less than 1, it is easily seen that the level of expenditure $\hat{x}$ chosen by the firm will be less than that which minimizes expected social costs x^*, as shown in *Figure 5.5.*

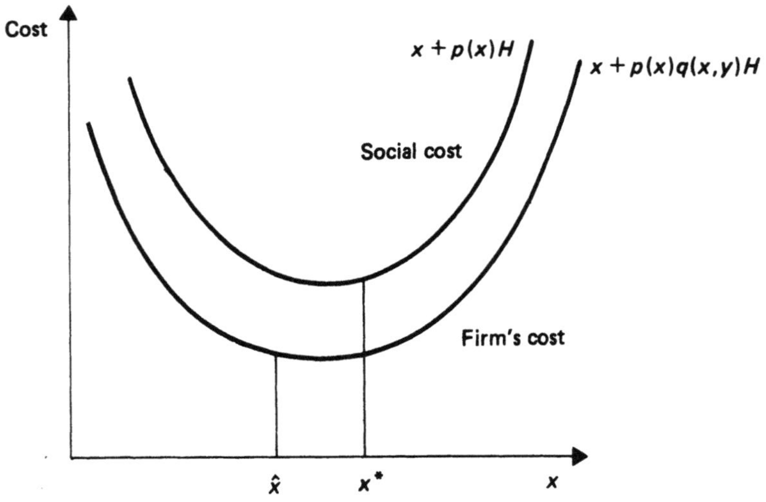

Figure 5.5. Firm's cost versus social cost.

Note that we have represented q as depending not only on y, but also on x. The probability of being held liable may depend in a court of law on whether "adequate" care was taken or not. Under strict liability, q would not depend on x but only on the level of monitoring y. This results in lower expenditures x for reducing the probability $p(x)$ of an accident.

This divergence between private (the firm's) cost and social (society's) cost can strongly affect the level of care taken by the firm and the resulting expected social cost. One would expect this divergence to be less the greater is the level y of regulatory expenditure on monitoring and litigating damages. However, these monitoring and litigation costs are themselves a part of the total social cost associated with waste management and are not freely available. We will not pursue here the problem of appropriately setting such expenditures (see Diver, 1980), but it should be clear that whatever this level is, there may still be significant departures between perceived private and social costs of safety and waste reduction measures undertaken by a firm.

What we pursue here is the issue of insurance and liability. The above model assumes that the firm meets all assessed damages from its own assets. This is unrealistic on two counts. First, the firm may not have sufficient assets to pay assessed damages, and second it may, freely or by law, insure itself. In *Figure 5.6*, we present a more complete model of the firm's decision alternatives.

Let us first consider the mechanics of *Figure 5.6*. The first branch of this decision tree reflects the firm's decision to buy insurance or not. Here we consider only two choices: buy insurance coverage of $\$I$ or buy no insurance. Actually, of course, the firm must decide how much coverage to purchase. The cost of insurance is denoted by r in premium per dollar coverage, so that the cost of $\$I$ of coverage is rI. The firm's assets are denoted A. Thus, if A is less than the net loss $H - I$, after insurance payments, then the firm (if required to pay H) would lose all its assets and some damage would go uncompensated by the firm. This explains the payoffs $x + rI + A$ of the branch "buy insurance I, accident occurs, firm held liable", which has a probability of $p(x)q(x,y)$ of occurring. For the same branch, if A is greater than $H - I$, then the firm would pay damages and final payoffs would be $x + rI + H - I$. The explanation of the other branches is similar [2]. A detailed discussion of the incentives embodied in "solving" the decision problem of *Figure 5.6* would take us beyond the scope of this chapter. However, the following points should be clear intuitively.

First, the issues of insurance (I), liability [$q(x,y)$] and care taken (x) are joint decisions. Second, the level of assets at risk (net of any insurance coverage) strongly influences the decisions which a profit-oriented (or cost-minimizing) firm will undertake. If we assume cost minimization, the lower these assets, the less the incentive for the firm

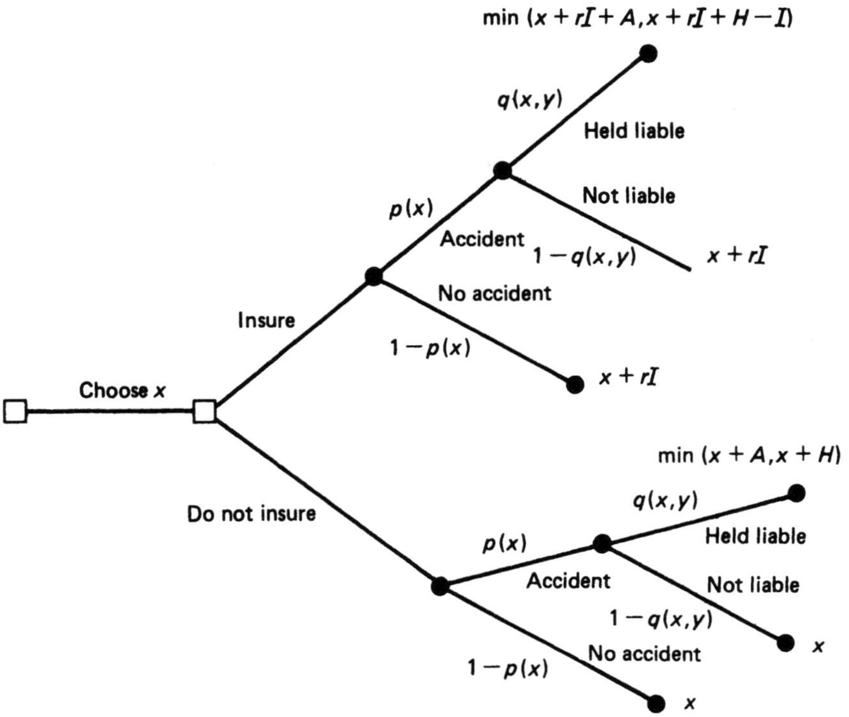

Figure 5.6. Firm's problem with insurance.

to invest in waste reduction or safety measures [3]. The policy conclusion of this analysis is that firms with small assets at risk may present a special problem in designing appropriate incentives for control/disposal of hazardous wastes. For such smaller firms, financial requirements under the Resource Conservation and Recovery Act (RCRA) and Superfund are just coming into effect in the USA. Appropriate levels of assets and insurance coverage remain a controversial topic (see Katzman, 1985).

Third, as noted above, if additional expenditures on safety or waste reduction reduce liability or assessed damages, the firm will have a stronger incentive to invest in such measures. Fourth, as has been noted by several recent commentators, large firms have not only assessed damages at stake in environmental impairment suits, but also reputations. To the extent that reputation losses are perceived to be substantial, the firm may perceive its net probable losses from an accident to be higher even than the total social costs. In this case, the firm could expend considerably more on waste reduction and safety measures than would be socially optimal.

Finally, the issue of compensation in the form of insurance premium subsidies or partial public payment of damages from accidents would have the joint effect of lessening the social inequity of uncompensated victims (due to inadequate assets or insurance coverage) as well as diminishing the incentives of the firm to undertake protective or waste reduction measures. Minimal insurance coverage laws would have the same equity/incentive effects.

Let us now consider some additional complexities not covered in the above analysis. We first note that the actual application of this general model to specific problems in waste reduction technology and transportation analysis involves detailed and firm-specific trade-offs in the allocation of a firm's scarce resources between output enhancement and waste/risk reduction. The interested reader is referred to Kleindorfer (1986) for a discussion of these issues.

A more serious simplification in the above model is the assumption that all damages are expressible in monetary terms. A richer view of the consequences of hazardous waste management is given in *Table 5.1*, where monetary (M), environmental (E), and adverse health effects (H) from an accident are related to four impacted groups.

Table 5.1 The consequences of hazardous waste accidents.

	Nature of damages caused by accident		
Source of compensation	*Monetary*	*Environmental*	*Health*
Firm	H_{FM}	H_{FE}	H_{FH}
Insurance	H_{IM}	H_{IE}	H_{IH}
Public and industrial funds	H_{PM}	H_{PE}	H_{PH}
Victims	H_{VM}	H_{VE}	H_{VH}

Here we show the disparate nature of potential losses from hazardous waste accidents as well as potential sources of compensation. The point of this table is that firms may only pay a fraction of actual damages, for example, owing to problems of assessing damages and proving causality. Remaining damages may have to be paid by insurance from public and industrial funds, or by victims. When victims pay, damages are uncompensated.

Given the nature of the firm as a profit-oriented organization, one would expect that only monetary, compensable consequences will be considered by the firm in its choice processes. These compensation considerations give rise to many related and unresolved policy issues concerning insurance in the hazardous materials area. These include the following:

(1) The problem of insuring nonsudden environmental impairment and health impairments. Given the latency periods for many such impairments, there are fundamental problems in assessing liability for an "accident", depending on when it happened, when it was noticed, and who the insurance and industry stakeholders were at these different moments of time. For example, the US insurance industry has vigorously complained against interpreting general commercial liability (GCL) coverage written in the 1960s and 1970s as implying coverage for nonsudden environmental impairments discovered in the 1980s (see Katzman, 1985). These risks, says the insurance industry, were never meant to be covered by the GCL policies in question.

(2) The problems of determining an appropriate historical data base for various coverage situations. The scientific uncertainties and ambiguities associated with hazardous materials and environmental impairment make it virtually impossible to assemble a "predictive data base" for assessing risks. This ambiguity strongly affects insurability and total available coverage for such risks, as we discuss further in the next section.

(3) The problem of small-asset firms and "the polluter pays" principle. Fair as this principle sounds, strict application would likely leave many victims uncompensated. Contrast the "polluter pays principle" with the "public pays principle". In the former, the firm faces complete liability for its actions, but this requires enforcement [4]. The resulting enforcement gap may be quite significant (see Diver, 1980), leaving many victims uncompensated, especially those with limited resources and access to the law. If the public pays, the problem of uncompensated victims is ameliorated, but now firms face even less of the social costs of potentially unsafe practices, and their incentives for undertaking efficient protective measures will be correspondingly less. This leads to the trade-off noted above between enforcement costs, incentives, and inequities of uncompensated victims.

(4) The problem of establishing fault and apportioning liability when several parties are involved in an environmental or health impairment case. The problems of establishing fault have led in the USA to the requirement of joint and several liability for nonsudden environmental impairment. This has, in turn, led to tremendous problems in the insurance industry in providing adequate coverage and to industry efforts to police its own clean-up and liability apportionment activities (see Conservation Foundation, 1984).

The essential point of the above discussion is the mediating influence of compensation, liability, and insurance on the firms' decisions concerning the volume of waste generated and transport means/modes/routes for disposal activities. A crucial question arising from this is where and how one locates disposal facilities. Concerning the approximate "where", this involves a trade-off between transportation costs and risks as firms go through the choice processes described above. The more difficult question is "how" to site such facilities, an issue we now examine in detail.

5.3. Compensation and Insurance Issues for Siting Hazardous Facilities

The siting of HMDFs presents a set of challenging problems owing to the conflicts among the interested stakeholders. Although there appears to be general agreement that properly designed and managed disposal facilities are desirable for society, few people are willing to sanction one in their own community. The perceived benefits from tax revenues and perhaps a few jobs pale in relation to the perceived risks. A public opinion survey, conducted by Resources for the Future in 1980, revealed that fewer than 10% of 1576 respondents were unconcerned as to how near their homes the disposal site for hazardous waste chemicals was located. Approximately 30% were willing to accept the facility within 10–14 miles of their residences. On the other hand, over 70% of the respondents favored the facility, as long as it was more than 100 miles from their residences (US Council on Environmental Quality, 1980).

The inability to site new facilities in the USA during the last three years (Bacow and Milkey, 1983) as well as in European countries, poses grave risks to society since the currently generated volume of hazardous materials exceeds the safe disposal capacity of existing facilities. In particular, the volume of hazardous materials generation is growing at a rate of 3% per year; at least 50–60 major new sites will be required over the next few years (US Environmental Protection Agency, 1980).

As has been pointed out in two excellent studies on the siting problem (O'Hare *et al.*, 1983; Morell and Magorian, 1982), there is a need to utilize policy tools, such as compensation and insurance, for helping to negotiate a feasible site, and to develop appropriate technology for storing the materials (e.g. incinerators, the deep well injection system). The need for a site will be affected by the type of waste reduction methods that industrial firms are undertaking, and its location will be partially determined by the costs of bringing the waste from the producer to the HMDF. Hence, there is a clear interrelationship between the production of waste, its transport, and the site selection process.

5.3.1. Key stakeholders

We will consider the role that compensation and insurance can play in helping to facilitate the negotiation process by first looking at the goals and objectives of key interested parties in the siting debate:

(1) *Waste generators*. Industrial firms are the principal sources of hazardous materials and, hence, most interested in finding a place to store them so they can continue their production of goods and services. They should be willing to incur part of the cost of siting a facility, particularly if they know that their own production processes will be threatened by their not having appropriate disposal mechanisms. Legally, waste generators in the USA are partially liable for costs of any accidents associated with an HMDF through rules of joint and several liability. These rules hold that defendants who were remotely connected with a hazardous chemical may be liable for a share of the damages, even if they were not directly associated with the handling of the waste themselves (Katzman, 1985).

(2) *Facility developer*. The developer has a financial incentive to compensate the host community for locating the waste treatment and disposal facilities in its backyard. These payments may partially be in the form of taxes; however, these revenues are normally relatively small unless the disposal facility is unusually large (Morell and Magorian, 1982, p 59). The developer will also be primarily responsible for property losses as well as adverse health and safety consequences due to the disposal facility. Hence, the developer will likely want to purchase insurance as protection against these potential losses. In the USA the RCRA requires all developers that store or dispose of hazardous materials to demonstrate financial responsibility to cover any accidents. This means that they must either obtain liability insurance, have enough funds to self-insure, or have some parent company guarantee them funds in case of an accident.

(3) *Host community*. As pointed out earlier, there are limited gains from siting a hazardous facility. The costs, on the other hand, are likely to be large, ranging from intangible psychological impacts on residents to negative environmental and health effects from the facility.

(4) *Insurance companies*. There has been an increasing concern by the insurance industry with the feasibility of offering coverage against potential losses from hazardous materials. There is limited statistical data on which to estimate the probability that there will be an accident from a hazardous facility. In addition, there is great uncertainty as to future court settlements regarding

liability from hazardous materials facilities, both in the magnitude of the awards and their timing (Huber, 1985). Following the Bhopal accident, there has been a withdrawal in pollution liability coverage and a large increase in rates. Today only two worldwide insurers provide coverage against large risks so that industrial firms, particularly the small ones, are scrambling to obtain coverage (Katzman, 1985).

(5) *Other residents*. Communities and residents who are not affected by the HMDF will benefit from the production of goods and services from the industrial concerns which generate these wastes as by-products. By having a facility located elsewhere they are provided with a cleaner and safer environment than if the wastes had been stored improperly.

We will restrict our discussion to the above stakeholders, since they are most directly involved in compensation and insurance arrangements [5].

5.3.2. Stages in the siting processes

In the negotiation process for choosing a site there are three distinct stages which need to be explicitly considered. These stages reflect different impacts that the waste disposal facility is likely to have on the community. Although we will be treating each of them separately below for analytic convenience, they will have to be viewed as part of a package when a final site is actually being selected.

Stage 1: Building the Facility

Unless taxes and employment benefits are significant, the disposal facility will be viewed as a "public bad" by any potential site. Hence, some type of ex ante compensation arrangement may be necessary to convince the community that the net benefits of the facility are greater than the expected costs.

The word "compensation" at this stage is frequently interpreted by the general public to mean a bribe or payment, particularly when the health and safety of people are involved. Society wants to preserve the belief that life is special; money may cheapen it, while laying bare the inequality of wealth (Calabresi and Bobbitt, 1978). For this reason, one should also consider changing the term "ex ante compensation" to "benefit-sharing" indicating that interested stakeholders, such as the developer, industry, or other communities who benefit from the facility, will provide either monetary payments or payments in kind to the host community. A number of successful examples of sharing gains from winners to potential losers suggests that this type of benefit-sharing may be feasible for future siting decisions:

(1) *Massachusetts Hazardous Facility Siting Authority Act.* This legislation, passed in July 1975, illustrates the use of monetary compensation for siting a regional resource recovery facility. One dollar per ton royalty would be paid to any city which agreed to host a regional resource facility. The town of Haverill initially agreed to play this role but, in the face of opposition, the City Council withdrew its offer. The town of North Andover was awarded the site in May 1977 after it virtually unanimously approved the project at a town meeting. By early 1981 several communities had committed their wastes to the proposed plant (O'Hare *et al.*, 1983).

(2) *Wes-Con Inc.*. This company converted two abandoned Titan missile silos in Idaho into small HMDFs. Although there was no public opposition, the developer offered in-kind compensation in the form of free disposal services, additional fire protection, and medical training (O'Hare *et al.*, 1983).

(3) *Gray Rocks power plant.* Environmental opposition halted construction of a coal-fired plant in Wyoming because of the potential damage to the surrounding environment. The suit was settled when the utility company agreed to set up a $7.5 million trust fund to preserve a stretch of the Platt River which was the habitat of several species of migratory birds, including the whooping crane (Lave and Romer, 1983). The coal plant was completed in 1981 and is fully operational today.

(4) *French nuclear power plants.* People living within 15 km of a nuclear power plant, who feel they are adversely affected, can apply to the local authorities for a reduction of 15–20% in electricity rates. Both businesses and households are eligible for this compensation (Kunreuther and Linnerooth, 1984).

(5) *LNG facility in Wilhelmshaven.* The FRG provided an additional subsidy for the construction of a recreational facility, because of the adverse effects that a liquefied natural gas (LNG) plant promised to have on the tourist industry in the area (Kunreuther *et al.*, 1983).

Stage 2: Living with the Site

An important consideration in the decision by a community to accept an HMDF is the impact that the facility will have on property values and future economic development. Residents will normally be concerned that the location of a waste treatment plant may discourage others from wanting to live in the community so that the market price of their property will fall and the community's tax base will be eroded.

If the developer is willing to provide some type of compensation to residents forced to sell their house at a price lower than its comparable market value elsewhere, then this may partially allay these economic fears. To our knowledge this arrangement has not been implemented in any community. One reason may be the difficulty of determining what a fair market price would be in the absence of an HMDF.

A principal reason that communities are concerned with the effects of an HMDF on their economic base has to do with the operation of the plant. Property values are likely to plummet if there are shown to be adverse health and safety impacts on workers and residents near the facility. Insurance can be a useful policy tool for encouraging firms to incur costs for improving plant design and undertaking protective measures (e.g. thicker liners for landfills, a better stack gas scrubber for an incinerator) if premiums are based on risk.

Regulations and specific standards can also be imposed with regular inspections to ensure that the plant is complying with the rules. Before a community embraces a waste disposal facility, it generally wants to be assured that the plant will be forced to meet these regulations or else be shut down. The following two examples illustrate these types of monitoring and control arrangements in practice:

(1) *Antonelli Corporation* (Providence, Rhode Island) constructed a facility to treat and store electroplating materials. The corporation allowed annual monitoring inspections by the citizens and city officials, as well as providing additional fire-fighting equipment to deal with any accident (Sanderson, 1984).
(2) *Wes-Con*. The developer informally negotiated with residents to permanently shut down the facilities if there were ever a fire.

In the USA, the RCRA program has taken a step in this direction by setting minimum standards with regard to the handling, transport, storage, and disposal of hazardous materials. Developers must locate, design, construct, and operate their facilities in conformance with Environmental Protection Agency requirements or equivalent state requirements. Federal and state regulatory officials have free access to inspect all these facilities at all reasonable times (Baram, 1982).

Stage 3: Accidents

Communities are naturally concerned with the risk of an accident from an HMDF storage site and its consequences to residents. There is a need for some type of guaranteed ex post compensation, by the facility developer, insurance industry, and/or the government, to cover the accident costs to victims. As we have already indicated, the ambiguity of the risk plus the long latency period associated with health consequences have made traditional insurance arrangements inadequate in this situation.

In addition, chronic health problems have multiple causes and may be influenced by genetic factors and nutritional status. It is difficult to determine whether hazardous substances produce specific effects and, hence, traditional toxic tort law may fail to provide adequate compensation for these damages (Trauberman, 1981).

Federal–Private Insurance. Some type of private–government insurance program may be necessary to cover the costs of accidents while at the same time offering protection to the operators of an HMDF. Specifically, a reinsurance guarantee by the government for losses above a certain magnitude may enable the private sector to cover the first layer of losses. The Federal Flood Insurance Program as well as the Price Anderson Act for protecting residents from nuclear power accidents have incorporated this feature.

The insurance industry will undoubtedly still be concerned with the uncertainty associated with claims payments owing to the latency problem. One way to deal with this problem is to offer "claims made" policies in contrast to standard occurrence policies. Under this new policy, an insurer's coverage obligation is triggered by the receipt of a written claim rather than when bodily injury occurs (e.g. exposure to asbestos dust). Hence, the insurer will be able to determine its losses in any given year, since only individuals with current policies can collect on their claims even if exposure had occurred in prior years.

Trust Funds. Some type of trust fund similar to the one established by the Black Lung Act may help provide compensation to those exposed to toxic substances from an HMDF. In the Black Lung Program miners who are totally disabled owing to pneumoconiosis (a type of respiratory cancer) and can demonstrate that the disease arose from employment in a US coal mine are entitled to compensation. The Act requires miners to obtain insurance but a trust fund is established to pay claims where a "responsible" mine operator cannot be identified or has not paid benefits.

The concept of Superfund established under the RCRA is modeled after the Black Lung Program. A fund is created to clean up existing environmental waste financed partially by the US Government but primarily by charges on the wastes generated by the chemical industry. In contrast to the black lung disease where causality is easily determined, there are problems as to who will be compensated from Superfund owing to the causality problems mentioned above (Baram, 1982).

Self-Insurance Funds. If private or government insurance is unavailable then special types of self-insurance arrangements may be necessary to protect industry and developers from suffering large losses from an accident. For example, the chemical industry may want to administer an insurance fund where premiums would be collected from the parties engaged in the storage and disposal of waste. A fund would then be established to pay for specific losses. The concept may be attractive in theory but has large-scale administrative problems. Some type of monitoring and control of facilities would be essential to base premiums on risk and to assure that those participating in the process were not behaving carelessly because they knew they were protected from losses through this industry–developer fund.

5.3.3. Integrating stakeholders and stages

The different interested parties and the three stages of the siting process can be integrated through a stakeholder–strategy matrix. *Table 5.2* illustrates the use of the matrix by depicting which interested parties are likely to be involved with a policy strategy associated with a particular stage in the siting process. For example, monetary or in-kind compensation in Stage 1 should involve the host community, other communities, the developer, the waste generator, and the state siting

Table 5.2 Stakeholder–strategy matrix for siting an HMDF.

	Stage 1: Locating the Site	*Stage 2: Living with the Site*		*Stage 3: Accident*
Stakeholder	*Monetary or in-kind compensation*	*Compensation for property value decreases*	*Environmental regulation*	*Insurance and ex post compensation*
Host community	•	•	•	•
Other communities	•			
Developer	•	•		•
Waste generator	•			•
State siting authority	•		•	
Federal government agencies			•	•
Insurance industry				•

authority. One can see from the matrix that the insurance industry will normally be associated only with Stage 3 of the process.

We have intentionally developed the stakeholder–strategy matrix in rather primitive form to illustrate its potential for analyzing the siting process rather than being very specific as to the roles that each stakeholder will play. The institutional arrangements for a particular country or state will define the available compensation and insurance options for consideration. A number of studies have revealed the importance of cultural differences in influencing the way siting decisions are made and the policy strategies which have been utilized [6].

We can illustrate the conditions necessary for determining which sites are feasible by constructing a relatively simple decision tree for any given stage under the assumption that there are n candidate sites under consideration for an HMDF. Consider Stage 1. *Figure 5.7* depicts the branches of the tree if there are m different events with event j having a probability p_{ij} that outcome C_{ij} will occur. An example of a particular event in Stage 1 is the construction of an HMDF that meets all specifications; another event might be the discovery that the HMDF is on an earthquake fault after it is completed. We will assume that there are no events which will yield any net benefits from the site itself so that each $C_{ij} < 0$.

In order for site i to be willing to accept the HMDF, some type of compensation or benefit (B_{ij}) must be provided if event j occurs. If a fixed ex ante compensation package is given to the community no matter what event occurs, then $B_{ij} = B_i$ for all j. We have adopted this assumption in *Figure 5.7*. For Stages 2 and 3 the amount of compensation will most likely depend on specific events. For example, if the developer pays residents for decreases in property values, then there will be a probability distribution associated with anticipated changes in market prices of property in the area.

The only feasible sites are those where the proposed benefit package provided by the other stakeholders is sufficiently large that the community is willing to accept the HMDF. In addition, the health and safety standards must be acceptable to both the site and the developer.

The above decision tree provides a framework for discussing the issues of risk perception and eliciting community preferences for accepting a site. Both of these issues are important in specifying a compensation and insurance package that has a chance of being implemented.

Perception of Risk

There is an extensive series of controlled experiments on public perception of risk that suggests that the figure C_{ij} associated with each

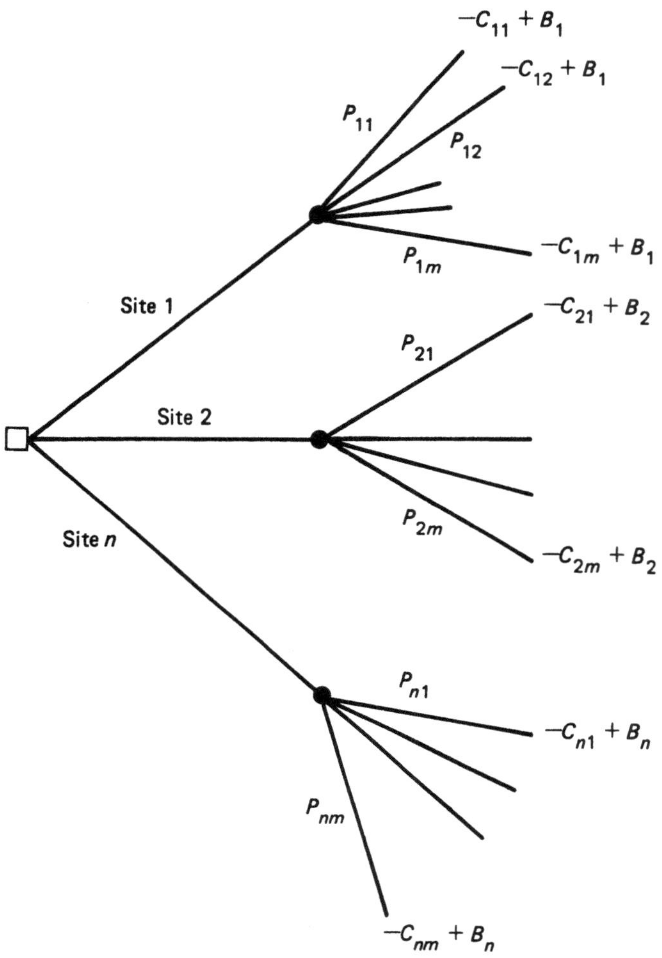

Figure 5.7. Decision tree for Stage 1 of siting process.

event j at site i has a number of attributes attached to it. Slovic *et al.* (1980) have analyzed extensive data on individuals' attitudes toward a number of risky activities ranging from skiing to living near a nuclear power plant. Two general factors appear to capture the public's rankings of different activities: the dread of the risk and whether the risk is knowable. Technologies perceived as uncontrollable, catastrophic, involuntary, and highly risky to future generations score high on the dread factor. Those that are not observable, unknown to those exposed, delayed, and new are viewed as unknown risks.

If a proposed technology scores high on both factors, as one would expect an HMDF to do, then the required amount of compensation will also be high. Events such as Love Canal, Seveso and Bhopal may raise the public's concern even though the technologies associated with these disasters were not the same as those of proposed hazardous waste facilities. One of the difficulties in siting new technologies is that there are differences in perceived risks by the relevant stakeholders. Industry and the developer may feel that an HMDF does not represent much of a risk because the technology is known to them and considered controllable, while the public has a different view. Von Winterfeldt and Edwards (1984) have noted that conflicts with respect to technology are due to differing values between the stakeholders.

A related challenge in siting HMDFs is estimating the probabilities associated with the different events. Given the limited historical data on accidents from an HMDF coupled with the difficulty of determining causal relationships between specific diseases (e.g. cancer) and specific toxic substances, there is considerable ambiguity associated with probabilities of losses being related to an HMDF. Extremely low probability events, such as the Bhopal accident, may fall in the realm of what Weinberg (1972) calls "transcientific" phenomena, which implies that there may be no practical basis for estimating the statistical chances and consequences of specific types of accident.

Ambiguity associated with losses has two principal effects. Insurance firms have relatively little interest in marketing coverage since they are unsure of the chances of incurring specific claims. For example, the insurance industry has been opposed to nuclear coverage claiming that the risk is not insurable because of the ambiguity associated with such losses (US Nuclear Regulatory Commission, 1983). Secondly, if there is ambiguity surrounding the probability of specific events, the potential host community may imagine the potential losses graphically rather than focusing on the probability dimension. An advantage of employing some type of benefit-sharing arrangements is that potential sites will be forced to concern themselves with tradeoffs between obtaining a certain payment now or having to incur possible losses in the future with some unknown probability.

Eliciting Preferences from Communities

There is a large literature which has emerged in economics that attempts to answer the following question: How can one design an approach for eliciting the true willingness of an individual or community to accept a facility which does potential harm to them but benefits others? If such a mechanism induces truth telling on the community's part it is called "incentive compatible" [7].

Economists have searched in vain for incentive compatible mechanisms which will maximize aggregate net benefits and balance the budget (i.e. the amount paid by developers and other communities covers the amount demanded by the host community). Problems of equity and fairness complicate the matter further [8]. The institutional arrangements associated with the siting process add a further dimension to designing appropriate compensation mechanisms. We will illustrate this latter point by briefly describing two facility siting laws, each of which implies a different set of negotiation procedures.

(1) *Massachusetts Hazardous Waste Facility Siting Act.* Under this legislation passed in 1980 the developer negotiates a siting agreement with the host community, offering compensation if necessary to satisfy the residents' concerns. The community itself cannot exclude an HMDF unless they can demonstrate that these facilities pose special risks. Compensation is also provided to communities likely to be affected by an HMDF in adjacent districts. Arbitration is offered to break any deadlocks (O'Hare *et al.*, 1983).

Under this arrangement the developer has an incentive to offer the lowest possible compensation to the host community as well as its neighbors while the communities have an incentive to demand as large a payment as possible. To the extent that there are a few communities vying for the facility, some type of auction might ensue with the lowest bid hosting the HMDF. Even here there is no incentive for a community to tell the truth if it knows that its bid is likely to be considerably below that of any of the other possible sites. To date there has not been any HMDF located under the Siting Act, although the town of Taunton was on the verge of accepting a facility in 1984.

(2) *Ontario Waste Management Corporation (OWMC).* This Crown Agency established in 1981 has the responsibility for determining the types and quantities of waste in Ontario that are not receiving proper treatment, specifying potential sites for locating an HMDF that will provide maximum protection for health and the environment, as well as developing ways for Ontario industries to reduce the volume of waste requiring final treatment and disposal.

With respect to the siting procedure, the OWMC has identified eight possible sites within the Golden Horseshoe region around the western end of Lake Ontario. This region generates at least 70% of the province's liquid industrial waste. Although compensation is not explicitly part of the OWMC siting process, it may have some potential value in facilitating the negotiation process between these eight sites.

We have been investigating an allocation mechanism whereby each site submits a bid indicating how much they would require to host the facility (Kunreuther *et al.*, 1986). The site which provides the lowest bid is the "winner"; all the other communities would have to pay 1/7 of their bid to the OWMC [9]. This procedure is guaranteed to yield a budget surplus so that the regional siting authority would be able to use any funds not given to the host community to reward residents along the approach routes to the site, all of whom will be demanding a share of the pie.

To limit the portion of the population requesting some form of compensation it is useful to construct the approach routes in a concentrated band so that the potential losers are well defined. If there are industries associated with each site, as there may be in Ontario, then the payments could be made by them rather than using the tax revenues of the communities. One other theoretical advantage of the proposed mechanism is the inability of any potential site to form a coalition with any other candidate area.

Both the Massachusetts and Ontario siting plans recognize the challenging process of finding a suitable location for an HMDF. They remind us of the importance of integrating compensation and insurance policies into a broader framework. More specifically, there is a need at the outset to screen all of the potential sites from an environmental and economic perspective, eliminating any that do not meet specific health and safety criteria and may be inefficient with respect to the transportation of wastes from industries in the region. Following this step one must gain an understanding of the values of each of the interested parties by detailing their perceived costs and benefits of different sites. This action can be taken using some type of value tree analysis as developed by von Winterfeldt and Edwards. At this point, conflicts between the stakeholders will be uncovered and specific benefit- and risk-sharing arrangements can be explored such as compensation and insurance. Specific standards for design and operation of the HMDF can be stipulated with appropriate monitoring and control procedures by a governmental agency instituted.

5.3.4. Normative criteria for a desirable siting process

To conclude this section we will specify four normative criteria in determining a siting procedure for any given state or country. These criteria are couched in relative rather than absolute terms since each society will determine its own appropriate targets. Since the status quo usually serves as a reference point, there may be a tendency to stay close to existing procedures.

Criterion 1: Degree of Openness

The prevailing viewpoint in the USA and Canada is that the siting process should be sufficiently open that different viewpoints can be heard and that the public can actively participate in the final siting decision. Both the Massachusetts legislation and the Ontario procedure provide for extensive public participation through meetings, information exchanges, and sessions with individuals and groups. Other siting procedures have had a similar character [10]. An illustration of a successful siting process which involved active participation was the Alberta, Canada, experience where over 75% of the local residents voted in favor of hosting an HMDF. One reason for the large support base was active involvement of the public at all stages of the process, beginning with the determination of siting criteria, review of developer proposals, and inputs into the final site analysis and environmental impact assessment (McGlennon, 1983).

Criterion 2: Nature of Deadlines

How rigid should time schedules be with respect to final decisions on the siting of an HMDF? To our knowledge there have not been any specific deadlines regarding the choice of a final location within any region or country. For this reason there have been few examples of successful siting decisions in recent years.

We feel that specific deadlines are very important for facilitating the determination of a feasible site. The necessity of either significantly reducing waste generation or constructing safe disposal facilities in the near future has forced this action. Fixed schedules also reduce the incentives for other interested parties to use information as a delaying tactic.

Criterion 3: Specificity of Contractual Arrangements

How detailed and well specified should contractual siting arrangements be? We feel that these arrangements should be explicitly specified at the time a host community is chosen so all stakeholders know their responsibilities at each of the three stages of the siting process. These specifications would provide a firm basis for negotiations between stakeholders since the expected gains and losses would be clearly delineated.

Criterion 4: Nature of Compensation and Insurance

The philosophical underpinnings guiding societal decisions will influence the role that compensation and insurance play in the siting process. For example, a libertarian system stresses the importance of individual freedom and any project that is likely to harm a single person would not be approved unless that person was sufficiently compensated so that (s)he felt as well off after the siting decision as before.

Under a utilitarian system, an HMDF would be approved if the aggregate net benefits were positive even if some people would be made worse off than under the status quo. We favor the explicit use of compensation and insurance mechanisms for guiding siting decisions with the caveat that they be utilized in combination with other policy mechanisms, particularly regulatory standards for health and safety of the HMDF.

5.4. Concluding Remarks and Future Research

Up to this point we have focused on hazardous waste management within a single jurisdiction. Recently, interstate and international problems have also received much attention [Cain, 1983; Gusman *et al.*, 1980; the Organization for Economic Cooperation and Development (OECD), 1985]. The issues arising in multijurisdictional settings include the following:

(1) Classification of wastes (see Wynne, 1984, Ch 5) strongly influences how wastes are regulated and controlled among several jurisdictions. Given the various purposes which classification schemes serve, it is not surprising that it has taken considerable time to generate international agreement on a common (if rough) classification scheme for regulating and monitoring international and transborder flows of hazardous wastes.
(2) Documentation and reporting of transborder flows is essential to maintain control of hazardous waste shipments. Recent agreements on reporting formats and procedures in the OECD (1985) provide a good example of the nature and magnitude of regulatory problems for transborder flows of hazardous materials.
(3) Financial responsibility and liability laws require international agreement as well. For example, if such laws are meant to provide standards and incentives for carriers, they will be ineffective if mere registry in another jurisdiction, with less stringent regulations, were an exemption to the requirement.
(4) Interstate and international cooperation in coordinating decisions on infrastructure development is necessary. For example, each jurisdiction might develop its own infrastructure for low-level toxic wastes but cooperate in constructing a single facility for disposal of highly toxic wastes. To date, problems of coordinating and conflicting regulations have hindered such international and interstate cooperative ventures (see Cain, 1983; OECD, 1985).
(5) A related problem is assuring sufficient demand for collection and disposal facilities. The point is that there are large fixed costs for such facilities. These fixed costs imply that a high volume of wastes must be processed in order to break even. They also lead to a number of sustainability problems, since high per-unit charges to users provide them with an incentive to look elsewhere

(e.g. neighboring states) for disposal opportunities. The fall-off in demand can be expected to be highest for high disposal-cost (e.g. high-hazard) wastes. The resulting "cream-skimming" and demand reductions then require either substantial subsidies by the state or significantly higher monitoring and enforcement costs to force industry to use such high-cost facilities. It is precisely this scenario which has unfolded in Bavaria and Hessen in the FRG (see Wynne, 1984, Ch 5) and in Sweden recently. This problem requires international or interstate coordination for its resolution.

In addition to the above international issues, several important research questions arise from our foregoing analysis. Concerning firms' choices of waste reduction and transport means, the key issue for future research is to determine how firms actually value liabilities and risks in these decision areas. The evaluation processes, as embodied in choices such as the one illustrated by *Figure 5.6*, are quite complex, even for "hyper-rational" firms. The question is, therefore, how do and will firms respond to the policy options on insurance and compensation discussed in this chapter (OECD, 1985).

With respect to siting decisions, there is a need to determine ways in which one can overcome the stigma associated with the use of compensation as a way of sharing gains from winners to potential losers. Siting is a political issue and there needs to be ways that local communities can convince their constituencies of the benefits of having an HMDF. Experimenting with in-kind compensation arrangements such as providing a community with improved health facilities may be helpful in this regard. Trade-offs between lives lost and lives saved can then be made more explicit than if just money were given to the community.

Concerning research on siting procedures, a key area of interest is bargaining and collective choice research. Both of these areas are well developed theoretically and some work has begun to focus specifically on bargaining in siting problems (Kleindorfer, 1985; Kunreuther *et al.*, 1986). Field and experimental validation of these theoretical concepts is incomplete in several important respects relative to HMDF siting problems. First, validation is required for theoretical predictions of bargaining outcomes in problems involving the existence of risk with possible compensation from winners to potential losers. Second, theories of bargaining typically assume that unanimous consent is required for an overall collective action; otherwise the status quo or some default options (e.g. restriction of industrial output) obtains. However, in siting there are typically multiple jurisdictions, each of which has enactment or veto power over only a part of the overall solution. This is clearly a different sort of problem from that dealt with in traditional theories of bargaining and more research on this topic is needed.

Regarding insurance arrangements, there is a need for some creative policies to deal with the current uncertainties associated with environmental hazards. This requires more well-defined liability arrangements, a better appreciation by the courts as to the impacts their decisions are having on the stakeholders concerned with hazardous waste, and a well-defined role for government in dealing with catastrophic losses. The potential for other industry initiatives to complement actions such as Clean Sites Inc. should also be explored. In particular, some type of industry-wide insurance program with rates based on risk deserves consideration given the reluctance of the insurance industry to offer pollution liability coverage.

Notes

[1] For simplicity we deal first only with monetary costs. Clearly a more realistic approach would measure consequences on at least three distinct dimensions: monetary, environmental, and health. We consider this issue further elsewhere in the chapter.

[2] For the mathematically inclined reader, we may note that the total expected costs for the firm facing the decision problem embodied in *Figure 5.6* are

$$\mathrm{COST} = x + rI + p(x)q(x,y) \min\,[\max(H - I, 0), A]\ .$$

[3] This could give rise to incentives, possibly to be thwarted by judicial precedent, for larger firms to set up minimum-asset subsidiaries to shield the assets of the larger firm from damage assessment. In the *Amoco Cadiz* case, which involved the transport of crude oil, the total assets of the transport firm was the market value of one (debt-ridden) tanker.

[4] This discussion essentially mirrors the analytical presentation associated with *Figure 5.5*, where the enforcement gap represents the diluting effect of $q(x,y)$ on firm incentives.

[5] Other stakeholders involved in the siting process are environmental and public interest groups, local politicians and planners, as well as contractors and labor: for more detail, see Morell and Magorian (1982, Ch 4).

[6] See Kunreuther *et al.* (1983) for a comparison of the siting of LNG facilities in four countries. The differences in locating HMDFs across countries are discussed in Wynne (1984) and Gusman *et al.* (1980).

[7] For a detailed discussion on incentive compatible mechanisms for this problem, see Raiffa (1982, Ch 22).

[8] There is a large body of work in recent years designed to elicit willingness to pay and willingness to accept values for public and private goods: see Brookshire *et al.* (1985) for a summary. Recently, Smith *et al.* (1985) completed a questionnaire to determine an individual's evaluation of the expected reduction of risks of hazardous materials. On the basis of the field survey in the Boston area, the authors tentatively conclude that one may be able to elicit households' benefits for

reduction of hazardous materials risks. If this preliminary finding is substantiated then one may be able to compare these estimates with those obtained from procedures, such as the one described in the text.

[9] The rule is that communities not selected pay $1/(n-1)$ of their bid if n communities are in contention for a site. In Ontario, as we noted, there are eight candidate sites.

[10] See Andrews and Pierson (1983) for a comparison of state approaches.

References

Andrews, R. and Pierson, T. (1983), *Hazardous Waste Facility Siting: A Comparison of State Approaches*, Mimeo, Institute for Environmental Studies (The University of North Carolina, Chapel Hill, NC).

Bacow, L. and Milkey, J. (1983), Responding to local opposition to hazardous waste facilities: The Massachusetts approach, *Resolve*, Winter/Spring, 1, 4, 8.

Baram, M. (1982), *Alternatives to Regulation* (D.C. Heath, Lexington, MA).

Bradshaw, M. (1985), *A Step Toward Certain and Long-Range Stability*, Address presented at Annual Meeting of Insurance Services Office, Inc., New York, NY, January 16.

Brookshire, D.S., Coursey, D.L., and Schulze, W.D. (1985), *Experiments in the Solicitation of Private and Public Values: An Overview*, Prepared for the Public Choice Society Meetings, New Orleans, LA.

Brusset, H. and Rochevolles, G. (1979), General survey of the quantities of solid waste and sludge produced in Europe and their processing cost, in *Recycling Berlin* (E. Freitag, Berlin).

Cain, J.T. (1983), Routes and roadblocks: State controls on hazardous waste imports, *Natural Resources Journal*, **23**, 767–793.

Calabresi, G. and Bobbitt, P. (1978), *Tragic Choices* (Morton and Co., New York, NY).

Conservation Foundation, The (1984), *Clean Sites and Private Action* (Collins Lithography & Printing Company, Inc., Baltimore, MD).

Diver, C. (1980), A theory of regulatory enforcement, *Public Policy*, **28**, 257–299.

Gusman, S., von Moltke, K., Irwin, F., and Whitehead, C. (1980), *Public Policy for Chemicals, National and International Issues* (The Conservation Foundation, Washington, DC).

Huber, P. (1985), Safety and the second best: The hazards of public risk management in the courts, *Columbia Law Review*, **85**, 277–337.

Katzman, M. T. (1985), *Chemical Catastrophies: Regulating Environmental Risk Through Pollution Liability Insurance* (Richard Itwin, Homewood, Ill).

Kleindorfer, P. (1985), *Negotiation and Bargaining for Noxious Facility Siting: An Experimental Design*, WP 85-02-06, Department of Decision Sciences (University of Pennsylvania, Philadelphia, PA).

Kleindorfer, P. (1986), Compensation and negotiation in the siting of hazardous waste facilities, *Science of the Total Environment*, **51**, 197–208.

Kunreuther, H., Kleindorfer, P., Krez, P., and Yaksick R. (1986), The role of compensation for siting hazardous facilities: theory and experimental design, *Journal of Environmental Economics and Management* (in press).

Kunreuther, H. and Linnerooth, J. (1984), Low probability accidents, *Risk Analysis*, **4**(2), 143–152.
Kunreuther, H., Linnerooth, J., Lathrop, J., Atz; H., Macgill, S., Mandl, C., Schwarz, M., and Thompson, M. (1983), *Risk Analysis and Decision Processes: The Siting of Liquefied Energy Gas Facilities in Four Countries* (Springer-Verlag Publishers, Berlin).
Lave, L. and Romer, T. (1983), Specifying risk goals: inherent problems with democratic institutions, *Risk Analysis*, **3**, 217–227.
Martens, B. (1981), Regulation of the transportation of hazardous materials: a critique and a proposal, *Harvard Environmental Law Review*, **5**, 345–387.
McGlennon, J.A.S. (1983), The Alberta experience ... hazardous wastes? – Maybe in my backyard, *The Environmental Forum*, February, 23–25.
Morell, D. and Magorian, C. (1982), *Siting Hazardous Waste Facilities, Local Opposition and the Myth of Preemption* (Ballinger, Cambridge, MA).
OECD (1985), *Proceedings of the Conference on Hazardous Waste, Paris, June, 1984* (Basler Publishing, Basel).
O'Hare, M., Bacow, L., and Sanderson R. (1983), *Facility Siting and Public Opposition* (Van Nostrand-Reinhold, Inc., New York, NY).
Raiffa, H. (1982), *The Art and Science of Negotiation* (The Belknap Press, Cambridge, MA).
Sanderson, R. (1984), *Compensation and Negotiation Primer*, Mimeo.
Shavell, S. (1984), A model of the optimal use of liability and safety regulation, *Rand Journal of Economics*, **15**, 271–280.
Slovic, P., Fischhoff, B., and Lichtenstein, S. (1980), Facts vs. fears: Understanding perceived risk in societal risk assessment: How safe is safe enough?, in R. Schwing and W. Albers (Eds) (Plenum Publishing Co., New York).
Smith, K., Desvousges, W.H., and Freeman, A.M., III (1985), *Valuing Changes in Hazardous Waste Risks: A Contingent Valuation Analysis*, Draft Interim Report (2 vols), prepared for US Environmental Protection Agency.
Trauberman, J. (1981), Compensating victims of toxic substance pollution: An analysis of existing federal statutes, *Harvard Environmental Law Review*, **5**, 1–29.
US Council on Environmental Quality (1980), *Public Opinion on Environmental Issues: Results of a National Public Opinion Survey* (US Government Printing Office, Washington, DC).
US Environmental Protection Agency (1980), *Hazardous Waste Generation and Commercial Hazardous Waste Management Capacity*, Report SW-894 (US Government Printing Office, Washington, DC).
US Nuclear Regulatory Commission (1983), *The Price Anderson Act: The Third Decade Report to Congress*, December (US Government Printing Office, Washington, DC).
von Winterfeldt, D. and Edwards, W. (1984), Patterns of conflict about risky technologies, *Risk Analysis*, **4**(1), 55–68.
Weinberg, A. (1972), Science and Trans-science, *Minerva*, **10**, 209–222.
Wynne, B. (1984), *Hazardous Waste Policy Management: Institutional Dimensions*, Working Paper WP-84-41 (International Institute for Applied Systems Analysis, Laxenburg, Austria).

5.D. Discussion

J. Nichols

This discussion of insurance and compensation as policy instruments presents a cogent argument in favor of applying these tools to facilitate the public dialogue on hazardous waste management. The authors are to be congratulated on articulating the complex problems faced by decision-makers in government and industry to assure a viable and safe infrastructure for the manufacture, transport, and use of hazardous materials. They also point to inherent uncertainties in assessing risk and to a related lack of public confidence in government's and industry's abilities to manage such risks associated with the technologies that comprise this infrastructure. Recognition of this is critical to a full understanding of both the utility and the value of "tools" such as insurance and compensation as policy instruments. The fact remains that the availability of economic and noneconomic incentives to, say, potential host communities of hazardous waste treatment facilities has proved ineffective in addressing the "not in my backyard" syndrome that has frustrated siting attempts in the USA and increasingly in other countries.

In my own state of Massachusetts a siting law does in fact recognize the significance of compensation and insurance as policy tools to ensure an infrastructure of safe waste management. The Hazardous Waste Facility Siting Act of 1980 requires insurance well above the Federal Environmental Protection Agency (EPA) requirements and calls upon developers to negotiate compensation with a host community. It is from the perspective of a public sector practitioner with responsibilities related to the implementation of that Act that this response is written.

Theoretically, these instruments should facilitate policy implementation. Unfortunately, in practice, they fall short of the mark. As it becomes increasingly difficult for industrial firms to acquire insurance for long-term liability from pollution, public policymakers will be forced to rethink regulations that have yet to be tested and to seek other means to foster appropriate and sound waste management practices. With respect to compensation, the issue is even thornier. The public perception of risk associated with the treatment or storage of hazardous wastes is so grave that it is virtually impossible to arrive at a dialogue stage with respect to compensation.

Consider the Massachusetts experience. Despite a well-defined process for identifying impact and for negotiating measures to reduce impacts and compensate a host community for accepting a facility, five attempts to develop facilities have been abandoned. The experience has been humbling – and has given rise to a broader perspective on what it will take to foster sound waste management and to site much needed facilities. The fact is that we have been unable to really test these tools because we remain at the starting gate with respect to hazardous waste treatment facilities.

This is not to say that these tools have failed us. Indeed, we have not even reached the point where they can be tried out. Rather, the problem rests in the broader context of society's understanding of the need for such facilities. Kleindorfer and Kunreuther stress this in the introduction:

> The key in this effort is to balance the benefits associated with the use of hazardous materials in producing products and services consumed by society with the costs associated with understanding and controlling their potentially negative side effects.

Society – the public at large – needs to understand this in order to move forward. And public policymakers need to incorporate such notions in the development and application of their policies. From a practical stance, integration of these tools into an array of regulatory and market forces must be accompanied by continuing efforts to communicate with the public to foster an appreciation of the notion that waste management is the concern of all of us. Industry, too, must simultaneously be prepared to demonstrate that effective measures are being taken to reduce the amount of wastes being produced at their source of generation.

5.D.1. Compensation, liability, and insurance issues for waste generators and transporters

Relief from insurance and liability regulations can provide incentives for appropriate waste management practices, but only to a limited degree. Economic, regulatory, and liability concerns together are forcing industrial, commercial, and institutional waste managers to examine and, in some cases, to reshape their hazardous waste management practices. In many cases these practices are moving away from an end-of-pipe treatment approach to a waste reduction, reuse, and recycling approach. In Massachusetts, where strict financial responsibilities are applied to waste generators and managers, the cost of insurance is regarded as part of the cost of doing business. On the other hand, it has been noted that insurance is not likely to cover the costs of major accidents or damages resulting from nonsudden environmental incidents. In the end, a firm's choice of how much waste to produce and what means it chooses to transport and dispose of the waste will depend primarily upon two factors: the state's enforcement capabilities (swift and effective) and economics. For private companies, the bottom line is profit: What can they do to earn the most?

Massachusetts is one state to implement strict financial responsibility regulations. The purpose of these provisions is to ensure that adequate funds are in fact available to close a facility and to provide postclosure care, if necessary. Our financial regulations are in fact much stricter than the federal requirements in several respects:

(1) Sudden/accidental insurance limits required are $3 million per incident and a $6 million annual aggregate (EPA: $1 million and $2 million, respectively).
(2) Non-sudden gradual pollution insurance limits are $5 million per accident and a $10 million annual aggregate (EPA: $3 million and $6 million).
(3) Cost estimates and funding for closure may not be lumped with similar requirements for out-of-state facilities owned by the same company.
(4) Neither the "financial test" (exempting firms with large net worth) nor corporate guarantee is included as an option for either closure or postclosure funding or as a substitute for third party liability insurance.

These regulations may prove problematic in the long term, especially with respect to the nonsudden gradual pollution insurance limit. As insurance companies withdraw from offering this coverage, policy makers will have to consider other options to assure communities that there will be adequate protection.

One fruitful avenue to consider in insurance and liability regulation is to consider waste generator group systems with incentives for companies that institute appropriate waste management systems. A management designed to minimize environmental impacts and potential liabilities would attempt to maximize waste reduction and to reuse or recycle those wastes that cannot be eliminated.

5.D.2. Compensation and insurance issues for siting

Compensation and insurance are clearly issues in siting facilities to treat hazardous wastes. Any community willing to consider hosting such a facility would certainly want to be assured of adequate insurance to cover the costs of damages or losses from operations or closures. Moreover, such a community could legitimately argue for compensation to accept risks generated by society at large – in as much as all of society relies on the products that generate the wastes.

To assess where we are today in hazardous waste management and to determine the effectiveness of tools like compensation and insurance we need to remember the social impact of the revelations of Love Canal and the Valley of the Drums on society; and we need to remember the fact that we were unprepared to deal with these calamities. The shock effect of those revelations crippled the public's confidence in industry's and government's ability to cope with this issue. This means that the public will be wary of *any* tools designed to facilitate siting: there is much to be done to win public confidence before such tools can be effectively applied.

When we first started working on this issue, we did not know a lot about it – except that we had to do something to avoid the mistakes of the past. We did not know where the generators were; we did not know what was in their waste streams, or how much waste there was, or how hazardous it might be; and we did not know where it was going. Our institutional structures for dealing with enormous data collection needs were minimal. Governments had virtually no institutional capability whatsoever to firmly anchor necessary actions. We had to set up the institutional structures, multiply enforcement capabilities, develop planning and analysis teams, and address the very legitimate fears of the public. We had to move quickly. And we had to do so in a climate of fear, mistrust, and acrimony.

A great deal has been accomplished in a short span of time. First, we know a great deal more about the hazardous waste problem. There is no question that it is a large problem. Second, we know that it must be dealt with in various ways – that there is no single solution. Third, institutional mechanisms for dealing with it are now in place.

In Massachusetts, the Dukakis Administration put a comprehensive plan for hazardous waste management into place in its first days. It is a program that addresses both the problems of the past and the needs of the future. The program encompasses five elements:

(1) The identification and clean-up of contaminated hazardous waste sites: the state's Superfund legislation.
(2) The promulgation and enforcement of strict regulations for waste management: Phase II regulations are in place and Phase III will be ready shortly.
(3) The reduction of hazardous waste generated at the source: we are now developing one of the nation's first source reduction programs.
(4) Siting of treatment facilities to meet industry's needs.
(5) Establishment of sound public transportation and information programs to involve citizens in resolving this problem at regional and local levels.

Despite this sound multifaceted approach and a siting law that builds in mechanisms to ensure compensation and insurance, our ability to site facilities to treat hazardous waste has been seriously hampered.

At present, compensation under Massachusetts law can be negotiated between the developer and potential host community. This has not proved fruitful for one basic reason: in our siting process, no one has arrived at this point in negotiations. All attempts have been stopped by public opposition *before* they have reached this important point.

Thus, the fundamental question we must ask ourselves is not whether or not these are good tools, but, rather, how they fit into a comprehensive waste management strategy.

I must stress how important this is to overcoming the major obstacle to siting – local opposition. Tangible demonstration of the link between economic development and safe disposal is essential. And one way industry can help is through a commitment to assist economic development in communities that accept safe treatment facilities.

Finally, and most important – before tools such as insurance and compensation can be applied effectively – all those concerned with hazardous waste management must work to achieve better communications on the issues of safety, uncertainty, and economics.

A decade ago no national initiatives had been taken to deal with waste management, and hazardous waste activities were inadequately regulated. Today federal and state legislation aims to protect public health through "cradle to grave" regulation of hazardous wastes from the point of generation through storage and transportation to treatment or disposal. But hazardous waste is an issue that will not be resolved only by law and regulation. Those of us who have worked on this issue over the past few years recognize its scientific complexity and political volatility only too well. Solutions will not come easily – nor will they come quickly. But come they must. Our society has a responsibility to deal directly, affirmatively, and openly with this issue. Chemicals touch virtually every facet of our lives. New products and processes stemming from advanced chemical technologies have greatly improved the quality of our life over the past 50 years. But these same technologies generate wastes that threaten our environment and public health. We need to better understand how these wastes are generated and how we can reduce them.

We need to learn how we can deal with those that are integral to the manufacturing processes of companies that benefit society by providing jobs and paying taxes. And we need to be able to better communicate regarding what we know and what we do not know with the public at large in order to establish a climate in which instruments such as compensation and insurance can be effectively applied.

CHAPTER 6

Bargaining and Negotiation in Hazardous Material Management

M. O'Hare [1]

6.1. Introduction

Management of hazardous materials is interesting to people other than engineers because:

(1) People disagree about the likelihood of accidents.
(2) People disagree about the severity of consequences.
(3) Different people experience different benefits from risky activities even without accidents.
(4) People have varying abilities to influence the practices that impose risks.
(5) Different people are likely to suffer different amounts of injury when accidents do occur.

For a number of reasons, these differences persist through the decision-making processes that affect hazardous materials handling. If they could be eliminated, the same decision would be favored by everyone concerned with a particular case; however, until we learn how to eliminate them we will have to manage conflicts.

Risk management conflicts rooted in fact and value differences have attracted the attention of analysts and policymakers impatient with the poor performance of traditional "resolution" mechanisms like command and control regulation, and intrigued by the pervasive success of negotiated settlements in other kinds of disputes. Environmental disputes, which conceptually include hazardous material policy conflicts as a proper subset, have occasionally been settled by negotiation and bargaining in fact, and are increasingly perceived as being more amenable to the making of deals than to litigation or regulation [2].

The case for moving the regulatory process towards more, and more explicit, negotiation has three parts. First, arguments from equity principles, especially a concept of procedural equity, purport to show that individual autonomy, as manifested in the right to barter "goods" broadly conceived, is a virtue and one that is advanced by encouraging negotiation between the parties affected by decisions. This argument appeals to people who think individual differences important and respectable. (It is subject to attack on grounds that autonomous negotiation systematically excludes or underrepresents particular classes of interested parties and also that the political process, whatever its faults, not only advances the virtue of social cohesion and public responsibility, but also better protects individual autonomy.)

Second, arguments from efficiency point out the impossibility of a legitimate government authority ever being able to process the information needed to find "correct" solutions to all significant market failures and inequities. (Even in socialist countries, where the government subsumes the economy and where official policy holds that all economic activity intrinsically generates externalities, the impossibility of duplicating the productive and personal choice-making capacity of citizens and firms in a parallel administrative machinery has generated experiments in devolving economic decisions to nongovernmental forums.) Efficiency also favors practices that vary with the tastes of particular affected parties and the opportunities of particular situations: negotiated regulation can do this better than command-and-control rulemaking.

Third, arguments from pragmatism are made to show that, partly because people understand the impossibility of pure political regulation, regulatory standards will always be negotiated (as will regulatory enforcement actions) and the inequities and inefficiencies we suffer are less when the inevitable negotiations are explicit and legitimated rather than apologized for and hidden.

When two people meet, each of whom has something the other wants, they are usually seized by one of the most powerful drives in human psychology (a drive that governments can suppress only with the greatest difficulty!): the urge to make a deal. Rather than asking how negotiation might be employed for hazardous materials, it would seem sensible to ask why it is not the conventional mode of action already. Two responses to this question apply. The first is to note that, in a sense, all hazardous materials decisions are made by one sort of negotiation or another, as are all decisions involving more than one actor. The second is to review the important obstacles to explicit negotiations inherent in the hazardous materials problem: for each, I will suggest some implied opportunities to encourage negotiated policymaking. Illustrations where appropriate will be drawn from the context of hazardous materials transportation, where explicitly negotiated policy is a rarity [3]. Most of these obstacles are deeply rooted between potential negotiating parties' models of reality or of how reality ought to be described.

While the formal hazardous materials decision-making environment is usually not explicitly a negotiating forum, both implicit and explicit negotiations are under way. Statute-making is, of course, intrinsically a negotiation. Regulatory rulemaking involves negotiations between experts inside an agency and between regulators and overhead agencies, like the Office of Management and Budget. When proposed regulations are presented for comment, the agency will balance conflicting claims of both fact and value presented by intervenors. That intervenors' claims are rarely offered as quid pro quo alternatives should not obscure the balancing involved in responding to them: "We might relax the 90-day-notice requirement for shipments, but if we do we'll have to give the environmentalists more frequent truck inspections or a larger enforcement budget request."

Even enforcement is carried on in an atmosphere of continuous implicit negotiation. The plant manager bargains with the inspector over what constitutes a violation and what is merely an opportunity to improve practice. She bargains with the inspector's supervisor over the actions required to attain compliance, and she may find herself negotiating with the district attorney over the size of fine that can be expected for a guilty plea (see Bardach and Kagan, 1982).

Despite the foregoing, health and safety regulation does not usually evidence the formal machinery of, say, labor negotiation. Dealmaking often seems to have insinuated itself imperfectly into a process that "ideally" operates in another fashion entirely. Several characteristics of the hazardous materials problem, or of how we think of it, obstruct negotiated settlements. In the following pages I will discuss these approximately in the order they would be confronted in an ongoing negotiation. This is not, however, the order in which they take effect. Instead, most of these obstacles are anticipated by participants: if they look formidable, the participants will not come to the table in the first place.

All the following impediments, then, must be considered as affecting incentives to negotiate in the first place, and not just the process of negotiation itself. Most disputes offer several means of settlement, and negotiation is unique among these in requiring unanimous consent to its use by the parties. Anyone can usually withdraw to another field of battle if it promises better results: people only negotiate if they expect to do better that way than by any other means. A party to a dispute evaluates negotiation by comparing its likely outcome to his

"BATNA" ("Best alternative to a negotiated agreement") (Fisher and Ury, 1981). Importantly, the comparison is between the perceived negotiated outcome and the perceived BATNA, which may or may not be the "real" alternatives. Parties who think they can better advance their own interests through the ballot box or rulemaking public hearings or the courts or civil disobedience cannot be expected to enter a negotiating session in good faith. Parties who think *other* parties foresee better alternatives to negotiation cannot be expected to ignore the likelihood that the negotiations are a sham, likely to be vitiated if they do not come out the way the other party wishes.

Bringing to the table parties who can negotiate requires changing the relative perceptions of negotiated outcomes and BATNAs. This can be accomplished not only by diffusing knowledge of successful negotiations in environmental and safety disputes (disputes in which plausibly "right" sets of parties are demonstrably satisfied with the outcome), but also by making the predictable outcomes of negotiation failures less attractive.

To do this in practice is difficult. For example, the Massachusetts Hazardous Waste Siting Act provides that if negotiations between a local government and a waste processor over a facility proposal are found to have failed by an administrative body, an arbitrated solution will be put in place (O'Hare *et al.*, 1983). The expectation was that an arbitrated solution would look less attractive in prospect than what the parties could work out together.

Unfortunately, the process established by the Act has not been pursued to a project yet, which shows the importance of perceptions in this kind of analysis. While the real BATNAs are probably better than what the various parties in four proposals have achieved, looking elsewhere for a site (from the developers' viewpoints) and the status quo ante (from the towns' viewpoints) obviously appeared preferable. The developers could not expect enough profit from the projects to justify making the threat of arbitration real by fairly expensive persistence in the process, and in any case could not be certain that the siting process – not yet proved by success – would really work. The towns sensed this caution and, again because the process is so novel, have so far been understandably dubious that a hazardous waste plant would ever be able to provide enough compensatory benefits as a result of negotiations to be worth having.

6.2. Impediments to Negotiation

6.2.1. Branching risk structure

The first task in negotiation is to decide what should be decided. Injury from hazardous materials can be avoided by changing behavior at any of several points in a handling process, and it is not usually obvious to all which should be the target of policy. In the case of hazardous materials transportation, for example, many conditions are, explicitly or not, on the table:

(1) A definition of the substance in question.
(2) What other substances may be transported with it.
(3) The routes that may be used.
(4) The packaging to be required.
(5) The terminals where the substance may be accumulated.
(6) The frequency and hours of permitted carriage.
(7) The liability rules that will apply.
(8) The persons who may carry the material.
(9) The parties to be notified in case of accident or on the occasion of each shipment.
(10) The procedures to be followed by shipper, carrier, and government in case of accident.
(11) The records that must be kept.

Such conditions affect sequential opportunities to avoid the injury a hazardous material is capable of causing, whether to people, the environment, or property. For example, since transportation is almost always a less controllable or predictable environment than storage or processing, risk is usually reducible by not transporting a hazardous material at all, or by moving it after a processing step that reduces its hazard. The next opportunity, if shipping is necessary for whatever reason, is choice of route or mode: rail and water shipment concentrates risk by increasing shipment quantity relative to truck transportation (but seems about as safe on the average per ton of material carried [4]). Avoiding populated areas will reduce injuries per accident, though, if longer routes are required, the likelihood of accident may be higher. Risk to a subpopulation can be reduced, though without effect on total social cost, by diverting hazardous shipments through someone else's town.

Every shipment experiences the ordinary risks of carriage: trucks bounce, barges roll, people drop packages, and so on. Some shipments experience unusual insult: airplanes meet clear air turbulence, ships encounter hurricanes, trucks have traffic accidents. And a few are subject to extraordinary stress: airplanes crash, ships founder, and trains derail into rivers and rupture tanks. Injury can be made less likely by packaging hazardous materials to withstand transportation risks of lesser or greater severity.

Finally, damage from materials that escape their protections is reduced by increasing the effectiveness of accident response. A wide variety of actions can be taken to accomplish this, from equipping fire departments to handle certain kinds of chemical fires to "hot lines" like CHEMTREC [5] that provide information on hazardous shipments through a 24-hour telephone network.

The importance of the foregoing stages is that hazardous materials transportation danger is a tree-like structure beginning with a shipment and branching out to different kinds of accidents in different places [6]. Along each branch, risk is a particular amount of damage multiplied by a chain of conditional probabilities: e.g. E(injury from fire in Safeville) = P(shipment of flammable) × P(route through Safeville given shipment) × P(accident given shipment through Safeville) × P(container breaks given accident in Safeville) × P(Safeville fire department makes a mistake given previous events) × amount of injury.

Reducing any probability in the chain provides a proportionate reduction in expected loss: if a material is trucked in containers that will survive a 100-m.p.h. collision and subsequent fire, not much additional safety can be obtained by training fire departments to deal with a spill.

The "best" probabilities to reduce are those where the safety gain per unit of resources invested is greatest. Unfortunately, (1) the influence of particular probabilities on the risks particular parties face, (2) the degree of control available to different parties over different probabilities, and (3) the understanding different parties have of the relationship of actions at different stages in the chain to their respective safety or other benefits are not distributed in the same way. Not only do different parts of the tree seem salient to different parties, but a single party is likely to find some probabilities interesting because his control is great but others interesting because reducing them would more effectively reduce his risk.

The result of this complicated set of emphases is to discourage some parties from becoming involved in the hope that others will act first, and to make simple policies, including many with poor efficiency, look attractive to others. It also turns the problem as a whole into a game quite different from direct bargaining, since the actions taken in one decision forum, such as national packaging standards for nuclear materials, will have value varying with the risk-averting decisions made downstream (Harrison and O'Keeffe, 1984). Many parties will find it preferable to wait for others to act rather than to commit themselves on the issue they can affect.

Opportunities

The complexity and the underlying asymmetries described above have been frequently observed. A common recommendation is for a systematic and holistic attack on the problem, involving extensive technical analysis of complete strategies rather than piecemeal corrections of smaller defects in the existing system [7]. In the abstract, it is hard to object to a set of hazardous materials practices determined by such a thorough analysis; the problem is that no one has described a process by which it could be attained. Trying to achieve it, furthermore, directly discourages any serious effort to deal with parts of the system. It may be wiser to endorse an attainable local optimum, representing the best we can do through piecewise accommodation, relative to an unattainable global optimum.

At the same time, it would be useful to reduce the number of decisions to be made by simplifying the regulatory environment so that as few agencies as possible supervise any particular hazardous materials activity. The obvious way to do this is to move control to higher levels of government: this is consistent also with the desire of some parties for uniform and geographically consistent regulations. Unfortunately, this displacement will systematically discourage the participation of such locally concentrated interests as neighbors of transportation routes, and suppress the expression of different individual aversions to hazardous materials risks. There is no simple solution to this complexity.

6.2.2. Unrepresentable constituencies

Negotiated social regulation will fail if it is seen by society as a whole, or by effective interest groups, not to have been generated by the "right" parties. Accordingly, negotiations can fail if essential interests are not represented. The failure can be latent – negotiations are never undertaken because they are expected to be ineffective – or post facto, when a deal negotiated by a few parties is delegitimized by others.

Parties can fail to negotiate for several reasons. First, some of the parties to hazardous materials disputes lack the ability to bind themselves, or (if associations) their members, to agreements. If such parties present themselves at the table, other parties will realize that they offer only predictions of contingent behavior – "if you put flashing lights on the truck, my members [probably] won't picket your terminal" – and not contracts. Other parties either have nothing to trade or have only the ability to trade in such large lumps that all-or-nothing demands are inevitable. For example, in an environment of competitive pricing and good substitutes, a shipper may be unable to offer any important protection beyond what his competition provides; safety advocates will correctly infer that they must accept conventional practice or forbid the particular carriage entirely.

Sometimes the problem is uncertainty over where property rights or authority lie: for example, just which aspects of hazardous materials transportation are under the respective control of local, state, or the federal governments is a matter of continuing and litigious confusion in the USA (Price *et al.*, 1983, Vol 1, pp 17–18). A government that does not know whether it has the power to permit or forbid cannot negotiate over its future use of such power; parties who each think they already have what both want (but only one can have) will be in court rather than at a bargaining table.

In many conflicts, important latent parties cannot properly be represented; their absence may predictably bias the outcome of negotiations enough to discourage their implicit allies from participating. For example, when a town considers excluding a hazardous shipment, neighboring towns are potential recipients of the risks. But no particular neighboring town is threatened enough to take action, so an important advocate for allowing the shipment is absent from the negotiation.

The result can be a series of contests, no one of which is of great consequence, but all of which (successive exclusion of routes) are so unevenly balanced because of the absence of important parties that they are resolved alike and therefore produce a major social cost: a net beneficial shipment is not made at all. As we have seen in the siting paralysis for hazardous waste treatment plants and prisons, a series of highly correlated decisions to put risks or costs "somewhere else" amounts, despite everyone's good intentions, to a decision to put them nowhere (O'Hare *et al.*, 1983).

The most important obstacle to the representation of particular interests in decisions governing a multitude of actions, like a rulemaking, cannot be overcome: I refer to the limits on the time and attention citizens can give to all issues that concern them. No one can make his voice heard, much less useful, about every problem whose solution will affect him, no matter how conscientiously he commits himself to any particular issue. This constraint is, at bottom, why we have political parties that aggregate positions at high levels of abstraction ("pro-environmental", or even "liberal") for public decisions that comprise many policies. It is also a reason for establishing executive regulatory agencies to apply a general political and economic philosophy to particular problems, in effect substituting the labor of a few experts for that of millions of generalists.

Opportunities

A wide variety of options can be recognized under this heading. Statutes can explicitly empower local governments or trade associations to commit themselves or their members to agreements. The scope of authority of governments can be clarified by statute more quickly, and probably better, than through the accretion of case law. Parties to a decision who have a right to affect events but lack the resources with which to make their presence consequential, such as low-income neighbors of routes or terminals, can be endowed by government ***or by other parties to a dispute*** with money or delegated authority or both.

The challenge in the context of hazardous materials is frequently to identify and empower specific, legitimate, representatives in a problem that is highly diffuse in space and time. It is much easier to create a quasi-government to negotiate over a terminal facility in a particular location than it is to enable the neighbors of a variety of 100-mile transportation routes to negotiate on conditions or location. It is easier to find the people who will care about a new prison than to distinguish among the multitude of parties claiming to represent future generations at risk from toxic materials with long latency periods.

In the end, it might be wise to embrace this class of obstacles rather than struggling with them. If the fundamental limits on time and attention I refer to are significant, it would often be more useful to strengthen the public trust in political and regulatory processes than to try to substitute explicit negotiation for them.

6.2.3. Incommensurable risk models

The foregoing obstacles are predictable on the basis of standard negotiation theory. Another, which I have come to think at least as important, is an epistemological conflict that usually appears disguised as a clash of values. The conflict is between two ways of thinking about uncertainty that I have elsewhere called "sophisticated" and "practical" (O'Hare, 1984). In the sophisticated view, the world presents hazards that vary widely and that can be avoided only at a cost; risk itself is unavoidable. Sophisticated observers know that people accept many risks for convenience, for money, or even for fun, and that some actions, like taking powerful medicines, are worthwhile in some circumstances and ill advised in others. They distinguish between very small probabilities and impossibility, and allow small probability to balance large costs. They take pains to guard against the errors in assessing risks introduced by such inferential "errors" as the well-known "representativeness" and "availability" heuristics [8].

The sophisticated approach to risk assessment is familiar to professionals in engineering and policy analysis. It is most visible when someone is analyzing the risks others will bear; indeed, it is not possible to analyze most important decisions intelligently without it.

The practical view divides the world into acceptable and unacceptable actions: potential policies are either safe or dangerous; if dangerous, they should be forbidden. While practical observers know that some people seem to take more risks than others, they see an important role of government as making drugs, skiing, coal mining, and appliances "acceptably safe". When using the practical approach, decision-makers embrace shortcuts, heuristics, and approximations: probabilities are rounded and costs coarsely categorized. This view is familiar not only to people inexperienced in formal analysis but also, through practice, to the same experts who use sophisticated analysis in their work. Indeed, it is not possible to accomplish all the tasks of everyday life without it: though one might take a sophisticated approach to any particular problem, no one has time to analyze a decision tree for more than a tiny percentage of the important choices we have to make. In fact, the sophisticated analyst must rely on practical methods for most of his professional decisions: even if we consider the regulations for shipment of a class of hazardous materials with careful attention to finite probabilities and costs, we will handle each individual shipment in that class by practical analysis. Note that a sophisticated analysis of a class of problems is itself a heuristic technique for simplifying any member of the class!

The difference in attitudes to risk portrayed above makes consequential discussion among people taking both approaches extremely difficult – often hostile. The mismatch will not go away, either, because both approaches are correct in a complicated way. The sophisticated approach is obviously necessary for thorough analysis of *any* particular large decision, but it is impossible to apply, because of legitimate competing demands for our time, to *all* or even *most* decisions. It is also irrational for people to use it in a particular decision that is of small importance compared to their work, family, and other interests; a hazardous materials proposal is "small" in this way for most members of the public affected by it. Thus, it is not possible to resolve the misunderstanding by teaching citizens statistical decision theory on a large scale, nor by teaching experts to deal with risk in a dichotomous way as "ordinary" people do.

The importance of this mismatch for negotiations is that the practical view makes a negotiated settlement of a risk conflict look like an immoral tolerance of an unacceptable risk. If a particular hazardous materials proposal is either safe enough or not safe enough, the benefits that might be obtained in compensation *or mitigation* of it are either unnecessary or insufficient. To put a proposal on the table as though it might be acceptable if packaged with a quid pro quo but not otherwise appears inconsistent with the practical view of risk, and a reasonable person who thinks he has to choose between the practical view and negotiation must cling to the former.

Opportunities

The common ground between sophisticated and practical analysts of a particular hazardous materials problem is that both ultimately face a real decision, and have made such decisions before. An aggressive search by policy analysts for risk descriptions that compare new policies to familiar, accepted degrees of risk, like the examples given by Slovic in Chapter 10 in this volume, encourages attention to the effects of different general policies on particular actions, rather than to the making of atomistic decisions. Whether we should allow a solvent recovery facility to transport its products through the streets is easier to discuss if all parties think about the relationship between that decision and our tolerance of gasoline tankers delivering to filling stations.

A good mediator will trust his parties to see the implications of such comparisons for themselves rather than trying to browbeat them into accepting a particular inference. The results of practical and sophisticated analysis tend to converge as the scope of policy decisions increases. More generally, it will be useful to design choice mechanisms to accommodate both views rather than implicitly pitting them against each other. One way to do this is to place the process of risk analysis in the hands of the parties to the dispute instead of trying to enshrine it as an objective or technical process. This has the effect of focusing negotiation on the actions to be taken in the future rather than on the views of the world parties subscribe to: I discuss below the importance of this focus.

6.2.4. Non-fungible goods

Assuming that the parties to a conflict have been identified and brought together, the issue to be negotiated properly defined, and a common language for discussing the future agreed to, an agreement almost always turns on an exchange of different goods. In commerce, one of these is usually money, because it can uncontroversially be transformed into many other goods at predictable rates: one side of the transaction can think of herself as receiving a known amount of "whatever she wants". Hazardous materials conflicts, however, involve many "goods" not typically traded or sold, and not easily transformable into others. A trucking company can agree with a municipality to pay for damage to roads caused by heavy vehicles, because the town knows it can buy "road repair" that will make the roads very much like "the roads as they are now". It is not so simple for the carrier to agree to pay for damage to health caused by chemical leaks: "people repair" often cannot be bought at any price, and does not always make people's bodies just like "the bodies people have now" even when available.

Hazardous materials negotiation will therefore be much more like barter than haggling over prices, with each party offering specific goods not easily transformable by the other. The right to inflict environmental damage may have to be traded for treatment plant landscaping. This specificity alone greatly complicates the negotiation. It will be further complicated by difficulties in distributing these nonfungible goods over individuals: money payments to a local treasury can be allocated across citizens quite flexibly, but even when the new park looks, for the town as a whole, like a fair exchange for some river pollution it may not serve the people most concerned about water quality.

Hazardous materials conflicts involve health and safety damage with low probabilities, and this most salient issue is especially difficult to bargain over. People sometimes act, and usually speak, as though health and safety risks are not exchangeable – because of "pricelessness", infinite value, or moral obligation – for anything. Reasonable computations of the losses suffered by people whose lives are shortened, and by their friends and families, do not seem compelling as measures of the cost of risk, but the use of money as a numeraire, while off-putting to many, does not seem to be the main reason for the difficulty.

Reasonable people, in my experience, find it difficult even to discuss exchanging health and safety protection for other health and safety protection, as in the familiar awkward classroom discussion of whether we should be spending $5 million per life protecting people from risk A when they could be saved from risk B for only $500 000 each at the current margins.

Opportunities

The difficulty of exchanging different kinds of nonmarket goods, especially risk reduction, is especially frustrating to analysts comfortable with what I called the "sophisticated" approach to risk management, who see the myriad ways in which people do exchange safety for money (stunt persons and fire fighters), for fun (skiers), for overall life-style benefits (living over the California earthquake fault), and even for trivial convenience (seat belt nonwearers).

Perhaps slow progress can be made if policy analysts and public officials repeatedly and nonstridently draw attention to the inevitability of making these difficult exchanges and to the advantages of doing it thoughtfully rather than carelessly. Specific negotiating opportunities should be handled with special sensitivity to the reluctance of most citizens to be seen bartering their own health for other benefits, much less the health of people they represent; similar goods should be compared whenever possible, and money compensation should be put on the table only when the issue is raised by recipients. Unfortunately, all these techniques have been tried, and have shown only modest success in facilitating explicit negotiations. In the next section, I consider whether "safety" or "risk avoidance" may be the wrong name for the good actually being traded in these cases.

6.2.5. Confusion of anxiety and risk

A paradox of conventional risk analysis as used in policymaking [9] is that the most salient cost considered is not visited upon "anyone": all the people who directly suffer death are already dead! How should we assess the cost of possible death? Many other costs, such as illness, are costly when they are imposed, but may also injure people in a meaningful way before anyone becomes sick. They may even hurt people who will not suffer them directly. If we look carefully at the strict behavioral definition of a cost – "that which people commit resources to avoid" – we can see evidence that the important costs of risky decisions are imposed on people before the untoward event occurs, and whether or not it does. These costs, which we might as well call "*anxiety*", may even include the "actual" cost with which we have conventionally tried to model risk-avoiding behavior (Schelling, 1968).

The importance of this possibility in the present context is the difficulty of conducting negotiations in which A thinks, and B says (but subconsciously does not believe), that a certain kind of "good" (safety) is the central problem. If the real issue is "worrying about safety", A will seem, to B, responsive only when proposing actions that coincidentally increase safety and decrease worrying, while other actions (such as greater use of warning signs) that increase both safety and anxiety will generate frustration and hostility.

That the cost of danger is something other than an expectation of quality-adjusted-life-years lost is suggested by a variety of fragmentary evidence, much of which is at least qualitatively consistent with treating unlikely injury as though it hurts people before, and whether or not, it happens. Risks are avoided as though they are multidimensional, and the dimensions measure perceptions rather than objective qualities (see Chapter 10 by Slovic in this volume).

One important dimension is the degree to which risk is controllable by the potential victim: in this context, risk analysts are bemused by the common preference on safety grounds for driving rather than flying, despite ample statistical evidence of the relative safety of the latter. What people who fear flying avoid is a period of anxiety, not a mere likelihood of death. This anxiety is avoidable when driving by the following analyses: "If I have a blowout, I will steer carefully towards the shoulder and brake gently. If I see an obstacle in the road, I will stop. If I feel sleepy, I will pull off and have a nap or coffee." Almost any scary driving scenario can thus be terminated with a psychologically plausible, if not statistically sound, resolution, and the mind turned to other things. In contrast, few such scenarios in an airplane can be "put away" by a plausible conjecture: "If an engine catches fire I will ...." The poor passenger must make do with "I hope the pilot knows how to ...", which is a feeble substitute as reassurance for the confident "I will ...."

Thoughtful psychologists have conjectured that the (statistically) exaggerated fear we display of certain accidents results from distortion of our estimates of actual occurrence rates, probably induced by the greater frequency with which newsworthy disasters are presented to us relative to banal injury [10]. Might it instead (or also) result from a recognition that accidents of the newsworthy sort will, because of frequent reminder, be costly to us directly, through the anxiety induced by their being forced on our attention? Perhaps I want multideath accidents controlled out of proportion to their expected number of deaths not because I think them more common than they are, but because I know that every one will be put before me again and again on the evening news, while people who die one at a time will do so without making me think about them.

Another paradox partially resolved by considering anxiety rather than events is public tolerance for major risks frequently confronted, such as the gasoline truck on the city street. A standard result in psychophysics, the Weber–Fechner law, finds response to vary with (the logarithm of) a change in stimulus. Ubiquitous risks present a fairly constant stimulus: gasoline trucks are part of everyday traffic, we use the hard and slippery bathroom more than once a day, dangerous occupations surround their practitioners most waking hours, and so on. What provokes worry is not, arguably, great danger, but a danger signal that rises greatly above background. What provokes risk reduction may therefore be the expectation that a certain risk will generate rare signals of hazard, like irregular news stories concerning hazardous shipments or indictments for careless waste disposal.

Opportunities

To the extent that anxiety rather than occurrence is the fundamental cost of risk, negotiating over any risky policy is a very difficult task. To begin with, explicit recognition of the proposition is likely to be insulting to participants using a practical risk model. Risk imposers, such as hazardous material carriers, could improve their tactics by carefully choosing only risk reduction mechanisms that are likely to reduce anxiety, but the policy landscape will be dense with moral traps.

If anxiety imposes real costs, for example, they must be deserving of balance against direct injury costs. Some practices, such as warnings, intrinsically increase the former to reduce the latter; should we look for ways to achieve great reductions in anxiety at a modest price in lives? Most people would say "no", but this answer is inconsistent with treating anxiety as a real cost. Could a regulatory agency announce a policy of trading real risk against anxiety and stay in business, even if that would increase welfare as behaviorally evidenced? Should it?

This may be an obstacle that must be deliberately muddled through rather than overcome by a direct attack. The challenge of anxiety cost is much broader than negotiated risk management. Research is needed to clarify the place of anxiety in the structure of social cost, and if it has the importance I conjecture, some new political theory will be required to apply the result in practice.

6.2.6. Confusing disagreement with failure

If the foregoing obstacles can be overcome, agreement can probably be reached. Unfortunately, common expectations of what such an "agreement" should be can present a last important obstacle to negotiations over hazardous materials. As everyone knows, contracting parties must agree on certain things: the price to be paid, exactly what goods will be transferred when, and precisely what actions the parties are obliged to take in the different futures that may be realized.

However, what makes contracts possible and worthwhile is disagreement between the parties – about values, about the facts, or about both (Leonard and Sebenius, 1983). If two people agree on the value of one good in terms of another, a trade will leave them no better off: people exchange A for B at a particular rate because, while the person gaining B thinks it worth more than the A he has to give up for it, the other party thinks it worth less. Contingent contracts, such as stock purchases or horse racing bets, depend on differences between two parties' beliefs about the state of the world: an investor will buy a security from another because the buyer thinks the company's future is brighter than the seller thinks it is.

What is widely ignored (by people who have no trouble in everyday life acting in accordance with the preceding paragraph) is the tolerance of a negotiating process for these differences. As a result, a doomed search for unnecessary and impossible agreement cripples negotiations both in practice and in anticipation.

This puzzling ignorance seems to have several sources. First, the formal machinery for managing hazardous materials and analogous conflicts is fraught with procedures that imply a need to agree on the facts and on objectives. Environmental impact statements are supposed to present an "objective" description of the future, which must therefore be persuasive to all; the same standard is held up for a variety of consulting reports and blue-ribbon studies (O'Hare, 1980). Regulatory hearings, patterned after trials, suggest a similar standard of discerning universal facts that "everyone should accept".

Social convention, which in most societies rewards real or feigned consensus, obscures the difference between the kinds of agreement distinguished above. It is rude for a purchaser to dwell on his continuing belief that the seller has been taken, and labor negotiations usually end with a smiling group press conference emphasizing the "agreement" reached. This convention makes it difficult for people to begin negotiations consciously expecting to finish with important differences of view. Political good manners also emphasize similarities of values among citizens: the most vigorous efforts at income redistribution, in which one group will benefit only at the expense of another, are civilized by frequent obeisance to highly abstract values (caricatured as "motherhood and apple pie") on which the parties can agree.

Finally, the public success of science in generating both widely accepted interpretations of reality and a means of persuasion to transmit them probably distorts our view of the results one can reasonably expect from "scientific" analysis of risky policies.

An implicit belief that the agreement sought by negotiation must include agreement on the facts or on values to be served seems to me to lie behind much of the disappointment and mistrust that quickly poisons many promising negotiating processes. As soon as it becomes clear that some of these agreements are not going to occur, a party without a clear understanding of what must be agreed on and what may remain unsettled will either fear being forced to say things he thinks false, or despair of the group's ability to ever make a deal. Such despair is probably anticipated subconsciously by potential negotiators in hazardous materials conflicts.

Opportunities

The persistence of differences between the parties to hazardous materials conflicts provides the most important reason to seek negotiated agreements in policymaking and the most promising opportunities to make them possible.

While I emphasized earlier the extreme difficulty of having all hazardous materials practice tailor-made to the tastes of particular groups of neighbors, and thus the difficulty of implementing negotiated rulemaking for general regulation in this area, it is also true that much hazardous materials activity is limited in space and time. It is in these cases that practice can be made to respond to the differences in circumstances, tastes, and perceptions of parties. The community surrounding a hazardous material terminal or facility usually experiences a much higher density of pound-miles of hazardous materials than other places, and the parties to negotiation over its conditions are few enough to bring together practically. Some hazardous materials movements are special enough, and use few enough routes, that their conditions can usefully be attended to on their own terms; an example is the transportation of a feedstock between a refinery and a few users.

Most of these relatively localized disputes are covered, uncomfortably, by the application of regulation designed for a large category of hazardous materials cases. Many could be handled by negotiated agreements that depend on the fact that the parties are, first, different from the parties to other similar disputes and second, different from each other in several ways, while not exactly opposed. The first kind of difference is what makes negotiated settlements not only efficient but also respectful of the particular qualities of particular people in a way that general rules cannot be. The second kind is what makes it possible to agree.

Such negotiations would be advanced by several practices. First, the differences between parties in ways of making inferences and in the particular types of outcomes that concern them should be recognized by putting analysis directly into the hands of the parties and not into a separate "objective" investigation. I have argued elsewhere that there is no study of a controversial matter that will be treated as though it is objective (O'Hare, 1980): it is better to equip people with the ability to obtain information they trust than to waste resources writing analyses no one will use except as propaganda or a punching bag.

Diffused information-gathering will provide the parties with what they want to know (there is not much incentive to buy tendentious analysis for your own consumption). It will, however, leave the parties with very different views of the facts of the dispute, and not much change their conflicting values. Accordingly, negotiations should be focused on actions subject to the parties' respective control, and not on facts or values. For example, it is useful to discuss whether shipments may be made by truck at night, but not helpful to "negotiate" whether such shipments are really more dangerous than in daylight. Conditional agreements are especially appropriate when facts are in dispute: a carrier, who expects his drivers to be careful, can contract with a town government, that expects them to be careless, to pay for a police escort for each shipment after being convicted of two traffic violations.

Value differences require an attempt to distinguish the essential from the peripheral qualities of activities. Some values are exact opposites, and negotiation will be a zero-sum game. People fundamentally opposed to nuclear electric generation will find little scope for agreement with the operator of a nuclear generating station: what makes him better off intrinsically makes them worse off. But many conflicts conceal a poor or imperfect opposition of values: people opposed to hazardous materials are often sympathetic with a shipper's economic prosperity and the jobs and taxes it generates. Can the shipper consolidate his operation by bringing a downstream process on site, thereby improving the local economy and avoiding a risky shipment?

The prospect for agreement is improved when alternatives are described as continua and sets rather than yes-or-no propositions, if only because "how much?" and "which of these?" keep the parties' attention on (1) the opportunity to have movement without complete surrender, and (2) the existence of possible actions different from the parties' original positions. A good way for a shipper to open negotiations with route neighbors is to approach them with a variety of options for reducing risk and ask them to help him choose among them. The implication, favorable to negotiation, is that they have knowledge and desires different from his but equally important.

It is important not only to include as many concerned parties as possible but also to be clear about whose interests are not being represented and therefore must be cared for in another process. The most common example in a hazardous materials negotiation will probably be people distant from the dispute physically who fear not only injury to people but also diffuse damage to the natural environment. In some cases these concerns may be represented in a particular dispute, but in others they will have to be managed through a rulemaking process that may or not be an explicit negotiation.

6.3. Implications

The foregoing discussion has attended little to negotiation obstacles easily corrected by public policy, many of which are noted, with useful advice, in the works cited in [2]. The obstacles presented here, by contrast, are either intrinsic qualities of the choice to be made, like its branching risk structure, or habits of mind and conventions of discourse that have positive value in many circumstances people face, like the practical view of risk. They are troublesome in the particular circumstance of negotiation over policies and practices that will help everyone somewhat but might hurt a few of us a lot.

The analysis that illuminates these problems is for the most part experience and common-sense based, which is to say it regards people as being not only "rational" economic utility-maximizers of the familiar type, but also as knowing enough to include their own time as a resource with a cost attached, and as being smart enough to maximize in complicated games with a complex relationship between particular goods and their own utilities.

I have been able to make only modest contributions to the "solution" of these problems. For a variety of reasons, negotiated agreements covering hazardous materials will be difficult to obtain compared to agreements in other social choice contexts. However, the obstacles cited here seem to suggest that they will be dealt with better if we develop procedures that are accommodated to them rather than trying to change human nature, or the fundamental structure of hazardous materials problems, to make them go away.

Notes

[1] The author very much appreciates comments from Gail Bingham, Howard Kunreuther, and the discussants. Remaining errors are his own. The research described here was partially supported through the Interdisciplinary Programs in Health, Harvard School of Public Health, by the Environmental Protection Agency (EPA) under contract no. CR807809. The conclusions presented here do not necessarily represent the position of the EPA or the US Government.

[2] The literature on environmental negotiation and on expanding the role of negotiation in conflict generally has become too extensive to summarize conveniently. To sample the field, see Bacow and Wheeler (1984), Sullivan (1984), and Raiffa (1983). For an extensive analysis of negotiated settlements in the area of facility siting, especially hazardous material disposal facilities, including a prescriptive discussion of statutory and informal methods of encouraging such negotiations, see O'Hare *et al.* (1983).

[3] Two useful general treatments of hazardous materials transportation issues are the collection of research papers in Price *et al.* (1983) and the practitioner-oriented work by Abbott *et al.* (1984).

[4] Aggregate data for hazardous materials transportation are difficult to come by and imprecise. This observation is taken from Abbott *et al.* (1984, pp 8, 26).

[5] Funded by the Chemical Manufacturer's Association, CHEMTREC provides information about chemical hazards and links to shippers for possible direct assistance.

[6] For a critical discussion of conventional practice and philosophy of risk assessment in hazardous materials transportation, see Philipson *et al.* (1983).

[7] See, for example, Price *et al.* (1983, Vol 1, pp 14–15) (recommendations) and Philipson *et al.* (1983).

[8] See Chapter 10 of this volume by Slovic. For a discussion of inferential heuristics that appear as errors when decisions are reviewed carefully one by one, but as efficient policy when a way of making many decisions must be chosen, see Nisbett and Ross (1980).

[9] See, for example, the review of risk analyses in Kunreuther *et al.* (1983).

[10] For a review, see Chapter 10 by Slovic in this volume.

References

Abbott, L., Bulanowski, G., Foster, B., and Jordan, J. (1984), *Hazardous Materials Transportation: A Legislator's Guide* (National Conference of State Legislatures, Denver, CO).

Bacow, L. and Wheeler, M. (1984), *Environmental Dispute Resolution* (Plenum Press, NY).

Bardach, E. and Kagan, R. (1982), *Going by the Book: The Problem of Regulating Unreasonableness* (Temple University Press, Philadelphia, PA).

Fisher, R. and Ury, W. (1981), *Getting to Yes* (Houghton Mifflin Co., Boston, MA).

Harrison, D. and O'Keeffe, M. (1984), *How Much is Prevention Worth?*, Unpublished manuscript (Kennedy School of Government, Harvard University).

Kunreuther, H., Linnerooth, J., Lathrop, J., Atz, H., Macgill, S., Mandl, C., Schwarz, M., and Thompson, M. (1983), *Risk Analysis and Decision Processes: The Siting of Liquefied Energy Gas Facilities in Four Countries* (Springer-Verlag Publishers, London).

Leonard, H. B. and Sebenius, J. (1983), *Differences and Gains from Trade*, Unpublished manuscript (Kennedy School of Government, Harvard University).

Nisbett, R. and Ross, L. (1980), *Human Inference: Strategies and Shortcomings of Social Judgment* (Prentice-Hall, Englewood Cliffs, NJ).

O'Hare, M. (1980), Information management and public choice, in J. Crecine (ed), *Research in Public Policy Analysis and Management* (JAI Press Greenwich, CT).

O'Hare, M. (1984), *Speaking Truth to No-one: Information Management in the Public Sector*, paper presented at the Association for Public Policy and Management meeting, October 1984.

O'Hare, M., Bacow, L., and Sanderson, R. (1983), *Facility Siting and Public Opposition* (Van Nostrand Reinhold, Inc., New York, NY).

Philipson, L.L., Napadensky, H.S., and Maxey, M.N. (1983), Hazardous materials transportation risk assessment, in D.L. Price, K.B. Johns, A.G. Clary, and J.W. Bain (eds), *Transportation of Hazardous Materials: Toward a National Strategy*, Vol. 2, p 43 (Transportation Research Board, National Academy of Sciences/National Research Council, Washington, DC).

Price, D.L., Johns, K.B., Clary, A.G., and Bain, J.W. (eds) (1983), *Transportation of Hazardous Materials: Toward a National Strategy* (Transportation Research Board, National Academy of Sciences/National Research Council, Washington, DC).

Raiffa, H. (1983), *The Art and Science of Negotiation* (Harvard University Press, Cambridge, MA).

Schelling, T.C. (1968), The life you save may be your own, in S.B. Chase (ed), *Problems in Public Expenditure Analysis* (Brookings Institution, Washington, DC).

Sullivan, T. (1984), *Resolving Development Disputes through Negotiations* (Plenum Press, NY).

6.D1. Discussion

J. C. Davies

I found the chapter by O'Hare important and interesting. It alternates between practical and theoretical observations, which sometimes makes for difficult reading, but, like a good book, it is even better on a second reading. I want to focus on just one of a number of stimulating questions raised by Professor O'Hare: how to treat public perceptions of risk. This is a question that arises in several different contexts in Chapter 6.

Professor O'Hare suggests substituting the "value of not worrying about death" for the traditional "value of a life" when calculating the importance of a risk. There is no question that worry and anxiety are definite societal costs. But, as Professor O'Hare notes, one implication of this substitution is that "shippers, carriers, and regulators should look for ways to make risks invisible.... It might even be appropriate to accept higher absolute risk as a trade for less visible risk.... A moral trap seems to yawn here; how can we steer between cynical deception and appropriate response to what appear to be people's fundamental values?" The issue is at least as old as Plato. We should recognize that the determination of risk may pit the views of philosopher kings (or their contemporary equivalents, scientific regulators) against Jeffersonian faith in the general public.

A different aspect of the same problem is presented in O'Hare's discussion of the "sophisticated" versus the "practical" way of thinking about risk. He implicitly concedes that dichotomous (i.e. zero-risk) thinking about risk is popular. But he also states that it is "inappropriate" in regulatory practice. Why is it popular? O'Hare and others have suggested some reasons, but they are not fully adequate. Why is it inappropriate? Because it is impossible to have zero risk, and thus the public is not being realistic.

Paul Slovic, Professor O'Hare, and others have shown that in fact the public perceives more subtle dimensions of risk than do the experts who tend to view risk solely in terms of deaths and injury. The public has ways of integrating different dimensions of a given risk that, if they could ever be made explicit, probably would be superior to any of the methods currently used by the experts. But, at the same time, the accuracy of O'Hare's portrayal of the public as unrealistic and simple-minded about risk cannot be denied.

More specifically with respect to O'Hare's chapter, it is difficult to reconcile, on the one hand, his views that we should pay obeisance to public fears in that they should in fact be the definition of risk and we should place the process of risk analysis "in the hands of the parties to the dispute" with, on the other hand, his portrayal of a public that demands zero risk and that defeats politicians who favor cost-benefit analysis.

It is difficult to reconcile these views but not, I think, impossible. And both views represent different aspects of reality, so that to highlight the difference is in no way a criticism of O'Hare's contribution. Each view of the public implies significant political and moral choices, so that reconciling them is a task of immense practical importance.

The task is not made simpler by the fact that it is just one facet of the overall issue of experts versus laymen in a technological society. As the Vietnam War issue made quite clear, at least in the USA the experts do not trust the public and the mistrust is reciprocated. This is one of the crucial issues of our time, and risk analysis is at the cutting edge. The many sophisticated techniques for analyzing and managing risk that have been described in this Conference are all doomed to failure if we cannot successfully grapple with the issues of trust, credibility, and the role of experts. This, it seems to me, is our central task.

For the past nine years, The Conservation Foundation has been experimenting with and doing research on alternative methods of dispute resolution, particularly mediation and what we call "policy dialogues". We see these techniques as one of many possible ways of addressing some of the challenges outlined above.

Our research has shown that mediation – the involvement of a neutral third party – has become an increasingly common method of settling environmental disputes. A research report that we recently completed documents 160 cases, over the past ten years, of environmental disputes in which mediation was utilized. The number of such cases has increased each year, and in recent years institutions have been formed devoted entirely to mediating environmental disputes.

At The Conservation Foundation we have conducted a number of policy dialogues. These are meetings among the interested parties directed at reaching agreement about specific environmental policy issues. They have mostly involved representatives of the business community, representatives of the major environmental groups, and a neutral chairman. They have dealt with such subjects as implementation of the Toxic Substances Control Act, development of a handbook for siting hazardous waste disposal facilities, disposal of radioactive waste, and energy pricing. Two such groups are currently active: one, initiated by several church groups, is developing guidelines for pesticide use in developing countries: the other is working on several proposed legislative changes to the Toxic Substances Control Act.

The example most relevant to the subject of this conference is the formation of Clean Sites, Inc. This was not really a dialogue group because there was no neutral mediator. But it was an effort involving both the chemical industry and some of the major environmental groups under the auspices of The Conservation Foundation which served as a neutral institutional home.

The result of the effort was agreement among the Chemical Manufacturers' Association, several major chemical firms, The Conservation Foundation and the National Wildlife Federation to create a new institution – Clean Sites, Inc. – to clean up hazardous waste sites on a voluntary basis. The new institution was created about a year ago and is now active in cleaning up more than 20 sites, all of them on the government's list of priority sites.

Our overall experience has shown that mediation and negotiation can be effective methods for dealing with environmental problems, including many types of hazardous waste problems. The methods are successful in that they have a high rate of success in achieving agreement among the interested parties and the agreements often provide better solutions than could otherwise have been achieved. The main problem in some of the efforts has been implementation. The results of some of our dialogue groups were ignored, for example, because government officials were reluctant to accept a solution which they had not crafted.

Mediation and negotiation are not panaceas. There are many situations where these techniques are not appropriate: there may be no deadline to spur agreement, one of the parties may benefit from lack of agreement, the resources available to the different parties may be so disparate that they cannot bargain as equals. But the experience to date has been very encouraging. At a minimum it shows that new methods for dispute resolution can be developed, can be applied, and can serve the public interest.

6.D2. Discussion

F. Dinglinger

Siting of hazardous waste disposal facilities has been the subject of a Parliamentary Investigation Committee in our state (Baden-Württemberg, FRG), and waste management has led to the formation of another Parliamentary Investigation Committee that is sitting currently.

In this delicate situation it is hard to say whether the siting and realization of a waste disposal facility is the result of a bargain. It is the result of a legal and administrative procedure based on scientific expertise and facts concerning all possible risks people can think of. This is the reason why the new industrial waste disposal site in Baden-Württemberg is one of very few in the FRG that could be realized in the last decade.

First, I want to point out the legal situation. In the FRG it is possible to get a waste disposal facility strictly according to legal regulations, which include public hearings and expert studies. In cases of conflict between different parties, it is the courts that settle the dispute, basing their decision on formal law and expert opinion. This is the reason why – in principle – there is little scope for negotiation and bargaining but plenty of scope for experts of all kinds on either side.

Of course nobody would actually try the legal procedure without first considering other ways. Looking back, one can see three stages:

(1) Looking for a site.
(2) Bargaining.
(3) Legal and administrative procedures.

6.D2.1. Looking for a site

You have to start thinking very early! If, for example, the site obviously does not have a 30 m layer of clay, as assumed, or if the Prime Minister's house is just adjacent to the site, you will probably try the next alternative. There are other obstacles that you can establish very easily: if, for example, the site is situated in the Prime Minister's constituency or situated in the constituency of another strong minister or secretary of state, or situated close to the local mayor's home, then the site will be no good even if it is an intrinsically suitable site.

Such a situation may provide the first frustration; then a good engineer will start looking out for objective requirements that have to be met. In doing so he or she may easily pass five years, and end up with an honorable PhD. It is true that such a systematic evaluation of benefits helps a lot – but the Prime Minister has not moved away in the meantime. Thus, it can be seen that very important decisions – possibly the most important ones – are made in the first stage, and very often without the public being aware of it.

6.D2.2. Bargaining

If one of 50 possible sites has finally been chosen, the negotiations can begin. It may happen, however, that the community's officials have in the meantime decided, legally, that within the borders of their community no waste disposal facilities are to be "allowed". That does not bother you, however: there are 49 other sites left. If you want to go any further here, the experts have to start their job of convincing people.

6.D2.3. Legal and administrative procedures

In the third stage, all the settlements become part of the final decision or, if that is not possible, are verified privately. The legal procedure ends up with or without a court decision. Such settlements may involve:

(1) A new pond for homeless frogs.
(2) A new fire station for the community.
(3) Two more fire engines because of the dangerous materials to come.
(4) Connection of the community to a distant water supply system.
(5) A new street or improvement of existing streets.
(6) A new sewage plant for the community to take the expected dangerous leachate.

6.D2.4. Conclusion

Let me summarize. The role of negotiations must not be overestimated in FRG. I would put more emphasis on the first stage, and especially on the scientific study. If this does not provide a strong basis for further arguments, there is no way of winning a favorable court decision. And after the experience of the last ten years, one is inclined to return to clear court decisions if bargaining and negotiations are no longer timesaving elements.

CHAPTER 7

Rethinking the Siting of Hazardous Waste Facilities

R.E. Kasperson [1]

7.1. Introduction

Siting conflicts are as much a part of organized human experience as politics itself. Traditionally such conflicts centered upon competition among places to obtain *desired* functions, as with the location of state capitals in early American political history or the siting of national capitals elsewhere (as with Brasilia or Islamabad), other times upon desired facilities (e.g. "clean" industries in residential suburbs), and sometimes for federal contracts or grant awards. More recently, disagreement and controversy have erupted over where to locate facilities or functions such as prisons, cemeteries, taverns, "adult" bookstores, or town dumps with *undesired* characteristics. With widespread urbanizational and technological development, with increasing pressure upon land, and with growing concern over environmental and health protection, siting of controversial facilities of all sizes and kinds has become increasingly difficult and has emerged as a national policy problem of major significance (Popper, 1983). Nowhere are the difficulties greater and the stakes higher than with the perplexing issue of how and where to locate hazardous waste treatment and disposal facilities.

Let the stakes not be underestimated. In April 1985 the US Congress Office of Technology Assessment (US Congress, OTA, 1985, p3) estimated that more than 10000 disposal sites for hazardous waste will require cleaning up at a cost which may eventually total $100 billion. Dealing with this enormous task and the ongoing generation of hazardous waste will require successfully siting hundreds of new facilities across the country. Failure to do so will involve substantial prices (Morell, 1984). Industry is understandably reluctant to cooperate in site clean-ups that require relocating the wastes in existing landfills likely to leak in the future, thereby impeding the Superfund program. The national resolve under the Resource Conservation and Recovery Act (RCRA), and similar state initiatives, to close existing landfills is thwarted by the unavailability of new and better sites. The lack of a credible program of hazardous waste management also detracts from incentives to reduce the volume of hazardous waste generated, thereby exacerbating the long-run waste problem. While the proposed ban on land disposal of wastes under the RCRA would be a critical step in an effective rational management program, numerous sites for waste treatment facilities will still be required. Meanwhile, public skepticism and distrust of those charged with the management task grows, while the remaining margin of good will and cooperation so necessary for a coherent siting program erodes still further.

Underlying the various attempts to wrestle with these problems are conceptualizations of the nature of "the" siting problem, assumptions as to how it should be addressed, and preferred strategies of institutional response – siting paradigms, if you will. The adequacy of each of these paradigms will do much to determine the eventual success of the various national and state programs now under way. The initial returns do not inspire confidence that ongoing programs are adequate to the task, yet remarkably little rethinking appears to be occurring. Something like a conventional wisdom has evolved in the scholarly and practitioner communities, and it deserves critical scrutiny.

Accordingly, the objectives of this paper are:

(1) To conceptualize the nature of the problems involved in siting hazardous waste facilities and compare it to currently held views.
(2) To examine the adequacy of major siting models underlying current national and state waste disposal programs in the light of objective (1).
(3) To outline an improved siting paradigm to guide new initiatives in hazardous waste management.

The discussion to follow addresses each of these objectives in turn. To begin, however, a brief review of the current waste siting situation sets the context.

7.2. The Current Situation

Although hazardous waste landfills have been sited for many decades, sometimes resulting in societal alarms (as in Love Canal or the Valley of the Drums), the legacy of past failures and associated remedial efforts are not of primary interest in this review. Rather the focus is upon recent and ongoing efforts to find appropriate waste sites through the considered siting strategies embodied in national and state programs.

In its 1985 review of the national hazardous waste problem, the US Congress OTA painted an ominous picture. Disposal of such wastes over a number of decades had been largely unregulated by the states or federal government; some 80263 sites existed with contaminated surface impoundments, 90% of which posed a potential threat of groundwater contamination; the long-term health threats of hazardous waste were uncertain and potentially serious; and economic costs of managing hazardous wastes were estimated at $4–5 billion and rising rapidly (US Congress, OTA, 1985, pp 9–11). Arthur D. Little analysts have estimated that the quantity of hazardous waste generated would increase at a 3% annual rate, the same rate of growth as the chemical industry. Further, they estimated that off-site treatment and disposal would be used for 80% of the waste, as compared with the then existing level of 20% [2]. In a 1985 review of the national hazardous waste program, the OTA found that, despite the expenditure of $1 billion from the Superfund, only 50% of the 538 sites on the priority list of the US Environmental Protection Agency (EPA) were receiving remedial clean-up attention (US Congress, OTA, 1985, p 3). Equally disturbing was the OTA's observation that "current remedial cleanups tend to be impermanent. Some sites get worse, and repeated costs are almost inevitable. Environmentally, risks are often transferred from one community to another and to future generations" (Shabecoff, 1985, p 31).

Both the national clean-up effort and the continuing generation of new wastes will require the finding of *hundreds* of new facility sites over the next several years. Failure to do so will impede the EPA's clean-up program, undermine efforts designed to reduce hazardous waste generation and to develop alternative control technologies, export this generation's risk into the future, and magnify the costs of eventual solution. Hazardous waste management occurs under the auspices of the RCRA of 1976 which was designed to produce "cradle to grave" control, primarily through land disposal over solid waste. But, because the Act is silent on the siting of facilities and the federal government thus has no authority to become involved, the responsibility of siting has, as with the case of low-level radioactive waste, fallen to the states (Bacow and Milkey, 1982, p 267).

The state response to this burden has been highly uneven. Approximately one-half of the states have established siting programs for new hazardous waste treatment, storage, and disposal facilities. The approaches show considerable variation, ranging from those that emphasize central authority and preemption to those (e.g. Massachusetts) that emphasize bargaining and negotiation. Whatever the program, state siting efforts encountered determined, and often vehement, local opposition. The resistance shows an impressive ability to unite "grandmothers and US Congressmen, factory workers and university scientists, those who never graduated from high schools and those with doctorates in ecology and physical sciences" (US Environmental Protection Agency, 1979, p iii). The vehemence of the opposition often stuns facility developers and regulators, as in one example where angry citizens were prepared to blow up a facility and there were reports of threats of death or physical harm to key individuals and their families (US Environmental Protection Agency, 1979, p iii). In the face of such volatile local response, it is not surprising that the siting record is rather dismal: some 32 siting efforts occurred nationally between 1979 and September 1984, with five treatment and two storage facilities approved (but not yet built) but only one waste disposal facility successfully sited and constructed, and that (in Maryland) was subsequently closed because it could not compete economically with previously sited facilities (Ryan, 1984, p 3).

Siting experience over the past decade for high-level and low-level radioactive waste and for more recent attempts with hazardous waste facilities clearly indicates a serious national problem – one that is shared to varying degrees by other countries. Why has hazardous waste facility siting run aground while other unwanted facilities continue to find host communities despite local reluctance?

7.3. The Nature of this Siting Problem

Although hazardous wastes facilities share a number of problems with other unwanted facilities, they present a different configuration of impact on these problems. They also reveal some risk and choice considerations not found in many other siting tasks. Five major issues contributing to siting difficulties are the lack of a systems approach, risk uncertainty, public perception of risk, inequity in costs and benefits, and institutional distrust.

7.3.1. The need for a systems approach

By its nature, hazardous waste management requires a systems approach. The waste production process is a complex one, involving numerous opportunities for management to reduce risks, to lower economic costs, and to recycle wastes to beneficial uses. All opportunities need to be weighed against one another to maximize health protection and economic efficiency, and to minimize the transfer of risks to future generations.

No less than waste management, facility siting is also a system activity. The deployment of a waste management system requires a network of waste processing, storage, and disposal facilities, interconnected by waste transportation links. The network may be designed in ways that increase equity, minimize risk, and lower costs, or, alternatively, produce the reverse effects. This is clearly apparent in *Figures 7.1* and *7.2*, which show the implications of a centralized versus a regional siting strategy for high-level radioactive waste repositories. The centralized system creates a national system of waste movement involving many nonnuclear states in risks, public concern, and regulatory burdens. The regional system, by contrast, minimizes such problems, thereby suggesting the importance of the underlying policy choice.

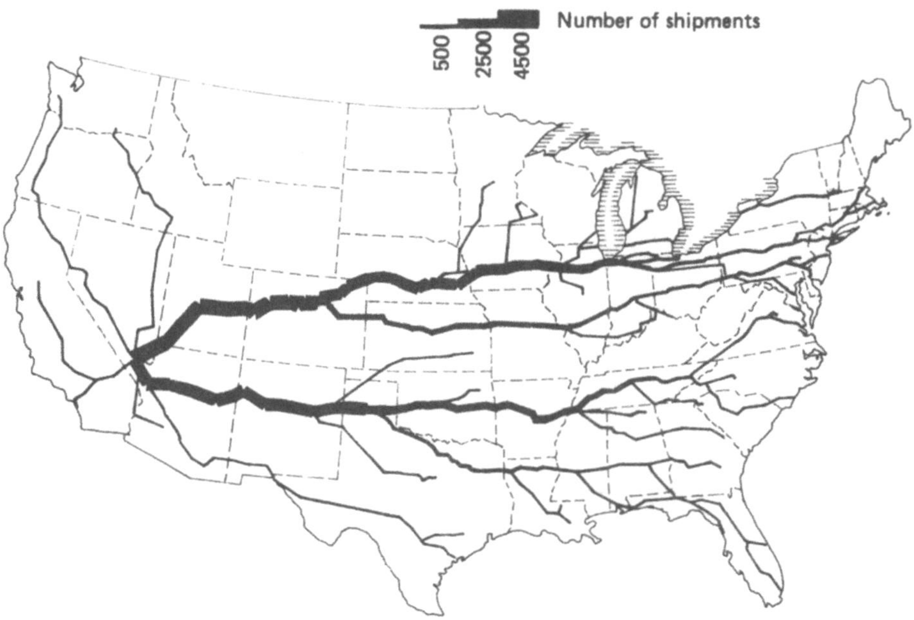

Figure 7.1. Projected annual spent-fuel shipments to a western storage site in 2004. Basis: truck shipments from all reactors (for demonstration purposes only). (*Source*: National Research Council, 1984, p 63.)

Unfortunately current hazardous waste facility strategies tend to be facility-specific. Most of the state laws governing hazardous nonradioactive waste facility siting are geared to the process for siting a given facility. Low-level radioactive waste facility siting, because of the scale of the institutional structure, addresses the siting of a single facility. Network and systems considerations receive little treatment in the Nuclear Waste Policy Act of 1982 and only belatedly are being

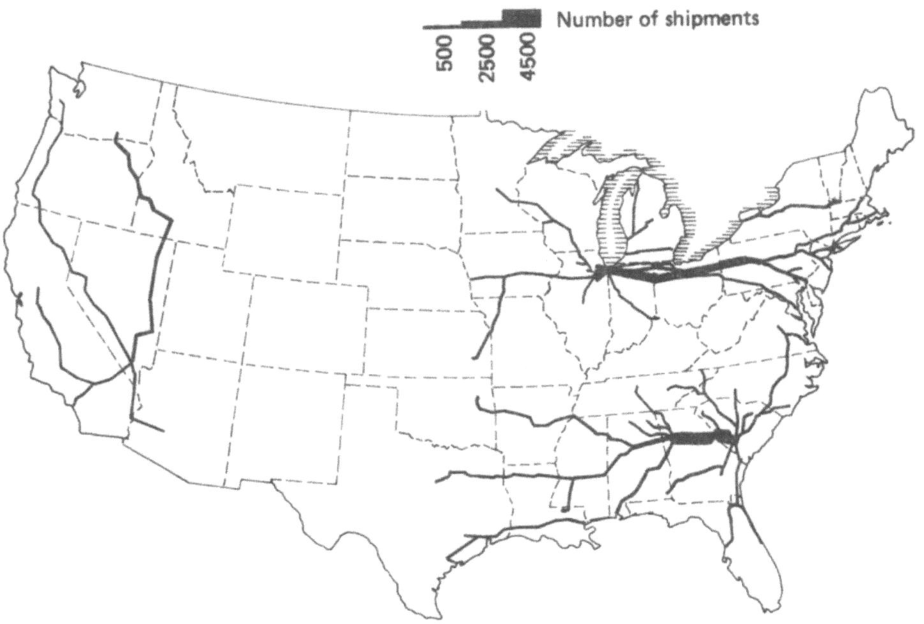

Figure 7.2. Projected annual spent-fuel shipments to regional storage sites in 2004. Basis: truck shipments from all reactors (for demonstration purposes only). (*Source*: National Research Council, 1984, p 70.)

addressed by the US Department of Energy, with regional distribution questions unresolved.

7.3.2. Risk uncertainty

Varying degrees of uncertainty characterize the health implications of waste facility siting. Generally there appears to be little chance of massive uncontrolled releases of hazardous wastes into the environment from well-designed disposal facilities (US Congress, OTA 1985, p 46).

On the other hand, residual risks unquestionably exist. The long time periods which characterize the waste disposal task, the limited experience with repository design and behavior, the necessary reliance upon computer models for simulating waste behavior, and limitations upon model validation and remaining gaps in scientific knowledge all suggest that uncertainties will remain. For land disposal (should it occur) of hazardous wastes, these uncertainties may be particularly substantial owing to limited knowledge of (US Congress, OTA, 1985, pp 22–23):

(1) The likely quantity and timing of releases of particular constituents.

(2) The rates of transport of released hazardous constituents through the environment and their rates of degradation in the environment.
(3) The extent of possible exposures of people and the environment to persistent hazardous constituents and their degradation products.
(4) The probability of damages.

Compounding these uncertainties is the fact that there is generally inadequate scientific knowledge to decide which locations are best for specific hazardous wastes (US Congress, OTA, 1985, p 20). All this suggests that:

(1) Given the magnitude of the siting task – with hundreds of facilities needed – some failures, and releases, must be expected.
(2) Authoritative statements linking the disposal of hazardous wastes at a particular location with associated long-term health and environmental effects are not possible.

Such risk uncertainties not only complicate the management task but exacerbate public concerns about the dangers posed by a particular suggested disposal facility. All this supports a policy initiative to ban the land disposal of hazardous waste.

7.3.3. Public perception of risk

Whatever the actual public health and environmental risks posed by new hazardous waste disposal facilities, they undoubtedly pale in comparison with what the public believes they are. There can be no doubt that members of the public perceive substantial dangers from such facilities and are intensely concerned about them. Indeed, community responses to hazardous waste threats take on characteristics which share some similarities (as well as several key differences) with contagious hysteria (Schwartz *et al.*, 1985). Intense concern is apparent experientially in the controversy that nearly always erupts whenever search activities are conducted for a hazardous waste facility. It is also apparent in the findings from a significant accumulation of pools, surveys, attitude studies, and psychometric research.

A 1980 national poll conducted by Robert Mitchell at Resources for the Future (US Council on Environmental Quality, 1980) found that only 10–12% of the US public would voluntarily live a mile or less from either a nuclear power plant or a hazardous waste disposal site as compared with 25% who would accept a coal-fired power plant and nearly

60% a ten-storey office building. Further, majority acceptance for the hazardous waste disposal site did not occur until 100 miles from the site despite assurances in the poll that the facility "would be built and operated according to government and environmental and safety regulations" and that "disposal could be done safely and the site regularly inspected for possible problems" (US Council on Environmental Quality, 1980, p 30). Finally, 10% of the respondents indicated that they would not voluntarily live at any distance from the site.

Highly consistent with these results are those from a recent study using a similar approach (Lindell and Earle, 1983) which assessed attitudes toward living near some eight different industrial facilities. Respondents rated each on 13 different risk dimensions and indicated the minimum distance they would be willing to live from each. The resulting attitude structure revealed three relatively distinct clusters (*Figure 7.3*). The least acceptable high-risk facility group included the nuclear waste and toxic chemical disposal facilities and the nuclear power plant. They were judged by respondents to pose a high threat to workers, the public, and future generations and to have risks which were less known and less preventable, catastrophic, with many deaths over their operating life, and dreaded by the respondents (Lindell and Earle, 1983, p 249).

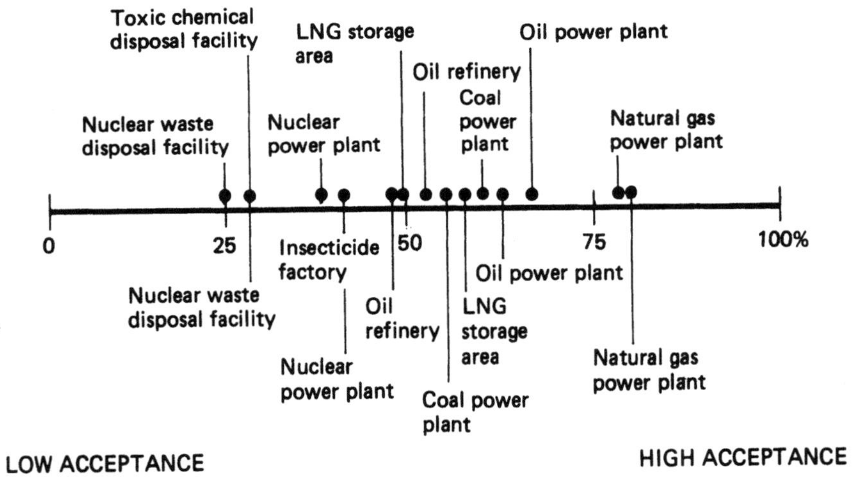

Figure 7.3. Acceptance scale for eight different facilities in two surveys. (*Source*: Upper data, Lindell and Earle, 1983, p 249; Lower data, Lindell *et al.*, 1978).

A 1983 Massachusetts survey of citizen attitudes to a hazardous waste treatment facility in each of five communities clearly indicated the importance of these risk and safety issues (Portney, 1983). Whereas the survey found considerable variation among response at the five sites, opposition was clearly linked to concerns about safety. Moreover, in testing some 11 different proposals designed to elicit attitudes to provisions necessary to change people from opposition to support, most opponents indicated that they would not change their minds. Among those indicating willingness to change, greater safety assurance and more safeguards were clearly most influential. A variety of financial incentives, by comparison, had considerably less impact. Further, opposition did not appear to be rooted in lack of information as the opponents tended to be somewhat more knowledgeable than the supporters.

Taxonomic work on hazards is not encouraging for public acceptance of hazardous waste facilities. The Clark University taxonomy (Hohenemser *et al.*, 1983) classifies hazards by some 12 biophysical attributes, resulting in some five major factors and a taxonomy of seven major classes. While hazardous nonradioactive wastes are not included, radioactive wastes score in the "multiple extreme hazards", sharing company with such other notable hazards as nuclear war (radiation), nuclear tests (fallout), nerve gas (accidents), pesticides (toxic effects), and recombinant DNA (Hohenemser *et al.*, 1983, Table 3, p 381). Since the taxonomy has proved quite successful in predicting public response, it suggests that the high concern over radioactive (and probably other hazardous) wastes is rooted in "real" properties of the hazards and is unlikely to disappear under the impact of fuller and more accurate information.

In their taxonomy of 162 technological controversies, von Winterfeldt and Edwards (1984) recognize three major classes: food/drug/consumer products, industrial development, and technological mysteries and value threats. Contrary to the assumption held by some that siting a hazardous waste disposal facility is akin to locating other large-scale facilities (such as dams, airports, or the Alaskan pipeline), it is apparent that hazardous waste facilities fall into the class of "technological mysteries and value threats". This group contains the most dramatic controversies, involving the potential for disaster or possessing side effects which are dreaded and which threaten social values. Such controversies have contents of debate which oscillate between factual disagreements and value disputes, receive widespread media coverage, and involve a broad spectrum of "stakeholders". This is quite apparent in a recent study of the siting of liquefied energy gas facilities in four countries in which it was apparent that expert calculation of safety risks did not resolve broad conflicts in social interests (Kunreuther *et al.*, 1983). In considering appropriate tools for conflict resolution for this class, von Winterfeldt and Edwards conclude that compensation, bargaining, and negotiation will, because of the shifting debate and the presence of moral considerations, be less effective than for other controversies, and call for the creation of institutional mechanisms that involve stakeholders committed to resolution of the issues (von Winterfeldt and Edwards, 1984, pp 67–68).

All this suggests that perceived risk is a central problem in hazardous waste facility siting, that public perceptions are rooted in "objective" characteristics of the risks, that the risk issues interact with related value conflicts, and that the underlying attitudes are likely to be persistent and difficult to change. In short, hazardous waste facility siting is one of the toughest technology problems to manage and should be expected to strain existing institutions and decision-making processes.

7.3.4. Equity and the ethics of risk imposition

Inequity, it is widely held, is a key underlying problem for hazardous waste facility siting. Indeed, for many it is *the* problem. Consider the following from the 1981 policy statement of the National Governors' Association:

> Once a site is identified, the community objects to being the dumping ground for the state or region. It opposes the proposed facility because the *benefits* will flow to the owner, operator, waste-generating industries, and the public at-large (which fears "midnight" dumping), while the *risks* will be concentrated locally – in their community. (National Governors' Association, 1981, p 134.)

Despite the wealth of allusion to locational equity, searching treatments are quite rare. The current empirical understanding of likely impact distributions from a hazardous waste facility at a particular site is quite limited. Reasons for this limited knowledge include:

(1) The relatively underdeveloped state of theory supporting analytic approaches to social impact assessment.
(2) The lack of comparative analysis using common methodology.
(3) The limited siting experience in recent years and the fact that some facilities will be first of a kind.
(4) The highly site-specific nature of social impacts.

In one of the few searching empirical analyses of equity at a hazardous waste site, Kates and Braine (1983) painted a complex picture of gains and losses over more than a dozen locations stretching across the entire USA, including Puerto Rico: benefits for some corporations, institutions of governments, and local residents; losses for others; and mixed balance sheets for still others (*Figure 7.4*). From the existing experience, considerations relevant to the calculations of impact distributions of hazardous waste sites are:

(1) The "special" impacts associated with the perceived risk and social conflict arising at hazardous waste sites may exceed the more "conventional" impacts customarily associated with locating large industrial facilities in rural communities.
(2) The most serious socioeconomic risks are also the most likely to be poorly understood.
(3) Many socioeconomic and (perhaps) health risks will become apparent only during the siting process or over the long term, and a number of them will be essentially irreversible in nature.
(4) Many socioeconomic risks, and especially those associated with special effects, will prove extremely resistant to quantification as a basis for calculating compensation.
(5) The residents of rural communities are among the most vulnerable members of society (Kasperson and Rubin, 1983; National Research Council, 1984; Seley, 1983; Murdock *et al.*, 1983).

In the case of hazardous waste siting, equalizing the distribution of harms and benefits may not be achievable in any real sense. The high perceived risk and the intense associated fear together with the small number of sites ("Why have we been victimized?") simply overwhelms any prospect of restoring the original conditions through the enlargement of benefits. Distrust that benefits will actually flow in timely fashion (especially given that they cannot be well predicted in advance) exacerbates the problem but is probably not decisive. Moreover, among at least a significant minority, substantial indication exists that benefits at any feasible time are unlikely to change committed resistance to a disposal facility.

In cases of risks which, as taxonomic research on risks has indicated, have attributes which elicit dread and intense fears, and simultaneously involve difficult value conflicts, fairness may well depend more fundamentally on the distribution of the risk (and particularly the *sharing* of the risk) and on the characteristics of the *process* that allocates risk than on the relationship between risks and benefits. If this is correct, fairness is best accomplished through strategies designed to reduce risks (even at substantial costs) and to produce widespread sharing of the risk rather than through strategies designed to convince some to take uncertain risks on behalf of others in exchange for compensation. Moreover, much attention will need to be given to the ethics of risk imposition, designed to take account of such questions as:

(1) Under what conditions and to what ends may risks be placed upon other persons?
(2) Who has the legitimate authority to make such decisions?

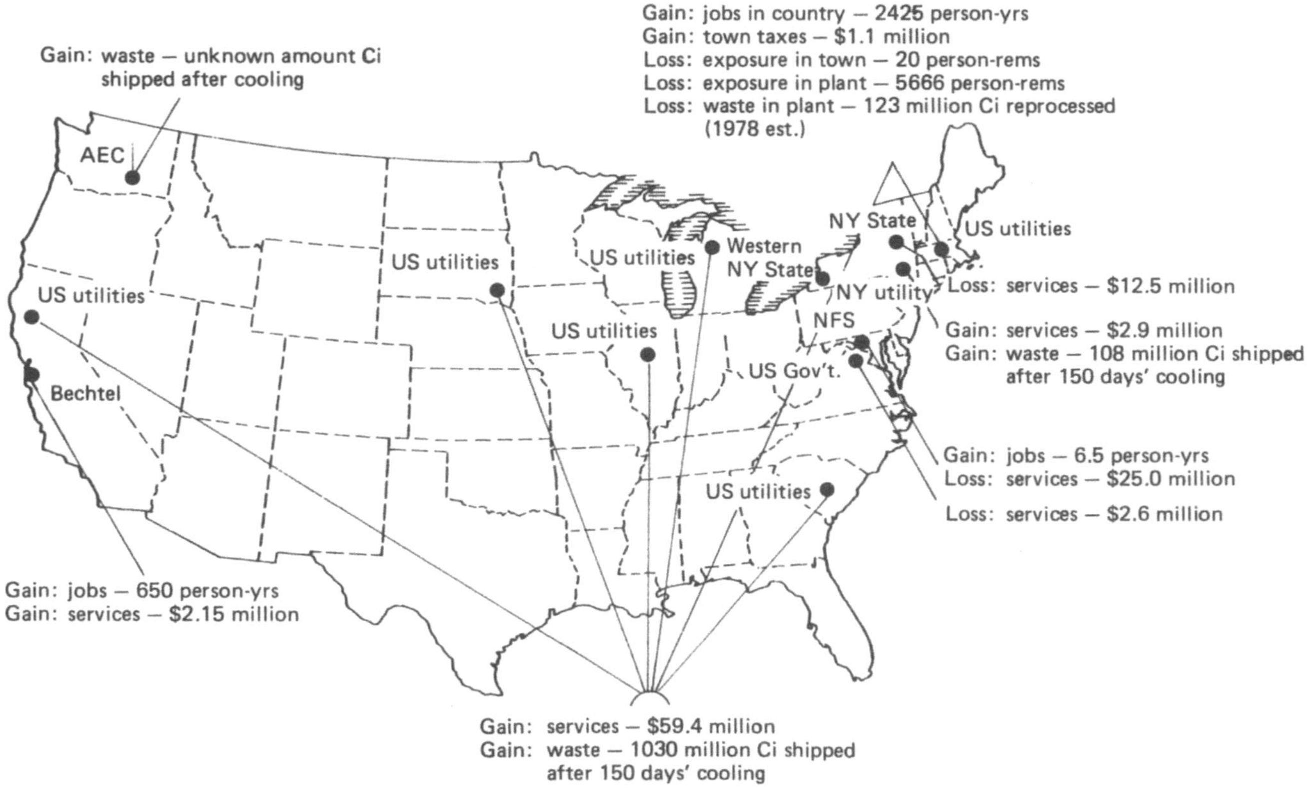

Figure 7.4. **Western New York nuclear service center impacts balance, 1960–1978. (*Source*: Kates and Braine, 1983, p 114.)**

(3) What responsibilities accrue to the risk imposer in such situations?
(4) What rights accrue to the risk bearer?
(5) How should the burden of proof be allocated?
(6) How may the process be designed to take account of differing needs and roles yet remain protective of basic values of the sanctity of human life, social well-being, and democratic principles?

7.3.5. Institutional distrust

The analysis of the siting problem to this juncture unmistakably points to the need for institutions capable of eliciting strong confidence that health and safety will not be compromised in the face of other needs and that fairness will be adhered to scrupulously. Unfortunately such confidence does not exist. Indeed, there is widespread distrust in the institutions responsible for siting decisions and for the assurance of safety at a particular site. This is not surprising, of course, for the ledger clearly shows that toxic wastes have been badly mismanaged over decades, radioactive waste disposal has been neglected, and regulatory agencies and industries have left a legacy of burden for future generations to assume. Should we be surprised that there is not a clamor to be the next site for a waste facility?

Although undoubtedly more marked in waste facility siting, the distrust of institutions is part of a very fundamental and long-term trend in US attitudes. This distrust is pervasive, ranging from social institutions, to the family, to the federal government, and most strikingly to industry. A 1976 poll, for example, revealed the following somber statistics concerning those groups in whom the public had "a great deal of confidence" on nuclear power issues: scientists, 58%; Nuclear Regulatory Commission, 39%; the heads of electric power companies, 19%; and companies that produce equipment for nuclear power plants, 12% (Harris and Associates, 1976, p 29). A 1980 survey of Wisconsin residents revealed that most respondents did not believe that government was moving fast enough to solve the waste disposal problem or was interested in what local citizens thought (Kelly, 1980). Similarly the 1983 Massachusetts survey of attitudes to hazardous waste facility siting revealed that a significant source of concern involved respondents' feelings that they could not trust the management of companies that operate treatment facilities and government regulators who oversee them or that proper procedures would be followed (Portney, 1983, p 36). In view of these data, it is not surprising that the OTA (US Congress, OTA, 1982, p 231) has concluded that "the greatest single obstacle that a successful waste management program must overcome is the severe erosion of public confidence in the Federal Government". The lack of credibility is likely even more profound for a private developer and potential operator of a disposal site.

7.3.6. Summary

At this juncture in the discussion it may be helpful to summarize, in the form of a set of six propositions, major findings concerning the nature of the siting problem:

> *Proposition 1*: The siting problem is not the same everywhere and may be expected to vary from place to place according to characteristics of the proposed facility, the siting process, and the host community.
>
> *Proposition 2*: While the siting problem is multidimensional, the central issue is the public perception of high risk associated with waste facilities and the intense fears which this perception of risk engenders.
>
> *Proposition 3*: Public perceptions of risk arise from some combination of misinformation about a modern waste facility, the attributes of the risks involved, the record of past waste mismanagement and neglect, and the memorability of particular failures, and they are likely to persist in the face of information designed to provide assurance or compensation to increase benefits.
>
> *Proposition 4*: Hazardous waste siting disputes belong to a class of technological controversies characterized by oscillation between factual disputes and value conflicts, widespread media coverage, and a broad structure of stakeholders – a situation calling for new institutional initiatives.
>
> *Proposition 5*: Fairness in hazardous waste facility siting likely depends more upon the distribution of the risk (and particularly risk sharing) and characteristics of the risk allocation process than upon the degree of association between the geographical distribution of risks and benefits.
>
> *Proposition 6*: Efforts to reduce local concern are impeded by the limited ability to communicate risk information effectively and by substantial public distrust in the institutions responsible for siting or for assuring public health and safety at the sites.

7.4. Siting Models in Critical Perspective

A large range of siting models are potentially available for locating hazardous waste facilities. Yet several of these have clearly dominated the approaches developed by states and the federal government over the past decade. Here the major options are set forth, with attention to the validity of conceptions of the siting problem in the light of the foregoing discussion.

7.4.1. Model 1: Locational opportunism

Historically, unwanted facilities have often been sited by a developer who has surveyed the various candidate sites and, once having met various substantive locational needs (available land, accessibility, physical site properties, etc.), has sought those places where the inclination or ability to resist is minimal. These are often communities that are rural and small, where unemployment is high and income low, and where connection to the centers of political power is weak. Residents of such places are more likely to trade safety or environmental quality for material gain – through jobs, increased tax revenues, and improved services. Places to be avoided are communities with high standards of living, for whose residents jobs have less appeal, where safety and environmental quality are highly valued, and which have an organized capacity to resist and ready access to political power. In short, a political marketplace allocates sites for unwanted facilities.

That locational opportunism has operated in hazardous waste facility siting is apparent in a recent study which compared towns with abandoned hazardous waste sites with similar towns lacking such sites in New Jersey. The results were unambiguous (Greenberg *et al.*, 1984). The communities which had escaped such sites had more affluent populations, and lower percentages of younger, older, black, and foreign-born people. In fact, correlation with socioeconomic status of communities was the most consistent relationship uncovered (Greenberg *et al.*, 1984, p 390).

Why should one object to such schemes, which deliver efficient market solutions? First, it is apparent that such procedures are objectionable on equity grounds, since burdens are disproportionately allocated to poor communities which usually share little in the benefits of waste generation, or upon places which are already so contaminated that additional health burdens are viewed passively. Second, the process of risk imposition is almost always objectionable: strategies of withholding information or creating intentional ambiguity tend to be pursued (Seley, 1983, p 34), capacity for participation tends to be minimal, and means of redress few. Finally, political opportunism in siting carries a potential for eroding the technical criteria necessary to assure the safety of present and future generations and the economic efficiency of the waste disposal system.

7.4.2. Model 2: Imposition by central authority

The underlying rationale for the exercise of centralized state authority in the selection of sites in local areas is that the general well-being of society requires the overriding of individual (or local) interests. This may be done with or without compensation arrangements for redressing inequities and with varying degrees of local participation. The actual selection process frequently includes safeguards that the decision be fair and unbiased, guided by technical criteria aimed at protecting health and safety. A modified version of this approach is a higher authority's overriding, usually according to specified conditions, of a lower authority's decision accepting or rejecting a site in its territory. For hazardous nonradiological waste siting, some 25 states apply some form of preemption to site facilities (Morell, 1984, p 560).

The key assumptions of site selection through imposition by central authority are:

(1) Local concern over risk can be abated through an unbiased technically oriented siting process using established means of public hearings and information dissemination.
(2) Higher authorities responsible for siting and the protection of public health possess sufficient credibility to command eventual local tolerance of the siting decision.
(3) Committed opponents will not succeed in producing lower government utilization of institutional means to resist the siting decision.

Although special cases may exist in which these assumptions hold, they certainly are not valid generally. As noted above, the public perception of risk evokes substantial fear of sites which is not amenable by institutions that command low trust and confidence. In the absence of special efforts to achieve fairness, the chosen site almost invariably views itself as victimized. The movement of the higher authority to preemptive actions to overcome the opposition usually succeeds instead in escalating its intensity and broadening its scope. For these reasons, as Morell (1984, p 560) points out, the power of central authority tends to be illusory. It is not surprising, then, that this approach (along with others) has failed to produce site successes.

7.4.3. Model 3: Bartered consent

The reaction to the evident problems with site imposition by central authority has produced a swing of the pendulum to a strategy of local acceptance for sites through intergovernmental bartering. The central problems of siting, in this view, are (1) the geographical dissociation of benefits and harms, and (2) the inability of the host area to share in the siting decision. This conception of the siting problem leads readily to a clear solution: provide compensation to the residents of a prospective host site and give them the means to bargain for the appropriate amount. Used in this way, compensation purportedly serves four purposes:

(1) It changes local motivation to oppose the facility "by reducing the costs each neighbor expects to suffer should the facility be built" (O'Hare *et al.*, 1983, p 70).
(2) It helps to redress inequity.
(3) It increases efficiency of facility planning because all costs and benefits are better accounted for (O'Hare *et al.*, 1983, p 70).
(4) It promotes "negotiation, as opposed to confrontation, in the resolution of siting decisions" (O'Hare *et al.*, 1983, p 74).

In the case of hazardous waste, eight states have enacted compensation plans coupled with state preemption whereas four others (Colorado, Massachusetts, Rhode Island, and Wisconsin) have coupled compensation with shared authority, the classic bartering approach. The case of Massachusetts is particularly interesting since it was closely informed by the conceptual work of O'Hare *et al.* (1983). The Massachusetts approach has a number of key ingredients: it gives the primary siting roles to the developer and the host community, requires a negotiated or arbitrated settlement between the two, includes impact mitigation and compensation to the host community as key features of the siting agreement, limits the basis on which the community may exclude the facility, and submits impasses between developer and host community to an arbitrator (Bacow and Milkey, 1983, pp 4–5).

Bartered consent as an approach to hazardous waste facility siting rests upon four key assumptions:

(1) The underlying problem which drives local opposition is the geographical dissociation of benefits and risks.
(2) Voluntary consent is achievable through sufficient provision of incentives and through direct bilateral negotiations that define the terms of community acceptance.
(3) The long-term impacts of the facility can be defined with sufficient precision to formulate an appropriate compensation package prior to siting.
(4) The developer and state regulatory bodies can command sufficient social trust to reassure local fears over the facility and to withstand a conflict-laden community consideration process.

All four assumptions are in doubt, if not outright wrong. As indicated above, the classic equity problem associated with large industrial facilities is not simply an accurate guide to that associated with hazardous waste facilities. Because perceived risk so dominates public response, risk minimization, risk allocation, and risk sharing questions become the dominant equity problems. Accordingly, and as growing experience confirms, the prospect of compensation does not effectively lower the degree of perceived risk and nor does it engender a propensity among local residents to "trade off" concerns. Rather, it tends to be viewed as a "bribe", exacerbating the risk sharing issue and increasing suspicion and distrust of the developer and state agencies. Negotiations, as Raiffa (1982, p 311) notes, depends upon a perception that the process is fair and that there is more to be gained by cooperation than by noncooperation. Such a reaction was particularly striking in a recent siting dispute in the Federal Republic of Germany (Kunreuther *et al.*, 1984, p 482). Further, although not widely recognized, it is also the case that locational opportunism, with its objectionable ethical elements, remains operative in the bartered consent approach because it is the developer (often in the private sector) who seeks out the potential sites. Finally, although the bartered consent model envisions a community social dynamic geared to growing community consensus on the terms necessary for residents to *accept* a facility, the dynamic that actually occurs is one which shows the classic features of social protest: local efforts to mobilize social resources and to identify institutional opportunities by which to *resist* the facility (Lipsky, 1968).

7.5. Fashioning an Improved Siting Paradigm

More effective approaches to hazardous waste facility siting depend upon a clear conceptualization of the siting problem, a sound ethical base to guide the design of siting strategies and to begin the recovery of social trust, and a set of policy tools by which the strategy can be realized. Each is considered in turn.

7.5.1. Conceptualizing the siting problem

Current approaches to siting are prone to predictable failures because of misconceptualizations of the siting problem. It is crucial to understand that this is a systems-level, not a facility-level, task. Waste management must be undertaken with an understanding of the relationships between disposal strategies and opportunities for reducing the generation of waste. It is also clear that what is at stake is not the deployment of a facility but a network of waste generation, waste processing, waste movement, and waste storage or disposal – a system which needs to be integrated with land-use planning. Relative risks associated with alternative designs need to be assessed and built into decisions on system design. High levels of perceived risk and associated public concern will accompany the deployment of this system – fears that are almost certain to persist in the face of assurances, technical studies of risk, and the offering of compensation. Equity will be centrally concerned with risk sharing and the process by which risk allocation occurs. Substantial distrust of the institutions responsible for siting facilities and overseeing safety will occur.

7.5.2. An ethical base for siting

Since hazardous waste facilities involve the imposition of uncertain and feared risks upon certain peoples for the general benefit of society, siting is inherently, and centrally, an ethical problem. Improved approaches that have potential for winning public trust must therefore build upon ethical principles.

Five such principles are proposed:

Principle 1: The general well-being of society requires that some individuals will have to bear risks on behalf of others.
Principle 2: Wherever feasible, such risks should be avoided rather than mitigated or ameliorated through compensation.
Principle 3: Reasonably unavoidable risks should be shared, rather than concentrated, in the population of beneficiaries.
Principle 4: The imposition of risk should be made as voluntary as reasonably achievable within the constraints of deploying sites in timely manner, and the burden of proof for site suitability should be on the developer.
Principle 5: Reasonably unavoidable risks should be accompanied by compensating benefits.

These principles provide a sound ethical base for siting strategies. Principle 1 makes clear that, from all we know, the location of hazardous waste facilities cannot be made a voluntary activity if conducted in a socially responsible way (thereby prohibiting, for example, locational opportunism). It also recognizes that the benefits of associated technological products and activity (e.g. chemical products, nuclear medicine) outweigh the risk associated with well-designed waste management programs. Principle 2, arguing from the widely accepted duty to avoid human harm, states the obligation (within technological and economic restraints) to avoid risks rather than to mitigate them (as through insurance, for example) or compensate for them once they have occurred. Principle 3 recognizes that, for reasons cited earlier, distributional equity depends principally upon broad sharing of the reasonably unavoidable risks among those who benefit from the technology or activity. Principle 4 addresses the ethical attributes of the process of risk imposition and, although recognizing that purely voluntary means are not possible, calls for the enlarging of the degree of voluntarism of risk assumption through, for example, full information (including uncertainty) to risk bearers, full participation in public proceedings, due process protection, and allocation of burden of proof to the site developer. Finally, Principle 5 sets forth the obligation, once risks have been reduced as much as is feasible, to compensate for the residual risks (to restore, to the degree possible, the original condition).

7.5.3. Policy tools

A large array of policy tools exist for erecting siting strategies based upon these ethical principles. A number of the more prominent are noted below.

Authority and the Systems Approach

The allocation of authority needs to be consistent with the systemic scale of hazardous waste management. Since hazardous waste facility siting is demonstrably contentious, sufficient concentration of authority must occur so that the actual deployment of sites in timely manner is possible. Widespread dispersion of authority in the face of a volatile siting process guarantees failure. But it is also essential that the level of the system be properly defined. The management of high-level radioactive waste is clearly a national-scale problem; the management of nonhazardous solid waste clearly is not.

Risk Reduction and Safety Assurance

The hazardous waste management system must be designed so that all elements, and stages, are sufficiently integrated so that risk reduction can be maximized. For hazardous wastes, this means a clear emphasis upon strategies and incentives designed to encourage reductions in waste generation and greater use of waste recycling and conversion, as called for by a recent National Research Council, Environmental Studies Board (1985) report. Specifically this should include banning or (at minimum) increasing the costs for land disposal sites to a level consistent with the total social costs of land disposal of wastes.

Institutional opportunities exist to enlarge the role of risk bearers in assuring their own protection. Local impact assessment committees can participate directly in the identification of relevant impacts, advice as to how they should be weighted, and strategies for avoiding and mitigating them. A formal local capability to monitor the facility and any potential releases linked to a means for corrective action can provide improved assurance of long-term health protection. Similarly, a direct local role in the design of the facility is a more appropriate sharing of authority than one centered on the siting decision. Postclosure trust funds can be developed to pay judgments arising from future harms caused by a facility owner or operator who is judgment-proof or otherwise not amenable to suit (Baram, 1982, p 215).

Risk Sharing

Several systems designs and institutional options exist for achieving a wider sharing of risk among the beneficiaries. First, the size and number of facilities can be altered to conform to a general plan of equity. Whereas a large multipurpose hazardous waste treatment facility may have economies of scale, several smaller more limited purpose facilities provide enhanced equity opportunities. Second, facilities may be regionally sited to make visible that all benefiting areas will share in the risk. Such siting strategies may also reduce the costs and risks of the waste transportation system. Finally, the siting strategy may also be arranged so that facilities begin operations simultaneously (rather than being staggered, as in the high-level radioactive waste program). Deployment of the overall waste system, of course, requires centralized planning but may be designed so that the network of facilities visibly demonstrates that each area will be expected to share in the waste burdens and risks (Morell, 1984).

The Role of Compensation

To conform with problem conceptualization and the ethical principles enumerated above, compensation would function as a means of providing distributional equity. Compensation may be of three major types:

(1) Means of reducing or mitigating anticipated adverse impacts of normal construction and operation of waste facilities through preventive or ameliorative actions.
(2) Payments for actual damage due to normal or abnormal events.
(3) Rewards for assuming burdens or risks for society as a whole (Carnes *et al.*, 1983, pp 330–333).

To these should probably be added the function, as perceived in the bartered consent model, of reducing local opposition.

Except for the last added function, all are appropriate purposes of compensation – with preference for the first because of its higher-order value of avoiding harm – and should be employed as policy tools. Explicit recognition is required that compensation levels are difficult to establish at the time of facility development because (1) effects often cannot be predicted in advance, (2) many effects are qualitative and difficult to measure, and (3) values are difficult to assign to human or environmental harm. Compensation should, in the view advanced here, be assigned by central authority through standardized procedures applied equally to all cases and should provide assurance that uncertainty and unforeseen events will not go uncompensated.

Risk Communication and Public Participation

The last set of policy tools relate to ethical principle 4, making the assumption of risk as voluntary as is reasonably achievable. The fact of the matter is that we do not know how best to do this, and programs need to be designed as research and demonstration efforts, with a substantial commitment of funds and command of expertise. A sound approach will recognize that:

(1) The developer has a conflict of interest in risk communication, so that risk and project communication should be vested in a more independent source (the author is partial to the League of Women Voters).
(2) Effective communication must take account of the social dynamic of how a community considers and debates the facility and its impacts.

A key objective of compensation arrangements should be the creation of a local technical capability, perhaps modeled after the Technical Advisory Committee to the Governor of New Mexico in the case of the siting of the Waste Isolation Pilot Plant in that state.

The paradigm outlined in this final section does not guarantee success, of course. If success were easy this would not be such a tough policy problem. But in the face of growing evidence that the current wisdoms are failing, new paradigms are needed.

Notes

[1] The author is indebted to Jeanne X. Kasperson and John Lundblad for comments and suggestions. I have also benefited from long-standing collaborative work on equity problems with risk with my additional colleagues Patrick Derr, Robert Goble, Dominic Golding, Jay Himmelstein, Robert W. Kates, and Mary Melville.

[2] See Arthur D. Little (1982). The projection was given by Dr Joan B. Berkowitz, Vice President at Arthur D. Little.

References

Arthur, D. Little (1982), *Transcript of Arthur D. Little Biennial Forum, Boston, MA, June 6–9* (Arthur D. Little, Inc., Acorn Park, Cambridge, MA).

Bacow, L. and Milkey, J.R. (1982), Overcoming local opposition to hazardous waste facilities: The Massachusetts approach, *Harvard Environmental Law Review*, **6** (2), 265–305.

Bacow, L.S. and Milkey, J.R. (1983), Responding to local opposition to hazardous waste facilities: The Massachusetts approach, *Resolve*, Winter/Spring, 3–8.

Baram, M.S. (1982), *Alternatives to Regulation* (Lexington Books, Lexington, MA).

Carnes, S.A., Copenhaver, E.V., Sorenson, J.H., Soderstrom, E.J., Reed, J.H., Bjornstad, D.J., and Peelle, E. (1983), Incentives and nuclear waste siting: prospects and constraints, *Energy Systems and Policy*, **7** (4), 323–351.

Greenberg, M.R., Anderson, R.F., and Rosenberger, K. (1984), Social and economic effects of hazardous waste management sites, *Hazardous Waste*, **1** (Fall), 387–396.

Harris and Associates (1976), *A Second Survey of Public and Leadership Attitudes toward Nuclear Power Development in the United States, Summary* (Ebasco Services, New York, NY).

Hohenemser, C., Kates, R.W., and Slovic, P. (1983), The nature of technological hazard, *Science*, **220**, 378–384.

Kasperson, R.E. and Rubin, B.L. (1983), Siting a radioactive waste repository: what role for equity?, in R.E. Kasperson (ed), *Equity Issues in Radioactive Waste Management*, pp 118–136 (Oelgeschlager, Gunn, and Hain, Cambridge, MA).

Kates, R.W. and Braine, B. (1983), Locus, equity, and the West Valley nuclear wastes, in R.E. Kasperson (ed), *Equity Issues in Radioactive Waste Mangement*, pp 94–117 (Oelgeschlager, Gunn, and Hain, Cambridge, MA).

Kelly, J.E. (1980), *Testimony on behalf of the State of Wisconsin regarding the Statement of Position of the United States Department of Energy in the Matter of the Proposed Rulemaking on the Storage and Disposal of Nuclear Wastes*, Docket No. PR50-51 (US Nuclear Regulatory Commission, Washington, DC).

Kunreuther, H., Linnerooth, J., Lathrop, J., Atz, H., Macgill, S., Mandl, C., Schwarz, M., and Thompson, M. (1983), *Risk Analysis and Decision Processes: The Siting of Liquefied Energy Gas Facilities in Four Countries* (Springer-Verlag Publishers, Berlin).

Kunreuther, H., Linnerooth, J., and Vaupel, J.W. (1984), A decision-process perspective on risk and policy analysis, *Management Science*, **30**, 475–485.

Lindell, M.K. and Earle, T.C. (1981), *Comparative Analyisis of Risk Characteristics of Nuclear Waste Repositories and Other Disposal Facilities*, BHARC-411/81/005 (Battelle HARC, Seattle, WA).

Lindell, M.K. and Earle, T.C. (1983), How close is close enough: public perceptions of the risks of industrial facilities, *Risk Analysis*, **3** (4), 245–253.

Lindell, M.K., Earle, T.C., Hebert, J.A., and Perry, R.W. (1978), ***Radioactive Wastes: Public Attitudes Toward Disposal Facilities***, BHARC-411-004 (Battelle HARC, Seattle, WA).

Lipsky, M. (1968), Protest as a political resource, ***American Political Science Review***, **62**, 1144–1158.

Morell, D. (1984), Siting and the politics of equity, ***Hazardous Waste***, **1** (4), 555–571.

Murdock, S.H., Leistritz, F.L., and Hamm, R.R. (eds) (1983), ***Nuclear Waste: Socioeconomic Dimensions of Long-term Storage*** (Westview Press, Boulder, CO).

National Governors' Association (1981), Siting hazardous waste facilities, ***The Environmental Professional***, **3**, 133–142.

National Research Council (1984), ***Social and Economic Aspects of Radioactive Waste Disposal*** (National Academy Press, Washington, DC).

National Research Council (1985), ***Reducing Hazardous Waste Generation: An Evaluation and a Call for Action*** (National Academy Press, Washington, DC).

O'Hare, M., Bacow, L., and Sanderson, D. (1983), ***Facility Siting and Public Opposition*** (Van Nostrand, New York).

Popper, F.J. (1983), LP/HC and LULUs: the political uses of risk analysis in land use planning, ***Risk Analysis***, **3** (4), 255–263.

Portney, K.E., (1983), ***Citizen Attitudes toward Hazardous Waste Facility Siting: Public Opinion in Five Massachusetts Communities***, Mimeo (Lincoln Filene Center for Citizenship Public Affairs, Tufts University, Medford, MA).

Raiffa, H. (1982), ***The Art and Science of Negotiations*** (Harvard University Press, Cambridge, MA).

Ryan, A.S. (1984), ***Approaches to Hazardous Waste Facility Siting in the United States***, Report to the Massachusetts Hazardous Waste Facility Site Safety Council, Boston, MA.

Schwartz, S.P., White, P.E., and Hughes, G. (1985), Environmental threats, communities, and hysteria, ***Journal of Public Health Policy***, **6** (1), 58–77.

Seley, J.E. (1983), ***The Politics of Public-Facility Planning*** (Lexington Books, Lexington, MA).

Shabecoff, P. (1985), Toxic waste threat termed far greater than U.S. estimates, ***New York Times*** (March 10), 1 and 31.

US Congress, OTA (1982), ***Managing the Nation's Commercial High-Level Radioactive Waste*** (US Government Printing Office, Washington, DC).

US Congress, OTA (1985), ***Superficial Strategy*** (US Government Printing Office, Washington, DC).

US Council on Environmental Quality (1980), ***Public Opinion on Environmental Issues*** (US Government Printing Office, Washington, DC).

US Environmental Protection Agency (1979), ***Draft Environmental Impact Statement for Subtitle C*** (US Environmental Protection Agency, Washington, DC).

von Winterfeldt, D. and Edwards, W. (1984), Patterns of conflict about risky technologies, ***Risk Analysis***, **4**, 55–68.

PART THREE

Risk Analysis

CHAPTER 8

The Inexact Science of Chemical Hazard Risk Assessment: A Description and Critical Evaluation of Available Methods

V.T. Covello and M. Merkhofer [1]

8.1. Introduction

In this chapter we describe and evaluate some of the more prominent methods of chemical hazard risk assessment. The methods are organized according to the component of the risk assessment process that they are meant to address (*Table 8.1*). Discussed first are methods designed for characterizing a source of risk, followed by methods useful for assessing exposures, methods for dose–response assessment, and, lastly, methods for risk estimation. Consideration is given both to methods used directly for the development of probability distributions over health and environmental consequences and to methods that provide information that would support this task. Methods useful for either continuous or discrete risks are also discussed.

8.2. Methods for Characterizing the Source of the Risk

Methods for characterizing the source of risk are directed at quantifying and describing the characteristics of chemical industry technologies, products, processes, or systems that have the potential for creating risk. Sources of risk (i.e. hazards) may involve chemical substances that are inherently flammable, corrosive, explosive, or toxic. For these chemical hazards to produce a risk, (1) chemicals must be

Table 8.1. Categories of chemical risk assessment methods.

Methods for characterizing the source of risk:	
	monitoring methods
	performance testing
	accident investigation
	statistical methods
	modeling methods
Methods for exposure assessment:	
	monitoring methods
	modeling methods
Methods for dose–response assessment:	
	animal research
	epidemiology
	tests on humans
	modeling methods
Methods for risk estimation:	
	methods adopting a perspective of classical statistics
	methods adopting a Bayesian perspective
	methods based on modeling

present in sufficient quantities or intensities, and (2) the natural or technological systems that contain or limit the hazard must be less than completely effective. If a chemical hazard is regarded as contained by some boundary that separates it from the general environment then risk-source characterization may be regarded as the study of the events or processes that may happen within that boundary that affect the level of risk.

Risk-source characterization has tended to focus on industrial or transportation accidents that may cause hazardous chemical substances or energies to be released into the environment (Health and Safety Executive, 1978; Rijnmond Public Authority, 1982; Lees, 1980; Bendixen and O'Neill, 1984). Routine activities (e.g. maintenance operations at a chemical plant) can, however, produce significant releases that can also be subject to characterization in a comprehensive chemical risk assessment.

Some examples can be provided to clarify the types of chemical hazard risk assessment methods that relate to risk-source characterization. If a chemical plant is regarded as providing a source of risk, then risk-source characterization will include the specification of the types, amounts, and timing of all toxic substances emitted into the air and water. If the concern is the failure of a chemical containment, risk-source characterization will include the probability of various failure modes and the volume of toxic substances that would be released under each mode. If the concern is a pesticide used in agriculture, risk-source characterization will include the quantities used, and when and where they are applied. The various subcategories of methods considered under risk-source characterization include monitoring methods, methods for performance testing and accident investigation, statistical methods, and modeling methods. Each subcategory is described below.

8.2.1. Monitoring methods for characterizing the source of risk

Monitoring methods are widely used to provide the basic data to support risk-source characterization. Monitoring refers to "the assembling of data sets, usually for the purpose of detecting some change in status of some variable of interest" (Hayne, 1984). Examples of monitoring activities for risk-source characterization include the collection of data on the operating parameters of a chemical plant's control systems and chemical analysis of residues from soil samples collected at a toxic waste disposal site.

Monitoring is usually a sampling process: rather than measure the status of some variable at all times or locations, selected measurements are taken. Issues that need to be addressed in designing a monitoring approach include:

(1) Decisions as to where and how to take representative samples.
(2) Decisions as to how to handle the samples en route to the laboratories.
(3) Decisions on which analytical methods to use, and investigation as to the presence or absence of certain conditions of interest.
(4) Decisions on which procedures and formats to use for aggregating and presenting the results of the analysis and for setting the degree of confidence in the data.
(5) Determination of how to interpret the monitoring data as to presence, quantity, transformation, migration, etc. of conditions of interest (Schweitzer, 1982).

Because monitoring focuses on current and past status, it is most useful in chemical hazard risk assessment for situations in which the regulatory options under consideration either will not significantly affect the risk source or will affect it in some well-defined way (e.g. eliminating it altogether). The first situation occurs if the regulatory decisions under consideration accomplish risk reduction through altering exposure or dose–response processes rather than the risk source itself. For example, if the decision involves restricting people from entering or residing near a hazardous site, data regarding the level of danger presented at the site (in terms of degree of toxicity, magnitude of potential energy stored, etc.) could be of direct use in the risk assessment, since regulatory actions would not be expected to affect this component of the risk chain. Monitoring is also of direct use to chemical hazard risk assessment if the regulatory action under consideration is directed at the risk source, but can be expected to alter the danger presented by it in a predictable way. For example, if regulation would require the installation of technology for purifying the emissions produced in some production process, and the technology is known to have some level of efficiency, then extrapolation using monitoring data might readily be used as the basis for assessments of risk with and without the proposed regulation.

Strengths

Monitoring provides a means for quantifying the current and historical status of the risk source. It thereby establishes a necessary base for identifying problems and for projecting the future. Most significantly, monitoring data may be used to help calibrate models, and this is its main value for chemical hazard risk assessment.

Limitations

An important limitation of monitoring is its focus on the past: it provides data on what has happened but cannot, in the absence of assumptions concerning the extrapolation of the past to the future, provide direct evidence of what will happen. For many chemical risks, the principal concern is not the continuation of the status quo, but a fundamental change that produces substantially greater risk. Monitoring can sometimes identify trends that foreshadow a significant increase in the risk posed by a hazardous chemical system. However, monitoring will be less useful if the principal cause of increased danger is a discrete event (e.g. an operator error or a terrorist act) that is not easily detectable from the behavior of the system prior to the event.

Another significant limitation is the usefulness of monitoring for rare events. If the source of risk is the extreme natural hazard which occurs once in 100 or 1000 years, then monitoring will provide too few observations to be of much use. Another limitation deals with the need to monitor a quantity that is readily measurable (a so-called indicator event) rather than a more difficult-to-measure quantity that may be of more direct interest. For example, while it might be desirable to monitor the transport of all hazardous waste, only those shipments properly documented and recorded will be available.

It is also important to note that monitoring can be an expensive and time-consuming activity. For example, determining the exact amounts and nature of the atmospheric emissions from a chemical plant can cost several hundreds of thousands of dollars. If, as is often the case, monitoring requires the use of complex measurement devices and depends on the collection of data over a long period of time, the development and implementation of a monitoring program can produce considerable capital and operating expense.

8.2.2. Performance testing and accident investigation

Monitoring systems are usually designed to collect multiple measurements of the status of the system while it is functioning in its normal environment. Sometimes it is useful to gather information by investigating the behavior of the system under stress, i.e. to subject the system deliberately to conditions that are of special concern.

Performance testing is a method for risk-source characterization that involves collecting data about a system under controlled conditions. In some situations, such as is the case with electrical or mechanical components that are produced in large quantities, sufficient data can be generated from performance testing to permit statistical methods (discussed below) to be used to derive conclusions concerning the operating and failure characteristics of similar components used in large chemical systems. In other situations, only limited tests can be performed, and engineering reliability and analysis is needed to estimate the characteristics of concern to risk assessment. In these latter cases, judgment plays a major role in deducing the implications of the results of performance tests.

Another method that rarely provides sufficient data for statistical analysis, but nevertheless can provide important information for chemical hazard risk assessment, is accident investigation. An important element of accident investigation is the determination of exactly what caused the system to fail. For example, although the cause of death in a chemical accident may have been contact with the chemical, the cause of the accident may have been an operator error coupled with a breakdown in the safety control system. Performance testing and accident investigation provide additional methods for augmenting information needed to understand and describe the source of risk.

Strengths

Performance testing and accident investigation can provide important information and understanding for chemical hazard risk assessment that are not readily obtainable from standard monitoring methods. In particular, methods based on investigation under unusual conditions, such as those causing failures and accidents, can allow the root causes of accidents to be better understood. Such methods provide important understanding, and aid the development and application of the statistical and modeling methods used in the development of risk estimates.

Limitations

A principal limitation in performance testing is the difficulty of determining and simulating the conditions that are likely to be of greatest concern in chemical hazard risk assessment. The applicability of the results depends on the assumption that the items or systems being tested and the conditions under which the tests are conducted are those relevant to the chemical risk under investigation.

In the case of accident investigation, a principal difficulty is the episodic nature of accidents, which denies investigators the opportunity to prepare their investigation in advance. Identifying the underlying causes can be difficult because the evidence required for such inferences is often obscured or lost during the accident. Controls may not exist, so the systematic comparison of factors in situations that did and did not produce accidents may not be possible. Furthermore, it is often difficult for the investigator to conduct the investigation in a timely and orderly way, since emergency treatment of those injured must take precedence over research.

8.2.3. Statistical methods for characterizing the source of risk

Because monitoring or performance testing often provides repeated measurements for a system creating a source of risk, statistical inference may be used to infer from past data what the future will hold. In cases where exposure and dose-response processes are very simple (e.g. an event that produces the same health consequences every time it occurs), statistical methods for characterizing the risk source are all that is needed for chemical hazard risk assessment. The most important example is accidents producing a fatality. If every time the accident occurs there is one individual present and the result of the accident is nearly always death, then it is not necessary to decompose the risk into a source, exposure process, and dose-response process: the frequency of the accident provides a measure of the risk of dying. A similar situation arises in the case of frequent accidents and common debilitating agents, as basic injury and fatality data may be available which can be analyzed directly to characterize the risk source. This type of analysis is often termed "actuarial risk assessment" or "historical risk assessment" (Vesely, 1984). The statistics that are usually estimated are the fatality rates (frequencies) for different population cross sections and for different causes. Actuarial data and other statistical records rarely identify the prerequisite causative factors, but they have the advantage of being relatively reliable.

In the more general case in which hazardous events produce different consequences, chemical hazard risk assessment requires an explicit accounting of exposure and dose–response processes. The statistical analysis of available data may, however, still be useful for characterizing the source of the risk. For example, monitoring data such as failure rates, temperature, and wind velocities can provide useful information for understanding the risk source.

The underlying assumption in statistical methods is that the variables of concern are random variables with probability distributions that can be derived from the collected data. The frequencies of risk-causing events are often assumed to be constant with time. The Poisson probability model for event occurrences, often used in statistical methods, makes this assumption. In some situations, however, frequencies are not constant. To model time behavior, various methods can be used, such as assuming a parametric function and estimating the values of the parameters from past data using standard statistical estimation techniques. If the underlying event produces varying characteristics or consequences, and adequate data are available, then the frequency of all events having a given range of consequences can be calculated. By choosing successively larger consequence values and calculating frequencies of events having consequences greater than these values, a plot of frequency versus consequences magnitude can be obtained. This type of curve is often constructed in engineering risk assessments and is called the "complementary cumulative distribution" or "risk profile". (This and other means for displaying risk estimates are discussed further in Section 8.6.1.)

Since statistical inference methods require a data base, applications to areas where data are limited require the use of constructed or surrogate data bases. For example, statistical inference is widely used in risk assessments dealing with the transportation of hazardous wastes (Philipson and Gasca, 1982). The data base supporting such an application is often less than ideal. Records of shipments of the hazardous material of interest are often not kept or are not accessible. Thus, estimates are based on samples of shipment data that are often of uncertain accuracy and validity and subject to considerable judgmental interpretation. For rare accidents, such as liquefied natural gas (LNG) tanker accidents, adequate data for a meaningful statistical inference may not exist. In such cases, surrogate samples are often used (e.g. oil tanker accidents as a surrogate for LNG tanker accidents).

A related problem for statistical methods occurs when the available data are for events with typical characteristics and the risk of concern relates to events having extreme characteristics. For example, the concern may be the frequencies of extremely high wind velocities when the bulk of the statistical data relates to winds of lesser velocities. Extrapolation methods, or "distribution techniques", as they are sometimes called, are a body of methods that estimate the frequency of severe events by smoothly extrapolating from less severe events which have occurred in the past. In applying such methods, the more severe events are assumed to be caused by the same physical mechanisms and processes which caused the less severe events.

Extreme value theory (Vesely, 1984) is sometimes applied if there are a large number of processes whose maximum (or minimum) states are involved in producing the risk and if there is not strong dependency among the processes. Extreme value theory has been used to predict frequencies of catastrophic floods, and similar methods have been used to predict maximum fire, maximum tornado, and maximum hurricane consequences in given time periods. The basic estimation process involves fitting a frequency versus consequence curve to past data. The function is specified except for one or more unknown parameters. Having observed past events with measured characteristics, the parameters are estimated and the resulting specified function is used to predict the frequencies of events with large consequences that have not yet occurred. Critical considerations for selecting the form of the function are the physical and engineering justification of the functional form, the assumption of smooth continuous behavior from less severe consequences to catastrophic consequences, and the availability of measurements of the past events.

To depict the uncertainty inherent in estimates derived from statistical data, probabilities or confidence intervals for the estimates can be computed. Since these and many other issues surrounding the use of statistical methods are independent of whether the methods are being applied to risk-source characterization or to some other component of the risk assessment process, more discussion of the strengths and limitations of statistical methods appears in Section 8.5.2.

8.2.4. Modeling methods for characterizing the source of risk

While even the most simple statistical methods for characterizing a source of risk are in effect "models", risk-source models can be quite sophisticated. For the case of continuous risks, models are frequently used to estimate the characteristics, amounts, and locations at which some toxic substance is emitted into the environment, and how these might change over time and location or with various controls. For example, to support a regional risk assessment of the effect of higher oil prices on emissions of sulfur oxides from electric power plants, a model might be developed to estimate the geographic distribution of emissions over time under different control scenarios. A variety of approaches are available for constructing such models (for an overview of basic analytic modeling approaches, see Gass and Sisson, 1975). For example, an econometric model (a model based on economic and empirical data) might be used to assess the effect of increased prices on high-sulfur coal due to a shift in use by the energy industry to oil and other low-sulfur energy sources (by comparing supply, demand, and price with interfuel competition, for example). Another approach might be to use engineering and process models based on judgment to estimate the effectiveness of new technologies in reducing harmful emissions from representative plants.

For discrete risks, such as accidents at chemical plants, models of the risk source are generally based on a specification of the circumstances and sequence of events that must occur for the accident to take place. For example, failure mode and effects analysis (FMEA) is often undertaken during the planning and construction of a facility to characterize hazards associated with possible system failures. The approach involves developing a functional block or logic diagram relating the various system processes and components, postulating possible failure models, identifying possible causes of failures, describing the consequences of the possible failures, and estimating the probabilities of each failure's occurrence.

Several modeling methods used in risk-source characterization involve the development of graphic structures. For example, the digraph, originally developed for safety analyses of chemical processing systems, is a deductive logic diagram that describes the interrelationships among process variables. The diagram displays the various system control loops, with the loops classified according to their ability to cancel a system disturbance of a given size.

Fault tree analysis (Gottfried, 1974; US Nuclear Regulatory Commission, 1981) is based on a specialized model that may be represented as a diagram of binary (yes–no) logic that traces backward in time the different ways that a particular event could occur. *Figure 8.1* illustrates the structure of a fault tree diagram. To construct a fault tree, the final, undesired (top-level) failure event (for example, off-site release of toxic substances from a chemical plant) is defined. Using a functional description of the system, the events that could logically cause the failure are identified. These events are related to lower-level events and the diagram is expanded until a level of basic events is reached – for example, the failure of individual subsystems or components (e.g. pressure relief systems, pumps, electrical components). The failure rates for the basic events are either assumed to be known with certainty or are assumed to be described by probability distributions derived from available failure-rate data. The uncertainty in the

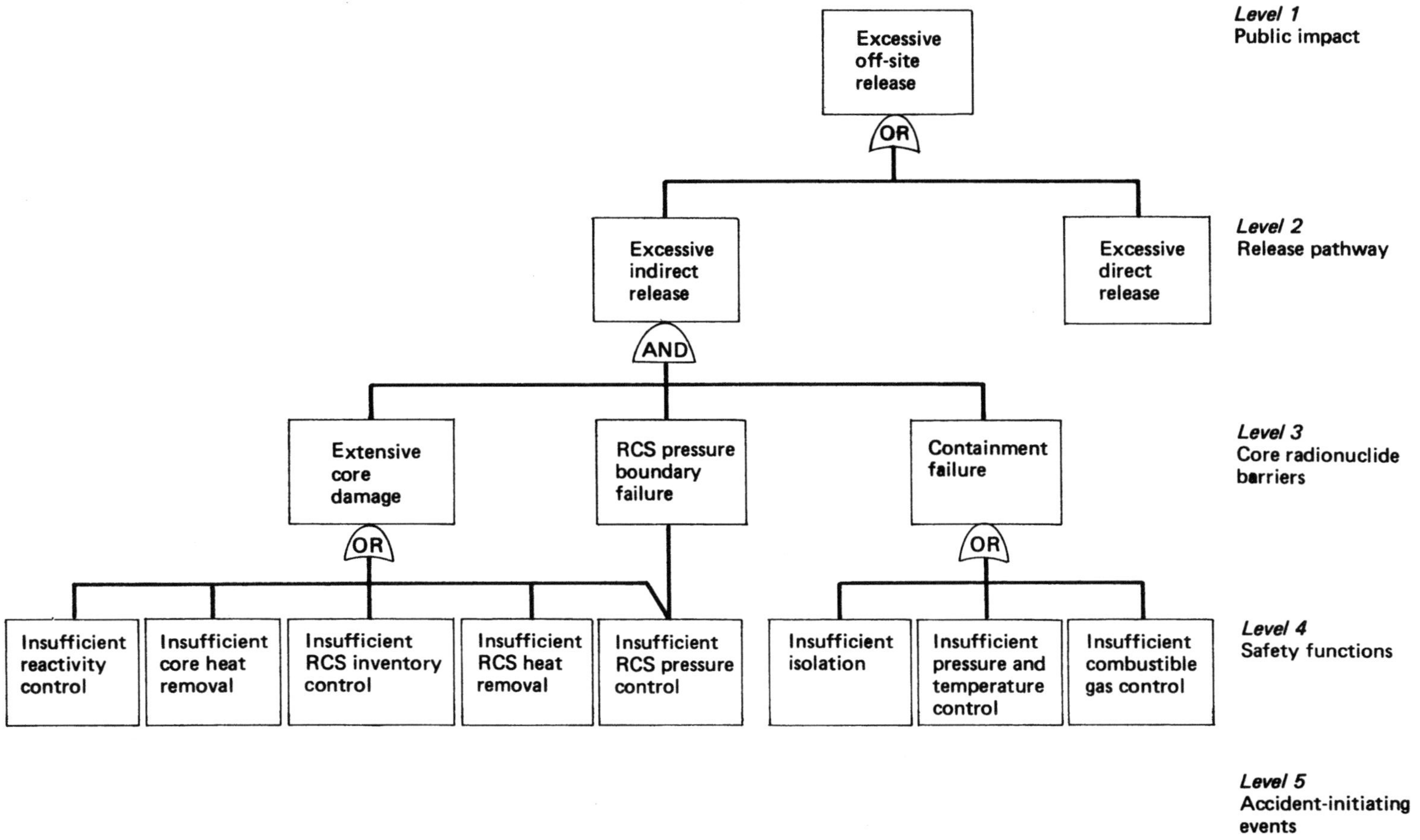

Figure 8.1. **Master logic for a fault tree diagram of a nuclear reactor: RCS, reactor coolant system. (*Source:* American Nuclear Society and Institute for Electrical and Electronic Engineers, 1983.)**

failure rate for the top-level event is derived from a calculation based on the equivalence between the fault tree diagram and a corresponding set of Boolean algebraic expressions for the manipulation of probabilities.

Event trees (see, for example, Rasmussen, 1975; McCormick, 1981) are also graphic tree structures. An event tree starts with some particular undesired initiating event (e.g. failure of a valve or pump) and projects all possible plant responses to that event. Each branch in a event tree, with associated probabilities, represents a possible state (often simply success or failure) for the plant's subsystems which would be called upon as the accident progresses. Such states might include the status of containment safety and mitigation systems. The probabilities associated with the individual branches may be obtained through statistical methods or fault trees, and a series of probability calculations permits the assessment of the probabilities of occurrence of the final plant accident states. Although event trees have been most prominently applied in the area of nuclear reactors, they have also been applied in risk assessments of chemical facilities, such as that at Canvey Island in the UK (Health and Safety Executive, 1978).

When fault trees and event trees are used to model accident sequences, additional models are needed to calculate the physical consequences of each accident sequence. This consequence calculation is in essence accomplished by determining the physical processes and phenomena that are associated with each accident sequence. In a risk-source characterization of an accident at an industrial facility, for example, containment analysis (US Nuclear Regulatory Commission, 1984) can be used to identify and assess a variety of containment-failure modes and to predict the amount of hazardous materials released to the environment under alternative accident sequences. This method combines the results of engineering analyses (based in part on performance testing) with models composed of fault trees and event trees.

Many of the issues concerning modeling methods are independent of whether the methods are applied to risk-source characterization or to some other component of the risk assessment process. One such issue is that, while it is often relatively easy to develop models of physical cause–effect processes, it is much more difficult to apply modeling when critical uncertainties relate to the behavior of humans. Important examples are the potential for operator error and sabotage. These are recognized to be significant sources of risk, yet the methods available for modeling these events are not as well developed as those for component and system failures (see, for example, Swain and Guttman, 1983). These and other general limitations of modeling are discussed in more detail in later sections of this chapter.

8.3. Methods for Exposure Assessment

Exposure assessment is the process of measuring or estimating the intensity, frequency, and duration of human or other exposures to a risk agent (National Research Council, 1983). Exposures may occur in a variety of ways: e.g. through direct external contact (produced, for example, in the industrial setting), air and water transport (as is the case with many industrial effluents), or ingestion through the food chain (such as occurs with mercury). The appropriate methods for exposure assessment depend on the pathways for exposure that are of greatest concern for the situation under study.

Gauging exposure is usually difficult because numerous factors must be considered and because information is often incomplete. Even for the special case of the worker population, only rarely is there adequate information about who is exposed and the degree of their exposure. Large chemical manufacturers, for example, generally keep records on workers exposed to potential hazards, but data on the level of exposure to particular chemicals usually are not precise. The situation with respect to the general population is often even worse, owing to lack of data on exposure levels. To regulate a pesticide used on citrus fruit, for example, officials would want to know not only how many oranges are consumed by a typical person, but also how many people are fond of eating large numbers of oranges, because the latter group would be the population at greatest risk.

If the chemical hazard is associated with food or consumer products, personal habits may have a large bearing on the nature of the exposure. For example, differences in food storage practices, food preparation, and dietary habits influence the fraction of a hazardous substance present in food that individuals actually ingest. Similarly, if the hazardous agent is absorbed when a consumer product is used, patterns of use affect exposure. A solvent whose vapor is potentially toxic, for example, may be used outdoors or in a small, poorly ventilated room. Thus, for risk assessments associated with consumer products, exposure assessment is concerned with understanding how the products may be used and the implications for estimating exposures.

If exposure is through the air or water, then consideration must be given to how the hazardous substance moves from its source through the environment and how it degrades or reacts with other substances. In this instance, exposure assessment has been defined as the determination of the concentration of toxic materials in space and time at the interface with target populations (Travis *et al.*, 1983). Pathways of exposure for consideration must include atmospheric and aquatic surface and groundwater transport and transformation.

Another important aspect of exposure assessment is the determination of which groups in the population may be exposed to a risk agent; some groups may be especially susceptible to adverse health effects. For example, in the case of toxic chemicals, pregnant women, very young and very old people, and persons with impaired health are particularly sensitive. Some regulatory laws (such as the United States Clean Air Act provision regarding promulgation of primary air quality standards) have been interpreted to require consideration of the impact of pollutants on "sensitive or susceptible individuals or groups".

Exposure to multiple chemical hazards often results in portions of the population becoming more sensitive to any single chemical hazard. Exposure to chemical hazards that act synergistically greatly complicates the chemical risk assessment because most assessments are conducted on individual chemical hazards. Synergisms also often make it necessary for exposure assessment to consider the activities that exposed individuals are engaged in and other substances to which they might be exposed. For example, strenuous activity often increases the magnitude or likelihood of adverse reactions to exposures to many air pollutants. Similarly, exposure to asbestos results in a cancer rate incidence that is much greater than indicated by carcinogenicity data on the substance individually (National Research Council, 1983). The number of sources also complicates exposure assessments. An individual can be exposed to a single chemical hazard from multiple sources. Exposure to lead, for example, can come from breathing air, eating food, and drinking water.

Thus, to be comprehensive, an exposure assessment must describe the levels of exposure and all conditions that might be needed to assess the effects of those exposures, including the magnitude, duration, schedule, and route of exposure; the size, nature and classes of the human populations exposed; and, of course, the uncertainties in all of these estimates.

Given the variety of pathways by which hazardous chemical substances can be transmitted, a comprehensive exposure assessment must account for the cumulative effects of simultaneous exposures via distinct media. Thus, multimedia exposure assessment is generally necessary to account for all significant exposure processes.

8.3.1. Monitoring methods for exposure assessment

As in the case of risk-source characterization, exposure assessment relies heavily on monitoring measurements. For any given risk assessment, monitoring activities can be used to provide a variety of relevant information. For example, to support the assessment of the risks posed by a site at which hazardous wastes have been disposed of, the air, soil, surface water, and groundwater contamination levels at various distances from the site may be monitored using conventional field measurement techniques. Remote techniques, such as aerial photography and multispectral overhead imagery, might also be used to delineate waste site problems. Ground penetration remote sensing technologies might be used to estimate the subsurface distribution of waste materials and to help target monitoring activities (Schweitzer, 1982). Biological monitoring (Gompertz, 1980) may be used to clarify possible food-chain problems, for example, by taking measurements dealing with food crops, livestock, or local fish. In addition, wild animals and indigenous vegetation might be sampled to find indications of local contamination.

Individual exposure monitoring can provide the most accurate information for exposure assessment. The method is employed in certain occupational settings and involves routine monitoring of exposed individuals (e.g. radiation workers, who carry dosimeters, film badges, or other radiation-sensitive devices) to measure their exposures to the risk agent in question. Site monitoring techniques also involve actual measurements of exposure levels. In this case the measurements are not specifically for individuals, but for locations. Site monitoring tends to be a more practical method when the numbers of people or areas of exposure are very large, and where the risk agent released is either routine (as in continuous risks) or where the release occurs over a sufficient period of time to allow a monitoring system to be established. An issue that is frequently relevant for site dose monitoring is the sizable variation of concentrations of a risk agent often found in media (such as air and water) over space and time. Thus, selection of a particular dose-measurement interval, or, more generally, the selection of particular sites or measurements, can have a strong impact on the numerical results obtained.

In addition to the direct monitoring of exposure levels, biological monitoring can provide useful information for exposure assessment by, for example, measuring the levels of chemical residues in human tissues, in food, or in tissues or food of specific animal species. Such data may be used to estimate exposures provided that there is some understanding of the preferential biological absorption and retention processes of substances in the particular organs or organisms (e.g. strontium in bone, iodine in the thyroid, and mercury in fish), together with knowledge of the physical decay processes of the hazardous material (e.g. radioactive decay or biodegradation). In addition to the use of chemical residue measurements in biological monitoring, some indication of exposure can also be obtained through measurements of individual physiology, such as the effect of cholinesterase inhibition on fertility, as with birds, or on growth, as with fish (Hayne, 1984). Population effects that may be measured for indicator species include altered population, numbers, or biomass; altered population processes, such as survival or productivity; and altered parameters, such as age, structure, or sex ratio. Biological community effects include changes in such parameters as species diversity and the ratio of food to consumer organisms.

Limitations

The most important limitation in monitoring for exposure assessment relates to its ability to provide the basic data needed to apply statistical methods. The logic of most statistical methods is based on an assumption of random samples: the data are assumed to provide an unbiased representation of the larger universe of values from which the measurements were taken. In many monitoring systems, costs and practicality demand that sampling be carried out at convenient sites and times. Thus, monitoring often producers an "index", not an unbiased estimate. Common examples include restricting sampling to daylight hours.

8.3.2. Models for exposure assessment

For most chemical hazard risk assessments, individual and site dose monitoring data are insufficient to characterize exposures sufficiently for the development of frequency distributions describing the doses actually received by individuals. Even if site exposure measurements are available and an individual's activities are highly structured (such as in the workplace), the risk agent may be present at various locations and occupational settings not covered by the site monitors. Furthermore, because individuals are likely to move between locations covered by monitors, computations require estimates of levels between monitors and models for the movement of individuals between sites. Thus, chemical hazard risk assessments generally require the use of models to estimate exposures.

Increasingly sophisticated environmental models are being used in exposure assessment to describe the environmental transport and conversion of chemical substances. These include models of the chemistry, transport, and deposition of acid rain; models of the photochemical processes associated with materials disposed of at sea; models of wind directions and velocities and precipitation patterns that might affect the transport of pollutants from normal or accidental industrial or power plant emissions; and models for the leaching and runoff of chemicals to streams. The various models differ significantly from one another in terms of their approximations and the assumptions used in characterizing the source (e.g. point source or area source, instantaneous or continuous release), the media of transport, and the manner of spreading and entrainment considered.

In the case of exposures produced through the transport of materials through air, transport modeling has reached a relatively high degree of sophistication. One of the more straightforward of the available approaches is to develop regression or time series models based on least-squares fitting to monitoring data. Another approach, which allows more flexibility for investigating alternative assumptions, is the Gaussian plume model. This model assumes that the plume from an emission source spreads laterally and vertically in accordance with a Gaussian statistical distribution. Although Gaussian models have limited applicability in instances of complex terrain and variable release rates, they have been used for many years to predict ground-level concentration where reasonably flat terrain exists and where an average, fairly stable, release rate can be assumed (Travis *et al.*, 1983). Another, more computationally demanding, approach is based on computing trajectories that released material might follow. The trajectories are typically composed of segments estimated for fixed intervals of time and based on historical wind data. Trajectory models are often used for long-range predictions of atmospheric concentrations. To represent rapid short-duration emissions, such as might result from containment failures or explosions, "puff" models have been developed. The models share many of the same principles as continuous release models, but the applications are quite different.

In addition to transport, many atmospheric models account for transformations of pollutants (e.g. photochemical transformations) and are useful in situations where transformation products may exhibit greater or less bioactivity than the parent pollutant. Chemical transformation processes are often approximated by first-order rate reactions. Frequently, air transport models will estimate long-term average concentrations, while short-term concentrations are assumed to follow some specified frequency distribution (usually lognormal).

Aquatic transport models, although not as highly developed as atmospheric models, are also frequently used in exposure assessment. In the case of surface water, such models will attempt to account for the major source of contaminants either as point sources (e.g. industrial wastes) or as a normal rainfall runoff process or storm water runoff to water bodies from urban and agricultural areas. Surface water transport may be an important factor in situations where a principal exposure pathway is through drinking water or where aquatic food is ingested. Steady-state models are most frequently employed to account for the transport of chemical pollutants in surface water. For example, a river may be represented as a series of completely mixed reaches in which steady-state contaminant concentrations are estimated based on dilution and physical/chemical removal of contaminants from volatization, net absorption, net settling, photolysis, microbial degradation, etc. (Travis *et al.*, 1983). To represent spills (e.g. oil spills from tankers) dynamic models are available.

Groundwater transport is important in risk assessments of toxic waste disposal sites. More generally, toxic materials can enter groundwater by deposition from the atmosphere onto the ground, with subsequent infiltration into the water table. In modeling groundwater flow it is important to account for the interaction of the contaminants with soil: in many cases, a large proportion of contaminants in groundwater will be absorbed into the soil, thus greatly retarding their movement relative to that of groundwater. Very simple models tend to be used to represent these processes: (e.g. a linear absorption model is often used to predict the relative concentrations of a compound in the soil).

Food-chain models are sometimes used in chemical hazard risk assessments where there is a concern that contaminants that have been released may enter food chains and pose health risks to man via ingestion of contaminated food. Aquatic food chains, for example, can be important in producing bioconcentration of harmful compounds. To estimate risks to man via the food-chain exposure pathway, intake rates for contaminants in the various food categories must be calculated based on the consumption rate of food. Three elements are required for such calculations:

(1) The concentration of the component in question in each item of the diet.
(2) The amount consumed of each item that contains the component.
(3) The frequency with which each item is consumed.

Generally, very simple models are used to represent such effects. Market basket analysis involves an analysis of contaminants in prepared foods and typically produces an average or representative intake. Per capita consumption is based on dividing the sum of annual production plus imports of a given food item by the number of people in the country. Since these methods fail to reveal special groups with high intake rates, reliance is sometimes placed on dietary surveys, in which consumers are asked to recall what foods they ate with what frequency and with what amounts (portion size) over a defined period of time (Food Safety Council, 1980).

To estimate populations at risk from a given pathway of exposure, it is necessary to determine who works or lives at various sites, and who breathes air, drinks water, eats food, or uses products coming from a particular place. This inference is generally painstaking, but usually straightforward. Demographic models developed to support this task may include models for the population distribution within a geographic area surrounding a source of potential risk (e.g. dams, power plants, aircraft flight paths), models of the age distribution of a given population, and models of mobility and activity patterns of sensitive population groups.

8.4. Methods for Dose–Response Assessment

Dose–response assessment methods are concerned with characterizing the relationship between the dose of the risk agent received and the health and other consequences to exposed populations. These consequences can include early fatalities and injuries, latent cancer fatalities, genetic effects, environmental degradation, and economic losses. Most of the effort devoted to developing methods for dose–response assessment have been directed at understanding human health rather than on the more general ecological consequences of exposures to hazardous chemical substances. The discussion in this section therefore emphasizes methods related to human health.

8.4.1. Animal research for dose–response assessment

Since ethical considerations generally preclude the deliberate exposure of humans to risk agents, animal tests provide the largest experimentally derived data base on relationships between doses of hazardous substances received and their subsequent impact on health. To conduct an animal study, the chemical in question is administered to a population of laboratory animals and various data on their health status are then collected both before and after death. Concurrently, a control group is established and treated in an identical manner, except that the animals in this group are not exposed to the chemical. Data collected while the animals are still alive include observations of factors such as size, weight, condition of external features (skin, hair, eyes, etc.), gross evidence of internal abnormalities (such as bone deformation), reproductive capacity (including fertility, reproductive frequency, litter size, changes in frequency of spontaneous abortion or live births, and other abnormalities), life span, behavior (food intake, sleep habits, aggressiveness, problem-solving ability, etc.), body functions, and genetic changes (as evidenced by subsequent generations) (Marcus, 1983). As the animals die or are killed during the course of the study, additional information is obtained through pathological examinations, which are conducted to determine the cause of death and the nature and extent of any morphological abnormalities (e.g. tumors) in internal organs or tissue samples. The measurements obtained for the test group are then compared and contrasted with those for the control group.

In assessing the results of animal research, an important distinction is made between administered dose and effective dose. Because animal studies often must rely on relatively high exposure levels to improve the sensitivity in the detection of a response, the reaction may differ significantly from the mechanisms expected to be important in human exposures. Therefore, the effective dose, i.e. the fraction of the administered dose that impacts the animal in ways thought to be similar to that expected in humans, must be determined.

Strengths

Animal tests offer the researcher a considerable degree of experimental control. Critical experimental factors, such as dose, can be varied in a systematic way to perform scientifically rigorous studies. The controlled laboratory setting permits isolating effects from a specific substance. Because animals can be permitted to get a disease and die, the full effects of the substance can be examined. In addition, experimental results can be replicated to confirm their validity or to ensure the elimination of extraneous factors.

Limitations

The major limitation of animal research is the comparability of the results of research conducted on animals to humans. The complexity of biological organisms and biochemical interactions creates a great many difficulties for inference. Because of the inherent differences in susceptibility among different species, differences in biological or biochemical pathways, and variabilities in the sensitivities of individual agents, exposures to the same hazardous substance may produce different effects in different species, or a significant effect in one species and no apparent effect in another. Although the most reliable information might be expected to be produced through the selection of animals that are the most biologically similar to man, the choice of animals is, for practical reasons, largely influenced by such factors as cost, ease of care, shortness of life span, ease of breeding, ease of handling in large numbers, and genetic homogeneity (i.e. inbred strains). This is why small rodents – rats, mice, hamsters – are most often used for these studies.

Properly scaling doses to obtain comparability between test animals and humans is much more difficult than simply accounting for relative body weights. Metabolisms, respiration rates, organ uptake, elimination, etc. also influence the effective dose. In situations where the mechanism of effect is not well understood, the appropriate scaling factors to obtain comparable dose levels will be extremely difficult, if not impossible, to determine.

Even if the proper scaling factors for dose are known, dosage levels must generally be set at levels different from those that are of greatest concern to humans. Usually it is necessary to increase the dose to ensure that a limited sample size will be capable of detecting an effect. Sometimes it is necessary to diminish the dose. This would be the case, for example, if the agent in the form administered is toxic to the test animal in ways not of direct interest to the study. It may also be necessary to reduce the dose level if the desired dose would prevent the adequate uptake of necessary nutrients or if it would cause the animal to reject it owing, for example, to its alteration of the taste of food.

8.4.2. Epidemiology for dose–response assessment

Epidemiology, the study of the distribution and determinants of disease in human populations, can be used to detect patterns of risk and establish statistically significant associations with risk agents. Epidemiology is also a powerful method for hazard identification. For example, 20 out of 26 agents causing cancer in humans were first identified using epidemiological evidence (Tomatis *et al.*, 1978).

Two types of epidemiological studies that play a major role in dose–response assessment are cohort studies and case–control studies. Cohort studies involve the comparison of groups of people (cohorts) who have different dispositions with regard to some risk factor. For example, the comparison may involve a group of individuals who have been exposed to a substance with a group of individuals who have not. Similarly, cohort studies may involve comparisons of several groups having distinctly different levels of exposure, or groups possessing different personal attributes or behavior. Following the selection of the groups, each is studied over time to observe and compare its health status. Any excess incidence of disease or disability is regarded as a potential response to the risk agent.

Case–control studies involve the comparison of people with a given disease (cases) with individuals without the disease (controls). The objective in case–control studies is to identify and compare differences in characteristics of the groups that might be expected to be risk factors. In this case, a relationship or pattern might be regarded as an indicator of a cause or contributor to the disease. In both types of epidemiologic studies, the results are analyzed using statistical methods to determine the nature and significance of any identified relationships.

In any epidemiologic study, proper experimental design requires that the groups be similar with respect to all characteristics except for the specific factors under investigation. Thus, in cohort studies the categorization of groups must not result in preselection according to state of health, and case–control studies should not produce a preselection according to exposure to any given risk agent. Cohort studies hope to randomize over noncontrolled factors; however, whenever a comparison is made that the study was not specifically designed to handle, care must be taken that the relevant factors are controlled or randomized (i.e. do not vary in a systematic way that could distort the results). For case–control studies, statistical control is used (i.e. selection may attempt to match individual cases and controls for comparability, or standard statistical adjustment procedures may be employed). Demographic variables (age, sex, race, etc.) are most commonly used for adjustment or matching.

Epidemiologic studies have successfully been used to identify:

(1) The relationship between duration and level of exposure and biological or biochemical health effects.
(2) Disease incidence or mortality differences between geographic and socioeconomic regions.
(3) Time trends in disease incidence and mortality associated with the introduction or removal of a specific agent.

Strengths

Epidemiological studies can be used to provide direct evidence of the nature of the relationship between various risk agents and the occurrence of human health effects. Well-conducted epidemiologic studies showing a positive association between a risk agent and disease provide the most convincing evidence about risks to humans. Human studies also eliminate many of the problems encountered in conducting animal studies, including those arising from interspecies extrapolation. Additionally, since the exposure is already in the human dose–response range, the choice of an extrapolation model may be less critical than for animal studies.

Limitations

Although epidemiology is the only means of assessing directly the risks of environmental agents on humans, the method has several limitations that are difficult to overcome. These include:

(1) *Difficulties in detecting relationships at low exposure levels.* Epidemiologic evidence of an environmental hazard is usually obtained from persons with high or intermediate levels of exposure. As in laboratory animal studies, detecting causal relationships at low exposure levels is difficult, since the observed

associations with disease are usually less pronounced; they may also have alternative explanations, including those related to chance, errors, biases, or confounding factors. To provide a valid basis for risk estimates, large numbers of human subjects are often needed, especially if the exposure is low or rare, or if the excess risk is small compared to that of the baseline incidence rate.

(2) *Difficulties in measuring lifetime risks from partial lifetime data*. In most epidemiologic studies, the groups of people studied are exposed for a limited period of their lives and are then followed up for periods up to, but rarely exceeding, 30 years. As a result, in many cases the exposure duration is short compared to the long latency period between exposure and the development of certain diseases such as cancer. Since cancer incidence increases with the length of exposure and age, epidemiologic studies usually only detect a fraction of the lifetime risks resulting from exposure. The problem is to estimate full-lifetime risks from partial-lifetime data. There is no generally agreed-upon solution to this problem. This complicates the detection of a relationship and makes it impossible to identify the risks to humans of agents newly introduced into the environment.

(3) *Difficulties in measuring exposure and estimating dose*. Epidemiologic studies, unlike animal laboratory studies, are not based on controlled experiments. The induction of the disease is not under the control of the researcher; instead, the investigator is limited to the study of disease cases occurring naturally or accidentally. As a result, it is difficult to obtain precise measurements of the duration and intensity of past exposures. Several specific factors compound or contribute to this difficulty. First, since the exposure of interest often cannot be measured directly, surrogate measures of uncertain reliability may be used (e.g. occupation, place of residence). Second, since exposure data are usually derived from historical records generated for other purposes or from the recollections of subjects, opportunities for either random or biased misclassification of exposure are frequently encountered. Third, appropriate study groups are often simply unavailable or inaccessible.

(4) *Difficulties in eliminating biases and excluding confounding factors*. It is often difficult to implicate specific risk agents when environmental hazards involve complex exposures to a variety of agents, the effects of which are difficult to disentangle. Moreover, in epidemiologic studies it is often difficult to adjust for unknown risk factors, since control can be introduced only when the risk factors are already recognized. Thus, when a particular factor is related to exposure and disease outcome, it may be

confounding and give the appearance of an association when in fact none exists, or it may inflate or decrease the magnitude of the association. In view of these difficulties, it is not surprising that epidemiologic data exist for only a small number of chemicals.

8.4.3. Modeling methods for health dose-response assessment

A dose–response model is a functional relationship between the dose of an agent or biological stimulus administered and the response of a test subject. Dose–response models are most prominent when the risk agent is a hazardous chemical, but dose–response models have also been developed to estimate injuries and other effects from accidents or exposures to other forms of kinetic energy.

Most often, dose–response models are obtained by selecting parameter values for some specified model form to "fit" the model to toxicological or epidemiologic data. The model form used may or may not be based on plausible cause–effect mechanisms. A dose–response measurement may be quantitative (a measurement of the magnitude of the effect), quantal (a determination of whether or not a specific effect is elicited), or time-to-response (measurement of the time required to produce a specific condition). In some situations, sufficient epidemiologic data exist to permit a dose–response estimate to be developed directly from statistical analysis of observations of exposure and health effects in humans. In these situations, regression analysis in one form or another is the usual technique for estimating a quantitative relationship. Because the data are generally insufficient to distinguish among alternative functional forms, simple relationships, such as a linear form, are usually fitted.

Even when epidemiologic information is available, however, extrapolations based on theoretical models and expert judgment are generally needed to obtain appropriate dose–response relationships. For example, extrapolations from the exposures observed in an epidemiologic study of workers receiving occupational exposures to lower exposures experienced by the general population are often necessary to make use of epidemiologic results. Adjustments may also be necessary to account for the fact that the general population includes some people, such as children, who may be more susceptible than the people in the sample from which epidemiologic data were developed.

Animal-to-human extrapolation is generally based on standardized dosage scales for making interspecies comparisons, such as milligrams per kilogram body weight per day. The logic for such conversions must depend on the circumstances. For example, it is sometimes observed that the effective dose, expressed in milligrams per day of direct-acting chemical substances, is proportional to the body surface area.

Because a rough approximation to body surface area is the body weight raised to the two-thirds power, it can be shown that in these cases the equivalent doses to humans are higher than those in animals by the ratio of human body weight to animal body weight raised to the one-third power.

Because exposure assessment often calculates media contaminations to which individuals are exposed, an important issue for dose–response assessment is the relationship between exposure and dose. If dose is interpreted as what is adsorbed or absorbed at a target organ from an exposure, then different individuals might be expected to experience different doses from the same exposures, owing to differences in physiology, personal habits, etc. This is of particular importance for lead and other cumulative toxicants. Pharmacokinetics, which deals with the absorption, distribution, metabolism, and elimination of compounds in humans and experimental mammals, is the necessary source for information for the validation of dose–response models, including animal-to-human extrapolations.

In general, dose–response models are based on some assumed functional form for the relationship between dose and response. Simple relationships that are used include linear, linear with a threshold, quadratic, or sigmoidal (S-shaped) curves. More elaborate threshold distribution models are based on the assumption that each individual in the population has his or her own threshold tolerance for the hazardous substance. Specification of a functional form for the distribution of tolerances then determines the shape of the dose–response curve. For example, because the proportion of responses in a toxicity test generally exhibits a sigmoid relationship when plotted against the logarithm of the dose level, the frequency distribution of tolerances within a population is often assumed to be lognormal – the so-called probit model. This model has primarily been used for estimating within the range of empirically observed responses (interpolation) rather than estimating outside the range (extrapolation). Modifications are generally employed to extrapolate responses at low dose levels from the responses induced experimentally at high dose levels. One modification, the Mantel–Bryan equation, assumes a conservative slope (a slope of 1, which is shallower than is generally seen in experimental data sets) in order to correct for the steeper dose–response relationships that homogeneous strains of laboratory animals are more apt to exhibit than the heterogeneous human populations for which the dose–response is generally extrapolated.

Another commonly used tolerance distribution model is the log logistic, or logit function. Like the probit, it reflects an assumption that the proportion of response increases with dose and exhibits a symmetric sigmoid shape, but it approaches the extreme (0% and 100% response) levels more gradually than the probit. In the normal range of the dose–response curve, the logit and probit models appear so similar that it is nearly impossible to distinguish them. In the range of low doses, however, the logit model may predict doses producing no observable effects that are more than an order of magnitude lower than those produced by the probit model.

Depending on the level of understanding available, more sophisticated dose–response models may be developed based on theoretical assumptions for the biological processes involved. These can include known or suspected processes that control the uptake or residence time of the hazardous substance, mechanisms of cellular or organ damage or dysfunction, mechanisms of repair, and so forth. A description of several such mechanistic dose–response models is provided below. The discussion follows that provided by Fares *et al.* (1983) and Park and Snee (1983).

One class of mechanistic dose–response models is based on the premise that a positive response is the result of the random occurrence of one or more biological events. The one-hit model, also known as the linear model (even though it is not linear, but concave), has often been used to estimate the production of tumors from exposures to radiation. The model assumes that only one interaction between the target site and the hazardous substance is sufficient to produce an adverse condition, and that the probability of an interaction is directly proportional to the degree of initial exposure to the substance. This assumption is made regardless of the age of the exposed organism or the pattern of the exposure. The dose–response curve implied by the one-hit model is almost linear at low doses, but it actually has a concave slope over the entire dose range. The model is insensitive to minor fluctuations in the data, but ignores no-effect levels and thresholds (doses low enough to assume a negligible probability of an adverse effect in the lifetime of a population). It is therefore not a good predictor at very low exposure levels and has generated results that have been invalidated by good epidemiologic data (Gehring *et al.*, 1979; Ramsey *et al.*, 1979; Reitz *et al.*, 1978).

The "multihit" model, also called the "gamma multihit" model, assumes that two or more "hits" of a substance are required to induce a response, and that the number of hits over time follows a Poisson distribution. The dose–response curve predicted by the multihit model at low doses may be linear, concave, or convex, depending on the values selected for its parameters. At higher doses, differences between the multihit and probit models are difficult to distinguish. One major problem with the multihit model is that it may estimate doses producing no observable effects that are much higher than the doses that produce an observed effect in laboratory experiments from which the evaluation was made. Furthermore, even when there is no evidence of a confounding background level of exposure, the model may estimate a background rate, so that a possibly significant increase in an adverse health condition may be dismissed as being consistent with the estimated background levels of a substance.

The multistage model, also called "the Armitage–Doll function", is based on the assumption that an adverse health effect originates as a predisposed cell which must undergo a series of mutational stages. In addition, it also assumes that the timing of transitions between stages is expressed by a probability function that is approximately proportional to the dose rate. The multistage model, like the one-hit model, is considered a conservative extrapolation function because the curve approaches linearity at low dose levels. It tends to work well in the experimental dose range, but at low doses it ignores changes in kinetics, metabolism, and other potentially confounding or modifying mechanisms.

Pharmacokinetic dose–response models are based on the principle that biological effects are the result of biochemical interactions between foreign substances (or metabolites) and parts of the body. These models generally assume so-called Michaelis–Menten nonlinear kinetics, which, like many of the dose–response models, produce estimated effects that are linear when extrapolated to low doses. Like the multihit and multistage models, pharmacokinetic models can describe convex or concave curvature in the dose–response relationship; but, unlike the former models, they do not assume that nonlinear behavior is similar at both high- and low-dose levels.

Limitations

The major problem associated with the use of dose–response models in risk assessment is the difficulty of determining which of the models is appropriate for a given situation. The question is of considerable practical importance if the risk assessment requires extrapolations to dose levels lower than those represented in the supporting animal or human data. For example, as illustrated by *Figure 8.2*, at responses between 2% and 98%, the probit, logit, multihit, and similar models produce dose–response curves that have very similar shapes. Thus, all of these models give essentially identical fits to the observed data. All are also biologically plausible. At low doses, however, the doses estimated by the various models to yield no observable effects (virtually safe doses) can differ by three to four orders of magnitude (Interdisciplinary Panel on Carcinogenicity, 1984). Of the various tolerance distribution and mechanistic models discussed in this section, the virtually safe dose estimates, in order of increasing estimates, are one-hit, multistage, logit, multihit, and probit, when the data suggest convexity. When the data suggest concavity, however, the order of the virtually safe dose estimates is reversed (Park and Snee, 1983).

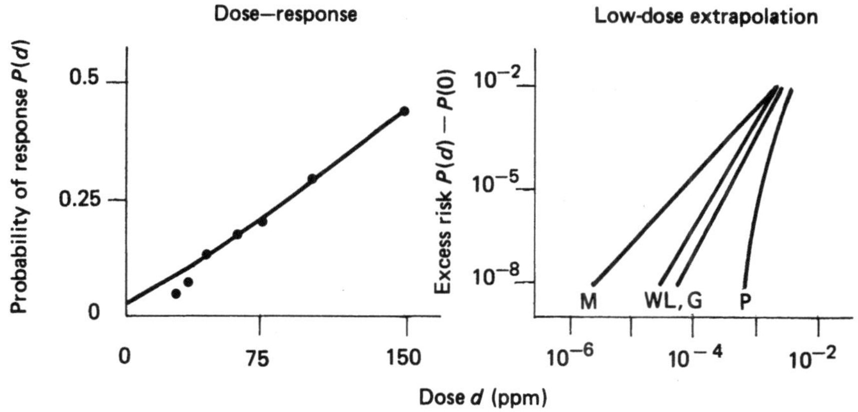

Figure 8.2. Models that fit the data equally well can produce very different results when extrapolated to low doses: M, multistage model; W, Weibull model; L, logit model; G, gamma multihit model; P, probit model. (*Source*: adapted from Krewski *et al.*, 1984.)

8.5. Methods for Risk Estimation

Risk estimation is the process of producing overall summary measures of the level of the health, safety, or environmental risk being assessed. This component of the risk assessment process has also been referred to as risk characterization (National Research Council, 1983). Because people's perceptions of the magnitude of the risk associated with an event depend on how likely they think the event is and how serious they feel its occurrence will be, risk must be regarded as (at minimum) a two-dimensional entity comprising the probability and magnitude of adverse consequences. Thus, although a variety of single-number summary statistics for the level of risk are often used (e.g. the probability of damage, the probability of an individual death, and the expected number of fatalities per year), displays of probability distributions over consequences are generally regarded as providing a more complete characterization of risk. Because single-number summary statistics can be computed from a probability distribution representation, probability distributions provide the most comprehensive means for summarizing the results of a risk assessment.

Despite their advantages, displays of probability distributions over consequences are in many cases the exception and not the rule: many chemical hazard risk assessments are conducted without explicit estimations of probability. For example, the risk assessor might simply acknowledge the existence of uncertainty but compute quantitative estimates of health and environmental consequences that are interpreted as "nominal" or "best estimate" values. Another approach is to specify a range of two or more alternative values for the consequences of concern (e.g. low, medium, and high values). If uncertainty is not a concern for decision-making – for example, if the consequences are almost certain to occur at the predicted level– then estimating probabilities may not be important. In the vast majority of risk problems of current concern, however, uncertainty is important. In these cases, it is frequently efficient to use simpler methods in the preliminary stages of a full-fledged risk assessment. Sensitivity analysis, for example, is extremely useful for the initial testing of the risk model. A complete chemical hazard risk assessment, however, should include estimates of both the probabilities and the magnitudes of possible consequences and suffers when estimates are collapsed into single-number summary statistics.

Although probability distributions over consequences provide a more complete characterization of risk than do single-number summary statistics, a variety of probability distributions are needed to describe the different aspects of risk that may be of concern. To reflect total (aggregate) societal risk, for example, the analyst might estimate a probability distribution over total adverse health and environmental effects (e.g., total number of fatalities). To represent individual risk, it might be desirable to estimate the probability of consequences to representative individuals (e.g. the probability of an individual dying). Other probabilistic representations of risk might also be important in certain situations. For example, for the purpose of aiding evaluations of the equity of the distribution of risk, it might be useful to compute measures of group risk consisting of probabilities and consequences to individuals grouped by occupation, geographic location, sex, race, etc. Developing useful measures of risk is a complicated problem for chemical hazard risk assessment. No single means for defining and displaying risk captures all of the aspects of risk that are of concern.

Generation of a probability distribution over environmental or health impacts requires the establishment of measures for quantifying the outcomes of concern and the determination of the uncertainty of these outcomes. Outcomes of concern are the health and other effects whose estimations are the objective of dose–response assessment. For example, health effects might be expressed as number of immediate and delayed fatalities, injuries of given severity levels, and health degradations of specific types or severities. Environmental effects might include reductions in abundance or production of commercial or game fish populations, reductions in timber yield and undesirable changes in forest composition, reductions in agricultural production, and reductions in wildlife populations.

Because the estimation of uncertainty in outcomes logically requires accounting for uncertainty in each component of the risk chain, methods for risk characterization involve integrating models for the risk source, exposure processes, and dose–response relationships. Ideally, the methods selected to account for each link in the risk chain ought to achieve compatibility in their costs and accuracies.

The scientific and technological uncertainty inherent in the risk situation can be reflected in risk-source, exposure, and dose–response models in several ways, depending on the nature of the uncertainty. Two possibilities are:

(1) The processes involved are sufficiently well understood that functional relationships among important variables can be presumed, but the values of some of the variables are not known (type 1).
(2) The physics, chemistry, biology, or other science and engineering aspects of the problem are so poorly understood that the functional relationships among important variables cannot be presumed to be known (type 2).

Morgan (1982) summarizes the available means for dealing with each of these cases. In the first situation, the functional relationships among the various important variables are known or knowable so that the risk assessor can build appropriate hazard, exposure, and dose–response models, using the various methods described in the previous sections. The problem of dealing with uncertainty then becomes one of how to characterize and propagate the uncertainty in the coefficients and variables through these models. The second situation is more difficult to handle. Typically, uncertainty in the form of appropriate models is addressed in one of three ways:

(1) By performing the analysis using the model form judged on the basis of current information to be most likely or the best of the available options. Uncertainty in the value of the model's coefficients is used as a crude way of capturing uncertainty in model form.
(2) By performing an order-of-magnitude bounding analysis that is designed to determine the extent to which changes in model form affect the assessment results so as to establish bounds on the range of possible answers.
(3) By performing a separate analysis using a variety of competing model forms, assigning probabilities to these alternative forms, and then combining the results probabilistically (this essentially converts a condition of type 2 above to one of type 1).

8.5.1. "Classical" versus "Bayesian" methods

Risk assessment methods for quantifying and propagating uncertainty through models differ according to the degree to which the methods reflect an objective or judgmental perspective of risk. The objective perspective of classical statistics views risk as a measurable property of the physical world. A method with a strong objective perspective will adopt a definition of probability related to the frequency with which events occur, will rely heavily on empirical data and empirically validated models, and will develop conclusions based on statistical inference. The outputs of a risk assessment adopting an objective perspective include both computed outcome values and probabilities, and measures of the uncertainties in these estimates.

The judgmental perspective, associated with the eighteenth-century mathematician, the Reverend Thomas Bayes, regards risks as a product of perceptions. A method with a strong judgmental perspective will adopt the Bayesian definition of probability: i.e. uncertainty is a degree of belief, and probabilities depend on the information, experience, theories, etc. of the individual. Bayesian risk assessment methods tend to make explicit use of expert judgment and theoretical models based on conjecture. Bayesians generally do not provide error bands or probability estimates to measure uncertainty in reported probabilities, because assigning a "degree of belief" to a "degree of belief" has little meaning. The "quality" of the outputs produced by a Bayesian analysis can, however, be reflected by a variety of computed measures, such as the sensitivity of those values to the range of judgments provided by different experts and measures that reflect the extent to which judgmental estimates might change in the face of new evidence.

Both the classical and the Bayesian views of risk assessment have merit, and there is no easy answer to the debate over which perspective is most appropriate for regulatory decision-making. Bayesians regard the models and assumptions inherent in the classical approach (linear models, normal probability distributions, independence among variables, etc.) to be arbitrary and as not reflecting the best information available for the analysis. They therefore attempt to incorporate all of the available data, including soft data and experience, in order to make maximum use of knowledge and expert judgment. Classicalists object that such an approach leaves too much to the idiosyncratic judgments of the analyst and produces analyses that cannot be fully confirmed or validated by others.

Although the classical and Bayesian perspectives represent polar points of view, the needs and constraints of applications generally force most risk assessments to adopt a perspective that lies somewhere between the objective and judgmental extremes. Assessments of chemical risks are generally derived indirectly using causal models designed to represent real-world systems. The validity of these models is determined both by empirical data and by their subjective credibility to experts. Nevertheless, the classical versus Bayesian distinction provides a useful framework for distinguishing risk assessment methods for quantifying and propagating uncertainty through models.

8.5.2. Methods adopting a classical perspective

Some methods attempt to develop probability distributions describing risk directly, without formally modeling the risk source, exposure processes, or dose–response processes. Such methods may be applicable when a sufficient data base is available for the outcomes of concern, and the objective is to estimate the risks as they exist – i.e. the risk assuming that no changes occur.

In this case, standard statistical analysis (e.g. as applied in epidemiologic studies) may be used to estimate probability distributions. Thus, for example, if over a number of years an average of 0.5% of a worker population has annually developed a certain form of cancer, and if no changes are made to the work environment, then the estimate of the inferred risk is that 0.5% per year will continue to develop this cancer. An appropriate confidence interval (or probability distribution) for such an estimate can also be developed.

More generally, statistical methods can be used to quantify uncertainty in the variables represented in the models developed to represent a risk source, exposure, or dose–response. The basic approach in such statistical methods, including those discussed in Sections 8.2.3 and 8.3.1, is to interpret empirical data as providing a sample from which "true" or "population" probability functions can be developed. These distributions are "named" probability distributions (e.g. the normal distribution), and they are specified by one or more parameters (e.g. a mean and variance). The "population" probability functions are developed through the estimation of their associated parameters.

The simplest case results if an event (such as an accident at a chemical plant) is assumed to occur with constant probability per opportunity, at any given opportunity, independent of whether or not it has occurred at any other opportunity. In this case, the number or frequency of occurrences of the event, n, for any given number of opportunities, N, is represented by a binomial probability distribution. The probability of occurrence per opportunity, p, is the parameter that, together with N, defines this distribution; and the empirically measured fraction of times that the event occurred, n/N, provides a point estimate for p. Confidence intervals for the estimate of p (based on an assumed probability distribution for p) may be easily calculated and are tabulated in statistics tables. Analogous methods exist for more complicated assumptions and for continuous variables.

Limitations

To be useful, statistical methods all require the existence of an adequate data base. Their application permits quantification of both the uncertainty associated with randomness reflected in the data base (e.g. the hourly variability of ambient air pollution concentrations) and the uncertainty associated with the limited sample size of the data base (the fact that samples provide only a limited indication of what might be learned if many more samples were taken). They do not reflect uncertainty over possible changes that may have occurred since the data were collected and cannot readily be updated to reflect changing information. Furthermore, the methods are predicated on a wide range of assumptions or underlying models for the processes by which outcomes occur.

Some of the pitfalls related to the assumptions commonly used are subtle. For example, it is customary in reliability estimates to use lognormal distributions to represent uncertainty in failure rates. The lognormal distribution is convenient mathematically, but, regardless of how its parameters are set, it has a "tail" that indicates finite probability of an actual failure rate that exceeds any finite value. In reality, a maximum failure rate exists for all items (e.g. a true failure rate on demand cannot be greater than unity). In most applications the presence of this tail can be regarded as a mathematical artifact that will not significantly distort the results of the analysis. However, as noted by Apostolakis and Kaplan (1981), there are situations in risk assessment where this common assumption can introduce errors into computed mean values and variances.

8.5.3. Methods based on the Bayesian perspective

The Bayesian view of risk assessment maintains that probabilities represent an individual's degree of belief about the world, not a specific property of the world (Lindley, 1970). Methods adopting a judgmental perspective for risk characterization rely heavily on the elicitation of probabilities from experts. Bayesians believe that relevant information for risk assessment is generally held by those individuals with the greatest knowledge and familiarity with the situation under study, and that this information is not entirely captured by the hard data of statistics. Hence, experts provide the logical source for admittedly subjective opinions and judgments regarding scientific and technical matters.

A variety of encoding procedures are available for eliciting judgmental probability estimates (see, for example, Stael von Holstein and Matheson, 1979). The basic types of encoding procedures are: probability methods, which require the subject to respond by specifying points on a probability scale corresponding to fixed values of the uncertain variable; value methods, which require the subject to respond by specifying points on the value scale while the probabilities remain fixed; and probability/value methods, which ask questions that must be answered on both scales simultaneously (the subject essentially describes points on a probability distribution).

Each of these encoding methods may be presented in either a direct or an indirect response mode. In the direct response mode, the subject is asked questions that require numbers as answers – for example, "What is the probability that the number of health effects is less than X?" Probabilities may be expressed directly or in terms of odds (e.g. 1:99 rather than a probability of 0.01). With the indirect method, the subject is asked to choose between two or more imaginary bets. Fixed-value methods are generally regarded as preferable to fixed-probability methods because they tend to produce more diffuse probability distributions and are, therefore, less likely to reflect overconfidence on the part of the experts (Morgan *et al.*, 1980). A widely used probability-encoding procedure is an indirect, fixed-value method involving a reference gamble. The reference gamble is frequently a series of intervals defined over the range of uncertainty (the "interval method") or a "probability wheel".

Spetzler and Stael von Holstein (1975) first described a detailed encoding process designed to help the analyst identify and reduce the effect of cognitive and motivational biases held by the subject. This process, which makes use of a probability wheel as well as other encoding devices, consists of five stages: motivating, structuring, conditioning, encoding, and verifying. As part of the process, the analyst describes to the subject how biases can arise and the implication of those biases for estimating probabilities. Once subjects have some understanding of the biases introduced by common thought processes, they are typically willing to take steps to avoid those biases so as to improve the internal consistency of their judgments. In addition to educating subjects, the analyst also actively tries to reduce biases. For example, probing is used to help identify hidden assumptions on which the assessment might be implicitly conditioned. This may suggest better disaggregations in which the probability of these conditions may be considered explicitly in separate assessments. The process also attempts to use known cognitive biases to counteract one another.

To address special problems that arise in probability assessment, several methods have been devised. For example, the Delphi method (Dalkey, 1968) is sometimes used as a means for promoting consensus when probability distributions are assessed from more than one individual. The method is designed to reduce the pressures associated with interpersonal interactions typical of group assessments. Several methods, such as the modified Churchman–Ackoff method and the normalized geometric mean method (Williams and Crawford, 1980), are available for resolving inconsistencies in probabilistic estimates. Other judgmental methods discussed in the literature include cross-impact analysis, study of precursor events, scenarios, surprise-free and canonical projections, authority forecasting, surveys of intentions and attitudes, divergence mapping, modes and mechanisms of changes, synetics, and contextual mapping (Mitchell, 1977). In general, such methods range from a casual assertion that "such and such will happen" (for whatever unstated reasons) to sophisticated computerized models in which judgmental weights are quantified and cross-correlated in complex ways and on many levels.

An oversimplification that has hampered the acceptance of even the most formal and systematic of the Bayesian methods is the belief that "objective" quantities are preferable to "judgmental" or "subjective" quantities. While Bayesian probabilities are "subjective" in the sense that they describe a state of knowledge rather than any property of the "real" world, they are "objective" (i.e. independent of the observer) in the important sense that two "idealized" individuals faced with the same total background of knowledge would, presumably, assign the same probabilities.

8.5.4. Methods based on modeling

Since models are developed to characterize the risk source, estimate exposures, and represent relationships between doses and responses, methods used for risk estimation involve combining these models to produce an overall model of the risk process. Measures of the uncertainties in the parameters of these models are then propagated through the overall model to obtain measures of the uncertainty over health and other consequences. *Figure 8.3* illustrates the process. Uncertainties may be quantified either by probability distributions or confidence bounds. If a Bayesian approach is taken, probability distributions will be encoded directly from experts; however, if the expert prefers and adequate data exists, some distributions may be generated from frequency data. In this way uncertainties due to variabilities of nature and due to lack of knowledge are combined to obtain quantitative measures of the overall uncertainty in outcomes. An approach more consistent with classical statistics might elect to retain a clear distinction between probabilities generated from data and those based on judgment. Such a method (e.g. Spencer *et al.*, 1985) might propagate frequency distributions generated from empirical data and then superimpose judgmental bounds to reflect additional uncertainties related to lack of knowledge.

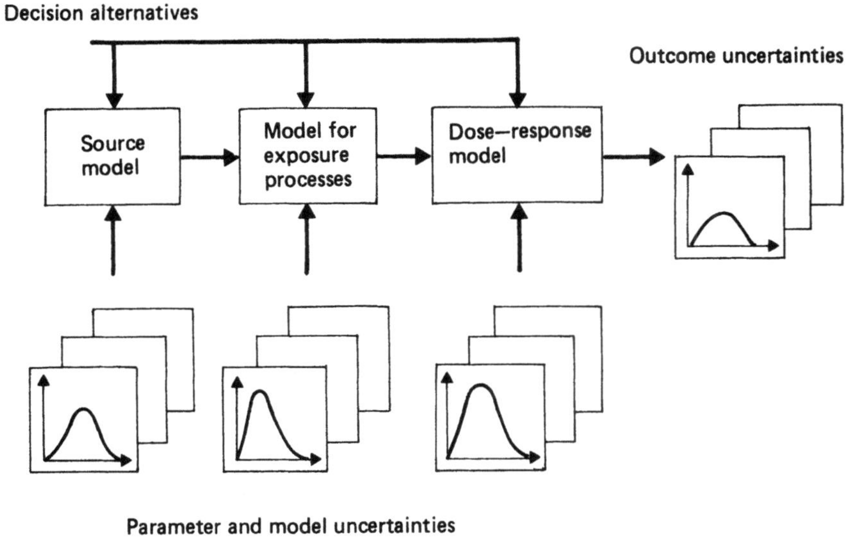

Figure 8.3. Combining models for the risk source, exposure, and dose–response to obtain a risk model.

The logic of combining and processing probability distributions depends on the situation. For example, a method commonly used for risks associated with the release of hazardous chemical substances involves calculating probability distributions for the physical effects suffered by representative individuals located within particular geographic regions surrounding the source, then combining these distributions with population density data to calculate a probability distribution over the number of various adverse health effects. The same data may also be used to compute the annual probability that each exposed individual may become a fatality or suffer a particular adverse health effect. The average over all individuals provides a measure of individual risk, whereas the average over various groups of individuals classified by location, occupation, health characteristics, etc., provides a measure of the risk faced by different population groups.

Several methods are available for converting the uncertainties about the inputs to a model (obtained through one of the methods described above) to the uncertainties that they induce on the outputs of interest. One method commonly used is probabilistic analysis. Here, probability distributions of the relevant input variables (derived from statistical data, fault trees, or event trees) are fed into a computer, and a simulation program selects values for each variable on the basis of their relative likelihood of occurrence. Through a sample of simulation runs of the combined risk model, the program estimates the distribution of overall outcomes. In Monte Carlo simulation, input values are sampled at random from their specified distributions, and each scenario is given equal weight. In other methods, values of particular importance (e.g. the extreme tails of probability distributions corresponding to quantities that might cause a catastrophe) are sampled more often and given reduced weight, so as to obtain improved resolution for important parts of the distributions. In Latin Hypercube Sampling, sample values are constrained to cover the distribution uniformly, one from each equal probability interval. This can increase the stability of the output probability distribution if the uncertainty is dominated by only a few sources.

Because the estimation of outcome distributions often requires a very large number of simulations, Monte Carlo and related forms of analysis can be costly if the risk model is complicated. Response-surface methods are sometimes used to overcome this problem. The idea here is to replace the complicated risk model with a simplified approximation called a response surface. Typically, the approximation is a linear combination of certain simple functions whose coefficients are determined by least-squares fitting. Because the response surface is much simpler to evaluate than the original model, it is less expensive to implement.

If the model is simple enough (including situations where response surfaces are used), outcome variables can occasionally be expressed as explicit analytic functions of the input variables. In this case, analytic methods (such as Taylor series expansions) can be used to express summary statistics (e.g. means and standard deviations) in terms of the parameters that define the distributions of the input variables. Thus, analytic methods provide another means for uncertainty analysis.

If the Bayesian perspective is taken, probability trees, rather than simulation methods, may be used for analysis because they permit the dynamic nature of events and probabilistic dependencies (characteristics to which Bayesians often attribute great importance) to be more easily addressed. Probability trees are generalizations of event trees: whereas the latter typically are composed of events with two possible outcomes (the event does or does not occur), the former may represent uncertainties with multiple outcomes whose probabilities of occurrence are quantified in a discrete probability distribution. Bayes' theorem [2] is used for "node flipping" (deriving posterior probabilities from prior and conditional probabilities in order to interchange probability nodes in the tree). A combinatorial scenario analysis is performed to compute the outcome of values for all combinations of uncertain values. The probability of each scenario is computed as the product of the appropriate input probabilities as represented in the tree. Tabulation and cumulation of the probabilities of the various endpoint consequences in the tree permits the development of probability distributions representing risk.

A major issue in the development of probability distributions using models is representing partial dependence or correlation among quantities. Accounting for correlation among continuous probability distributions can be mathematically difficult, except for certain standard named distributions, such as the normal and binormal, for which there exists much multivariate theory. If the probability distributions are assessed from experts, the encoding process for a given uncertainty may entail assessing separate distributions that are each conditioned on certain distinct values for the quantity on which the uncertainty depends.

Limitations

The major concern surrounding models is their accuracy or validity. Unfortunately, the validity (or invalidity) of a model is generally difficult or impossible to prove. Modeling uncertainty arises from the fact that the model is only a simplified version of the real world and must employ approximations to ease data requirements and computational burdens. Examples of the latter include limited disaggregation, such as the finite grid size for modeling spatially or temporally varying values, the use of discrete probability distributions to represent continuous distributions, and the finite number of runs used in a Monte Carlo simulation. Although in principle the effect of such approximations could be investigated by changing the level of detail (e.g. by decreasing the grid size or time steps), this usually would demand rather fundamental reprogramming of the model. Thus, it is relatively rare for this sort of sensitivity analysis to be conducted in a systematic way.

In practice, the validity with which a model is viewed rests as much on the acceptance of its assumptions by those most knowledgeable in the areas that it represents as on any quantitative test. It is, however, often possible for a model to be constructed so that it can simulate past history, and this approach is frequently employed as a means for validation. While models can often serve as good display and communication devices, detailed models generally require extensive examination and explanation to provide understanding.

Developing models generally requires a large amount of data (for those methods adopting an objective perspective) or a skilled and experienced modeler and considerable input from experts (if a Bayesian perspective is taken), or both. Computer costs for running models are typically small compared to development costs. Small probability trees may be analyzed by a hand calculator for a simple model, but the number of scenarios required increases exponentially with the number of uncertain quantities represented in the tree. Thus, the methods can become infeasible even for a computer if the probability tree is too large. Contrary to common perceptions, probabilistic simulation can be more efficient than analysis of probability trees for large models. Many simulation schemes have the added advantage that the modeling uncertainty due to the finite number of runs may be estimated statistically.

8.6. Outputs of Risk Assessment

There are two principal outputs of a chemical hazard risk assessment: (1) quantified measures of possible outcomes and their uncertainties (e.g. probability distributions, confidence intervals, judgment bounds) and appropriate summary statistics (e.g. expected values of risk), and (2) information on the relative importance of different sources of uncertainty.

8.6.1. Risk estimates and displays

As noted previously, probability distributions or other measures of uncertainty in risk should ideally be estimated for each of the potential health and other consequences that are of concern. These consequences may be estimated from a variety of perspectives to clarify the nature of the risk involved. For example, individual risk might be estimated from the probability distributions describing possible consequences to a typical or highly sensitive individual. An aggregated measure of risk is reflected by the probability distributions for consequences for the total population.

The probability distributions developed as the overall output of a risk assessment may be displayed in a variety of ways. A concise method of display is the cumulative distribution function (CDF) or complementary cumulative distribution function (CCDF). *Figures 8.4(a)* and *8.4(b)* illustrate a CDF and a CCDF, respectively. With the CDF the height of the curve denotes the probability that the actual value of the uncertain quantity will be less than or equal to any value along the horizontal axis. The CCDF is simply the complement of the CDF. With the

CCDF, the height of the curve at any point gives the probability that the actual value will be greater than or equal to any value on the horizontal axis.

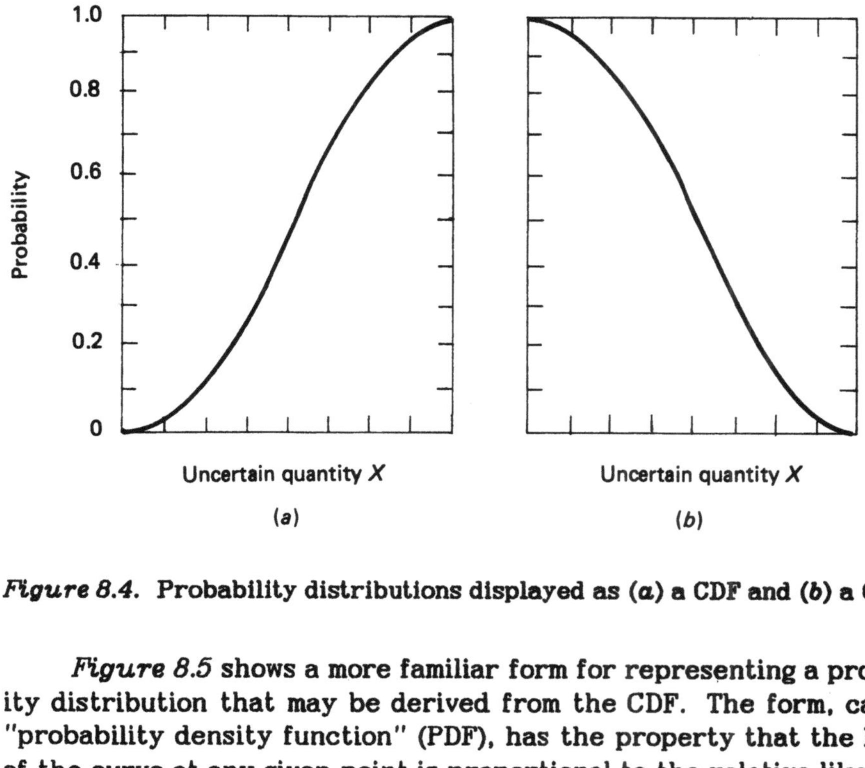

Figure 8.4. Probability distributions displayed as (*a*) a CDF and (*b*) a CCDF.

Figure 8.5 shows a more familiar form for representing a probability distribution that may be derived from the CDF. The form, called a "probability density function" (PDF), has the property that the height of the curve at any given point is proportional to the relative likelihood of the uncertain quantity having that value. Thus, in the PDF, the probability of a value between any range is represented by the area under the curve within that range. The relationship between the PDF and the CDF is such that the former is the first derivative of the latter: i.e. the height of the PDF at any point is proportional to the slope of the CDF at that point. Although the PDF has the disadvantage that it is more difficult to use to read off numerical probabilities, its shape provides an intuitive feel for the nature of the uncertainty involved.

Since it is often desirable to distinguish uncertainty due to the inherent random nature of the situation under study (e.g. the variable failure rate of components) from modeling and analysis uncertainties (not knowing the appropriate forms for models or the values of their parameters), displays such as *Figure 8.6* are often useful. The figure

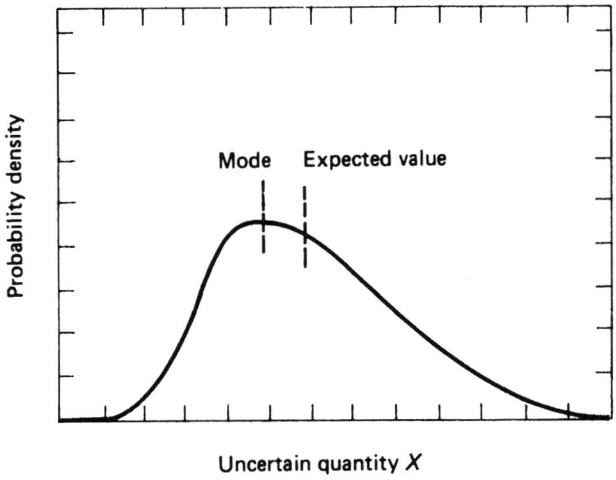

Figure 8.5. Probability distribution displayed as a PDF.

shows a series of CCDF curves, each generated according to a set of assumptions with associated (judgmental) probabilities. The numbers in the display represent cumulative probability. As indicated in *Figure 8.6*, such curves can easily be reduced to a form that makes explicit the state of knowledge or degree of belief behind each frequency curve. The conversion is based on the concept of a "cut curve". If a vertical line is drawn at any consequence level, then the intersection of this line with the family leads to a cumulative probability distribution for the probability of the actual frequency of occurrences of consequences with levels less than X_0. Such cut curves can also be drawn to decompose a family of risk curves into its various sources of risk.

A variety of summary statistics may be computed to summarize probability distributions. The expected value (or mean or average) is the sum (integral) of all values weighted by their probabilities. It is also the average value that would occur given a very large number of experiments if each were characterized by the given curve. Another popular summary statistic is the median, which is the point at which a vertical line would bisect the area under the PDF. The median is the point for which the odds are 50:50 that the variable lies above or below. Another useful statistic is the mode, which is defined as the value where the PDF is a maximum (and the CDF has the greatest slope). A symmetric (unimodal) PDF will have the mean, median, and mode occurring at the same value, while a skewed PDF will have the mean removed from the mode towards the skewed side of the PDF. The amount of uncertainty expressed by a PDF is often summarized by the standard deviation. The amount of skewness may be measured as the

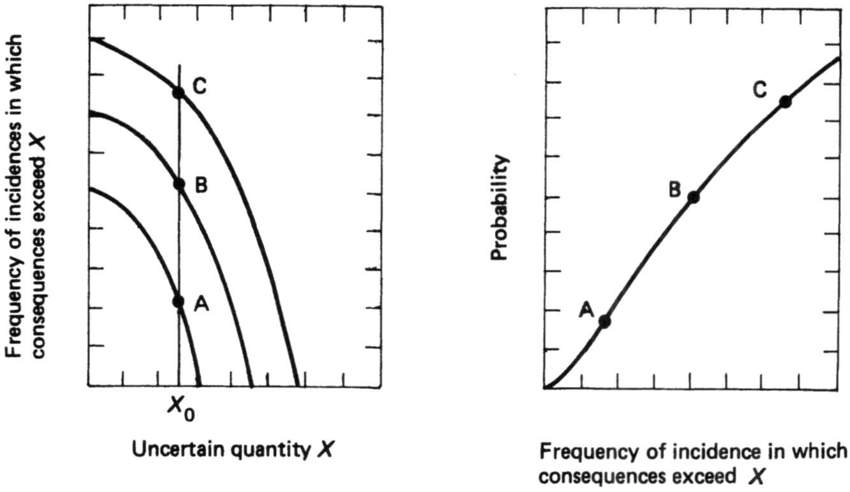

Figure 8.6. Using families of risk curves to represent both statistical uncertainty and uncertainty due to lack of knowledge.

difference between the mean and the mode values divided by the standard deviation.

While these and other standard statistical measures are frequently used to summarize important characteristics of probability distributions developed in risk assessments, special statistics are also often used. For example, one such statistic is the average risk per person, defined as the probability of death; another is the total accident rate, defined as the number of deaths in every 10^{-8} hours of exposure to risk. Keeney *et al.* (1979) suggest that, at a minimum, the analyst should compute the following measures to characterize risk in a way that is sensitive to social concerns:

(1) Total expected fatalities per year (to measure aggregate societal risk).
(2) Probabilities of fatality for each exposed individual (to permit comparison of the risk with other risks, such as the risk from smoking, driving a car, etc.).
(3) Probabilities of fatality for individuals grouped by occupation, geographic location, etc. (to allow equity issues to be addressed).
(4) Probabilities of exceeding specific numbers of fatalities per year (to allow sensitivity to catastrophe).

Care must be taken in presenting summary statistics as they may be misleading. For example, "maximum lifetime risks" are sometimes computed for assessments of industrial facilities. Such estimates may, in effect, assume that an individual is located at the fenceline of the

Table 8.2. Summary of risk assessment methods.

	Methods directed at:			
	Risk-source characterization[a]	*Exposure assessment*[b]	*Dose–response assessment*[c]	*Risk estimation*[d]
Methods relying primarily on experimental data and statistical inference ("classical" perspective)	Monitoring – equipment monitoring – environmental status monitoring Performance testing Accident investigation Statistical methods for risk-source characterization – statistical sampling – component failure analysis – extreme value theory Codified engineering methods	Monitoring for exposure assessment – biologic monitoring: food crops, livestock,, fish, wild animals, indigenous vegetation, etc. – remote geologic monitoring: aerial photography, multi-spectral overhead imagery – media contamination site dose monitoring: air, surface water, sediment, soil, groundwater – individual dose monitoring: dosimeters, film badges Calculation of dose – based on exposure time – coexisting or decay substances – material deposition in tissue	Short-term tests Molecular structure analysis Tests on humans Animal bioassay Epidemiology – cohort vs. case-control – retrospective vs. prospective Pharmacokinetics	Statistical analysis

Methods relying primarily on modeling ↓	Modeling methods for risk-source characterization – engineering failure analysis – simulation models: logic trees, event trees, fault trees – analytic models: industrial effluents, biological models for pests, containment models	Exposure modeling – air: analytic models, trajectory models, transformation models – surface water: dissolved oxygen models, etc. – groundwater: absorption models, travel time models – food-chain chemical migration models Population at risk models – census, sensitive groups, population estimation, trip generation models, etc.	Low-dose extrapolation models Animal-to-human extrapolation models Ecological effect models	Worst-case analysis Sensitivity analysis Confidence bounds Probability distributions: – Monte Carlo analysis – event tree analysis – probability tree analysis
Methods relying primarily on expert judgment ("Bayesian" perspective)				Probability encoding

[a]Measuring the degree of danger associated with the source of risk.
[b]Estimating the intensity, frequency duration, etc. of human and other exposures to the risk agent.
[c]Characterizing the relationship between the dose of the risk agent received and the health and other consequences to exposed populations.
[d]Developing overall measures of the level of risk.

"worst" facility continuously for 70 years, even though it may be unlikely that any such individual actually exists. Furthermore, such estimates may assume incorrectly that the last year of exposure (the 70th) contributes as much to the individual's health risk as earlier years of exposure. Since for effects such as cancer there is generally a long latency period between exposure and effect, such estimates may be unrealistically high.

8.6.2. Analysis of uncertainty

One of the main reasons for conducting a quantitative chemical hazard risk assessment is to be able to answer questions about the relative uncertainty contributed by various sources. Such information allows comparison of the practical importance of different sources of uncertainty, and, hence can guide the allocation of further analytical resources to reduce modeling uncertainty, or the design of research programs to reduce scientific uncertainty.

Given a risk model, there are several ways for analyzing and attributing sources of uncertainty. The simplest approach is point sensitivity analysis, in which the derivative of elasticity of outcome measures is computed with respect to each input evaluated at nominal input values. Because this method ignores the relative range of uncertainty in input quantities, more elaborate means are generally used. For example, parametric sensitivity analysis involves plotting or establishing a range of output values as each selected input quantity is varied from its extreme low to high values, while holding all others at nominal (or other) values. Another approach involves partitioning the uncertainty in the output according to a Gaussian approximation: the variance in the outcome distribution is divided among the inputs in proportion to the point derivative with respect to that input multiplied by its variance. Where the linear approximation is not too inaccurate, this method may be useful for initial screening to identify the major contributors to uncertainty.

Computing rank correlations between outputs and inputs is also used as a method for analyzing the relative contributions to uncertainty. In this method, the output and input values are ranked and the values are replaced by their ranks before correlation is computed in the usual statistical way. An advantage of correlating ranks rather than the values of the quantities is that it makes the results independent of the metric chosen to quantify the variables. Another advantage over the previously mentioned methods is that it does not require all other inputs to be held at their nominal value, but averages the effect as other quantities vary probabilistically over their entire range. The most comprehensive means for analyzing sources of uncertainty using the risk model is stochastic sensitivity analysis, in which changes in the probability distributions or their summary statistics (e.g. expected values and variances) are explored as various input values are fixed or varied across their range [3].

6. Summary

Table 8.2 summarizes the discussion of chemical hazard risk assessment methods by displaying some of the methods described in this chapter together with a number of other widely used risk assessment methods. The identified methods are arranged in such a way that their vertical and horizontal locations in the table reflect two of the dimensions for distinguishing risk assessment methods. Each method is placed in a column which indicates the element of the risk chain that it addresses. Whether the method is located near the top or bottom of the table reflects a judgment as to whether it adopts primarily a classical or Bayesian perspective in its view of risk.

As noted in the table, there are a great many chemical hazard risk assessment methods. The methods vary widely and illustrate the large number of analytic options available to the chemical hazard risk assessor. Selecting the methods that are most likely to be useful in any given situation requires experience, good judgment, and a careful consideration of several factors, including the aspect of risk assessment that they address, their input requirements, the outputs that they produce, and the underlying assumptions.

Notes

[1] The views expressed in this chapter are solely those of the authors and do not necessarily represent the views of their organizations.

[2] In extreme cases, Bayes' theorem is not of much practical value: if a new piece of information adds only insignificantly to existing knowledge, then it will not affect subjective probabilities much; if it is very much better than previous knowledge, then probabilities can often be calculated directly from the data. In the in-between cases, in which existing knowledge and new information must both be taken into account, Bayes' theorem provides a logical and consistent way for revising subjective probabilities: if $P(\mathrm{H})$ is the prior probability that hypothesis H is true, $P(\mathrm{D})$ is the probability that datum D will be observed, and $P(\mathrm{D}|\mathrm{H})$ is the conditional probability that D will be observed given that H is true, then the posterior probability that H is true given that D has been observed is:

$$P(\mathrm{H}\,|\,\mathrm{D}) = \frac{P(\mathrm{D}\,|\,\mathrm{H})P(\mathrm{H})}{P(\mathrm{D})} \quad .$$

[3] Such methods are closely related to computations of the expected value of perfect information. The method requires a loss function – i.e. a means for valuing risk – and, therefore, lies in the domain of risk evaluation as well as risk assessment. Although not strictly a risk assessment method, it offers a way of comparing the importance of different sources of uncertainty in terms of a common metric of value.

References

American Nuclear Society and Institute of Electrical and Electronic Engineers (1983), *PRA Procedures Guide – A Guide to the Performance of Probabilistic Risk Assessments for Nuclear Power Plants* (prepared for the US Nuclear Regulatory Commission, Washington, DC, as NUREG/CR-2300).

Apostolakis, G. and Kaplan, S. (1981), Pitfalls in risk calculations, *Reliability Engineering*, **2**, 135–145.

Bendixen, L. and O'Neill, J. (1984), Chemical plant risk assessment using HAZOP and fault tree methods, *Plant/Operations Progress*, **3**(3), 179–184.

Dalkey, N. (1968), *The Delphi Method: An Experimental Study of Group Opinion* (The Rand Corporation, Santa Monica, CA).

Fares, R.J., Burt, R.E., Britt, D.L., Kinsman, J.D., and Schoepke, S.A. (1983), *Preliminary Guidebook for Conducting a Regulatory Impact Analysis for Health, Safety, and Environmental Regulations*, August 22 (General Research Corporation, McLean, VA).

Food Safety Council (1980), *Proposed System for Food Safety Assessment, Final Report of the Committee of the Food Safety Council*, June (FSC, Washington, DC).

Gass, S.I. and Sisson, R.L. (1975), *A Guide to Models in Governmental Planning Operations* (Sauger Books, Potomac, MD).

Gehring, P.J., Watanabe, P.G., and Park, C.N. (1979), Risk of angiosarcoma in workers exposed to vinyl chloride as predicted from studies in rats, *Toxicology and Applied Pharmacology*, **49**, 15–21.

Gompertz, D. (1980), *Assessment of Risk by Biological Monitoring*, Paper presented at the National Health and Safety Conference, London.

Gottfried, P. (1974), Qualitative risk analysis: FTA and FMEA, *Professional Safety*, October, 48–52.

Hayne, D.W. (1984), *Development of Environmental Data Bases and Inventories*, Unpublished background paper prepared for the Conference on Long-Term Environmental Research and Development, The Council of Environmental Quality, 21–22 May.

Health and Safety Executive (1978), *Canvey: An Investigation of Potential Hazards from Operations in the Canvey Island/Thurrock Area* (HMSO, London).

Interdisciplinary Panel on Carcinogenicity (1984), Criteria for evidence of chemical carcinogenicity, *Science*, **225**, 682–687.

Keeney, R., Kulkarni, R., and Mair, K. (1979), A risk analysis of an LNG terminal, *Omega*, **7**, 191–205.

Krewsky, D., Brown, C., and Murdoch, D. (1984), Determining "safe" levels of exposure: safety factors or mathematical models?, *Fundamental and Applied Toxicology*, **4**, 383–394.

Lees, F.P. (1980), *Loss Prevention in the Process Industries* (Butterworths, London).

Lindley, D.V. (1970), *Introduction to Probability and Statistics from a Bayesian Viewpoint* (Cambridge University Press).

Marcus, G.H. (1983), *A Review of Risk Assessment Methodologies*, Committee Print No. 98-929 0, March, US Congress, House Committee on Science and Technology, Subcommittee on Science, Research and Technology (US Government Printing Office, Washington, DC).

McCormick, N.J. (1981), *Reliability and Risk Analysis* (Academic Press, New York, NY).

Mitchell, A. (1977), *Handbook of Forecasting Techniques, Part II: Description of 31 Techniques*, August, Report Submitted to the US Army Corps of Engineers Institute for Water Resources by the Stanford Research Institute, International, National Technical Information Service, Department of Commerce, Springfield, VA.

Morgan, M.G. (1982), Uncertainty and quantitative assessment, in J.V. Rodricks and R.G. Tordiff (eds), *Assessment and Management of Chemical Risks*, pp 113–129 (American Chemical Society, Washington, DC). (A more detailed discussion of the same ideas can be found in Morgan *et al.*, 1980.)

Morgan, M.G., Henrion, M., and Morris, S.C. (1980), *Expert Judgments for Policy Analysis: Report of an Invitational Workshop Held at Brookhaven National Laboratory, 1979 July 8–11*, Report BNL 51358 (Brookhaven National Laboratory, Upton, New York).

National Research Council (1983), *Risk Assessment in the Federal Government: Managing the Process*, Committee on the Institutional Means for Assessment of Risks to Public Health (National Academy Press, Washington, DC).

Park, C.N. and Snee, R.D. (1983), Quantitative risk assessment: state-of-the-art for carcinogenesis, *The American Statistician*, **34**(4), 427–441.

Philipson, L.L. and Gasca, J.D. (1982), *Risk Assessment Methodologies and Their Uncertainties: Vol. I, A Review of Risk Estimation Approaches*, Technical Report No. 80-1398, March (J.H. Wiggins Co., Redondo Beach, CA), 3–4.

Ramsey, J.C., Park, C.N., Ott, M.G., and Gehring, P.J. (1979), Carcinogenic risk assessment: ethylene dibromide, *Toxicology and Applied Pharmacology*, **47**, 411–414.

Rasmussen, N.C. (1975), Rasmussen on reactor safety, *IEEE Spectrum*, **12**, 46–55.

Reitz, R.H., Gehring, P.J., and Park, C.N. (1978), Carcinogenic risk estimation for chloroform: an alternative to EPA's procedures, *Food and Cosmetics Toxicology*, **16**, 511–514.

Rijnmond Public Authority (1982), *Risk Analysis of Six Potentially Hazardous Industrial Objects in the Rijnmond Area, A Pilot Study* (Reidel Publishing Co., Boston, MA).

Schweitzer, G.E. (1982), Monitoring to support risk assessments at hazardous waste sites, in F. Long and G. Schweitzer (eds), *Risk Assessment at Hazardous Waste Sites* (American Chemical Society, New York).

Spencer, F.W., Diegert, K.V., and Easterling, R.G. (1985), *Uncertainty Assessment in Probabilistic Risk Assessment*, Paper presented at the International American Nuclear Society/European Nuclear Society Topical Meeting, San Francisco, CA, 24–28 February.

Spetzler, C.S. and Stael von Holstein, C.-A.S. (1975), Probability encoding in decision analysis, *Management Science*, **22**, 340–358.

Stael von Holstein, C.-A. and Matheson, J. (1979), *A Manual for Encoding Probability Distributions*, SRI Project 7078 (SRI International, Menlo Park, CA).

Swain, A.D. and H.E. Guttman (1983), *Handbook of Human Reliability Analysis with an Emphasis on Nuclear Power Plan Operations*, NUREG/CR-1278 (US Nuclear Regulatory Commission, Washington, DC).

Tomatis, L., Agthe, C., Barach, H., Huff, J., Montesano, R., Saracci, R., Walker, E., and Wilbourn, J. (1978), Valuation of the carcinogenicity of chemicals: a review of the monograph program of the International Agency for Research in Cancer (1971–1977), *Cancer Research*, **38**, 877–885.

Travis, C.C., Baes, C.F., Barnthouse, L.W., Etnier, E.L., Holton, G.A., Murphy, B.D., Thompson, G.P., Suter, G.W., II, and Watson, A.P. (1983), *Exposure Assessment Methodology and Reference Environments for Synfuel Risk Analysis*, ESD Publication No. 2232 (Oak Ridge National Laboratory, Oak Ridge, TN).

US Nuclear Regulatory Commission (1981), *Fault Tree Handbook*, NUREG/CR-0492, January (US Nuclear Regulatory Commission, Washington, DC).

US Nuclear Regulatory Commission (1984), *Probabilistic Risk Assessment (PRA): Status Report and Guidance for Regulatory Application, Appendix A*, Report No. NUREG-1050 (Nuclear Regulatory Commission, Washington, DC).

Vesely, W.E. (1984), Engineering risk analysis, in P.F. Ricci, L. Sagan, and C. Whipple (eds), *Technological Risk Assessment*, pp 49–84 (Martinus Nijhoff Publishers, The Hague).

Williams, C. and Crawford, G. (1980), *Analysis of Subjective Judgment Matrices* (The Rand Corporation, Santa Monica, CA).

CHAPTER 9

Methods of Risk Analysis

N.C. Lind [1]

9.1. Methods of Risk Analysis

The risk associated with a hazard can be studied by many different approaches. Some methods serve to detect the presence of a hazard, others to judge the likelihood of undesirable events and their possible consequences. The methods range from qualitative to quantitative, and each may be useful for a particular purpose or a particular stage of the study of a project (design, construction, approval, or licensing). There is a need to develop risk analysis towards precision and objectivity. ***Risk analysis*** is here defined as the formal, quantitative evaluation of malperformance of a system.

Engineers have always had to consider the risks among all other consequences of their designs, and they are quite at home with risk analysis. Quantitative studies of risk were first employed in aeronautical engineering where, in the form of quantitative reliability analysis, it has been practiced for half a century. The reliability of multi-engine aircraft may, for example, be derived on the basis of single-engine failure rates. Some procedures of risk analysis are now routine, prescribed for certain complex, costly, or hazardous projects (e.g. nuclear power plants). Henley and Kumamoto (1981) have given a summary of the history of risk analysis up to the present decade.

Since risk analysis involves the forecasting of events, it is not strictly a scientific discipline. Nevertheless, it is not an arbitrary procedure or a free art form; it employs logical analysis and operates on data obtained by the scientific methods that are usual for the context.

But risk analysis has been called a "pseudoscience", perhaps because it calls for complex numerical procedures and employs scientific data, yet is not subject to verification by the scientific method. Indeed, a statement about the risk involved in a project is, like an engineer's prediction of lifetime maximum stresses in a girder, not verifiable. However, risk analysis is no more a pseudoscience than is bridge design, for instance. The proof of pragmatic disciplines is to be found in the practical performance of their products.

Moreover, risk analyses are scrutinized by critique from within the profession, by the clients, and by engineers and scientists in contiguous disciplines. This process is analogous to the way in which new knowledge is accepted in science, and high standards of skepticism are applied, ideally. The assessment of risks, by analysis or judgment, is an appropriate prerequisite for all prudent action, whether it is a matter of a pedestrian crossing the street or the siting of a hazardous facility.

The product of a risk analysis is a set of outcomes, each with an associated probability. Accordingly, there are requirements of consistency that must be satisfied in risk analysis: there is a right way to perform the analysis and it can be subject to rational dispute.

Risk analysis stops at this assignment of probabilities; it is not concerned with what values may be attached to the space of outcomes. It is value-free (unlike risk assessment).

Many different methods are available to analyze risk. Some examples are threat analysis, event trees, reliability analysis, and fault trees. These methods are not different means to the same end, although the same end result may sometimes be obtained by alternative methods. Each method is appropriate for particular purposes. It is useful to begin with a brief overview of this variety to provide a background for the discussion of fault trees, a popular and highly developed tool.

9.1.1. System reliability analysis

Some systems are so simple that an adequate mathematical description can be made in terms of a finite number, n, of basic random variables $X_1, X_2, \ldots, X_n$. The system has a finite number m of failure modes and fails in mode i if, and only if, a given failure function $F_i(X_1, X_2, \ldots, X_n)$ is negative or zero. The reliability of the system, then, is the probability of survival, and may be expressed as

$$\Pr\left[\bigcap_{i=1}^{m} F_i(X_1, \ldots, X_n) > 0\right] \qquad \forall\, i \in 1, 2, \ldots, m \tag{9.1}$$

A graphic representation of equation (9.1) is shown in *Figure 9.1*. There are two basic random variables, X_1 and X_2, and three failure modes. To fix ideas, X_1 may be the temperature and X_2 the humidity of a sensitive hazardous material mixture in storage. Mode 1 may be spoilage due to freezing, mode 2 may be fire, and mode 3 may be hygroscopic decomposition of one component. The survival event is an outcome in region S, which represents the intersection of the sets $F_1 > 0$, $F_2 > 0, \ldots, F_m > 0$. The boundary curve 1 is the locus of $F_1 = 0$ and represents limit state 1, i.e. failure in mode 1, and so on.

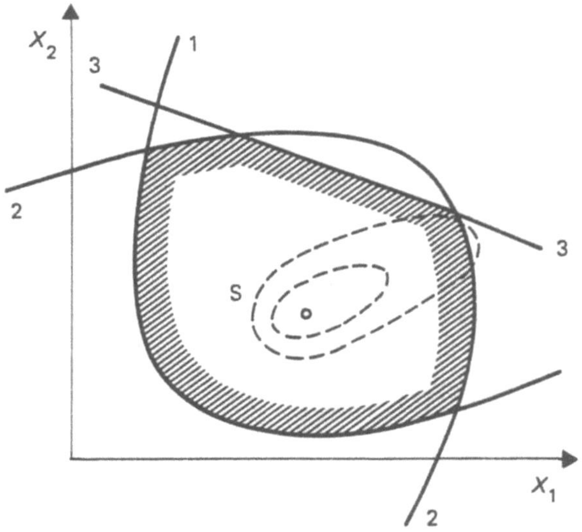

Figure 9.1. Reliability (schematic) of a system in two random variables X_1 and X_2 with three failure modes.

The state variables X_1 and X_2 usually vary as functions of time for a particular realization of the system. The point (X_1, X_2) in time traces a curve in the plane, oscillating around a mean or target state near the optimum. The probability of deviation from the target is assumed to be known and is indicated by the level curves of probability density shown by the broken lines. The reliability is calculated as the integral of the joint probability density of X_1 and X_2 over region S. Methods to calculate reliability have been surveyed by Madsen *et al.* (1986).

The modes of failure are sometimes related in a simple way to the elements of the system and to the basic random variables. For example, in a weakest-link system such as a chain, the strength of each link is a basic random variable, and each mode is a failure of one of the links. Many engineering systems are of this type, such as correctly made or proof-loaded structures: pressure vessels, dams, building structures, or airframes. These systems are subject mainly to the forces of nature and the influence of blind chance, while human influences (intervention, human error, or human factors) are neglected. The calculated reliability of such a system is conditional upon the absence of human influences. An example of the reliability analysis of a simple system is given at the end of this section.

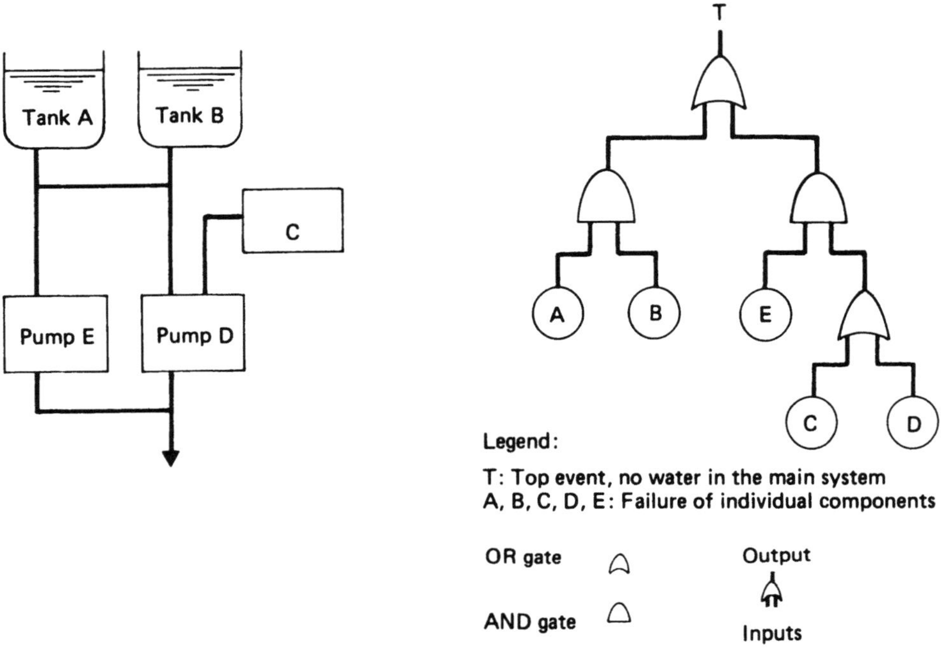

Figure 9.2. Water supply system and associated fault tree (after Paté-Cornell, 1984).

Figure 9.2 gives a more complex example. The failure, loss of cooling water, will occur if both tanks are empty and/or if both pumps fail to operate. Pump D needs a supply of power C to operate. There are five state variables x_A, x_B, ..., x_E. x_A equals 0 if tank A is empty and is otherwise equal to 1; x_C equals 0 if power to pump D is off and is otherwise equal to 1; and so on for the remaining basic random variables. The state of the system is represented by a point with coordinates (x_A, x_B, ..., x_E ^) in a five-dimensional space of binary variables. Only a finite number of states (32) are possible. Each of these states is reached with known probability of departure from the target state (1, 1, 1, 1, 1). There are three failure modes. For example, simultaneous failure of C and E, represented by the point (1, 1, 0, 1, 0) is one failure mode. Since the number of states of the system is finite, the probability of failure can be calculated by enumeration. For systems of the complexity met with in practice, efficient computer programs will normally be required to carry out the calculations.

9.1.2. Influence diagrams

A useful tool to gather and represent most of the information contained in fault trees and event trees in a compact form is a diagram listing all the variables of a system to display their relationships graphically. Intermediate variables are included together with parameters and input and output variables. The relationships can range from qualitative to quantitative.

9.1.3. Scenarios

Risk is of interest only in relation to systems that are subject to change by uncontrollable forces and subject also to our intervention. The important concept of a scenario [2] is appearing more and more in quite different fields, all concerned with forecasting, prediction, control, risk or management of resources. The scenario label indicates that the discourse is about a *model*, which is neither the reality of the future nor a complete description of the set of possible futures. The scenario concept reminds us that all analysis (of the past, present or future) applies to models and not directly to reality. This is of little consequence in everyday science, but it is very important in disciplines of prediction or control, such as risk analysis or engineering: every conclusion is based on a background set of scenarios.

Here is an example of a scenario: "There is a sudden temperature drop on a mild winter day, and ice forms on a bridge, with probability $P(1)$ per bridge per year. A tanker truck brakes, jackknifes, and overturns over the two-lane underpass, $P(2|1)$. A passenger car with N passengers, $P(N)$, $N = 1, 2, \ldots$ collides with the tank, $P(3|2)$. The tank ruptures, $P(\mathrm{R}|4)$ and the contents ignite, $P(5|4)$. There are $m\,(= 1, 2, \ldots)$ severe burns and $n\,(= 1, 2, \ldots)$ fatalities." This is actually a collection of scenarios each generated by assigning particular values to N, m, and n. Each particular scenario in this collection is possible but quite improbable, but the collective scenario is quite believable.

As Hawksley points out in his discussion at the end of this chapter, a scenario should be qualified by indication of the likelihood of occurrence. Thus, numerical values of the probabilities, or at least some indication of their magnitudes, are necessary. Moreover, the large number of possible outcomes of the scenario makes it necessary to consolidate the scenario (e.g. by assigning typical or expected values to N, m, and n) and let the result vicariously ***represent*** the entire collection of scenarios.

A scenario is defined for a system by assigning definite values to the set of variables of choice and chance. The scenario concept prompts the important question, "What if ...", and permits the partial study of the future by analytical methods that are similar to those used in the study of the past or present. The scenario is conditional upon the values assigned to the parameters – as is always the case in analysis – but the ensemble of scenarios can be used to make reliable predictions about the system under study. The scenario approach can serve to demonstrate that the operator has an appreciation of the risks of a facility or operation, and is incorporated, for example, into European Economic Community (EEC) legislation for the control of chemical hazards.

The decisive advantage of scenario thinking is that it liberates the analyst from the concern that the results have to be "true". Instead, attention can be concentrated on the rational analysis of the scenarios, permitting a freedom of imagination essential to good forecasting and design. The advantage is gained at the expense of a need for interpretation of the results – since there is no pretense of representing future reality directly. Risk analysis is often carried out without explicit interpretation of the results, and it is often criticized for not predicting what actually happened. The analyst is then likely to defend any discrepancies on grounds of insufficient data, being unaware of the model character of the analysis.

9.1.4. Threat analysis

Threat analysis is a recent and perhaps the most comprehensive formulation that encompasses risk analysis together with the analysis of crime, armed conflict, etc. It grew out of experiences in the area of computer security and has recently been presented in generalized form [3]. It is a technique to quantify hypothetical threats to any collection of assets. Although still in its infancy, threat analysis contributes a unified philosophy to the analysis of risk in its most general form. For many processes (e.g. nuclear power), the public perception of risk quite appropriately encompasses such hazards a sabotage, armed conflict, and proliferation, which risk analysis in the conventional sense often leaves out by assumption.

Threat analysis follows a sequence of seven stages:

(1) Enumeration of the assets.
(2) Evaluation of the assets.

(3) Listing of all scenarios for the threats.
(4) Determination of the loss expectancy rate for each event.
(5) Proposal of protective mechanisms.
(6) Recalculation of loss expectancy rates.
(7) Performance of a cost–benefit analysis for each protective mechanism.

The analysis is supported by computer procedures and reflects a more complete concept of analyzing risk not in isolation but in conjunction with a slate of protective mechanisms, presenting the sensitivity to each mechanism in a form that prepares for a rational selection. Indeed, there are many risks that we have no protective mechanism against. Such risks are of little interest and we choose to ignore them as a result [4].

Threats are classified into accidental and intentional threats, and the analysis unifies the approach to protection against malicious agents (war, crime, competition), natural hazards, and accidents.

The most important assumptions and limitations of threat analysis are as follows. There is no universal way to evaluate assets, and some intangible losses (e.g. "leakage of information") can be difficult to quantify. Although the threat analysis method yields a complete set of threats to all asset categories, no exhaustive method to generate a complete set of threat scenarios is known. Finally, there is no systematic way to prescribe mechanisms for protection. Threat analysis is not a panacea, but a conceptual framework for thoughtful management of a complex of risks.

9.1.5. Event trees

Change in a system is represented in terms of states, with events being transitions between states. The linear graph suggests itself to represent this kind of situation because it has the appropriate mathematical structure. A *tree* is a linear graph without loops. The event tree is oriented in a particular manner, such that only one branch enters each node. The event tree is a straightforward way to account for the various possible processes of change. Each edge (branch) in the event tree represents a possible state or trajectory in time, with time flowing in the direction of the orientation.

A probability is associated with each edge, conditional on the preceding edge. Each vertex (node) represents a possible transition to alternative states or trajectories and marks the introduction of a new state variable. The state variables are usually discrete-valued or discretized; however, they may be continuous, in which case the graph becomes an infinite manifold without much resulting complication. A scenario is a path from the initial state to a final state. There is a correspondence between scenarios and final states. The probability of the scenario and of the associated final state is simply the product of the probabilities of all the edges in the scenario [5].

9.1.6. Fault trees

In a fault tree the edges are oriented and represent binary events that can take two values (on or off, 1 or 0, true or false, and so on). Only one edge exits from each vertex, and its value is a Boolean function of the values of the events represented by the edges that flow into the vertex. The Boolean function represented by a vertex may be a union, and the vertex is then called an "or" gate. Or the function may be an intersection of the input events, in which case the vertex is called an "and" gate. Other logical functions, though less common, can also occur. Each type of gate has a specific symbol as shown in *Figure 9.2*. The tree represents the output, the top event, as a logical function of the input, called the basic events. The tree is used to calculate the probability of the top event as a function of the probabilities of the basic events [5, 6].

The fault tree is an appropriate representation of a system when the state of the system and of each component is a binary variable, such as electrical or mechanical components that are either on or off, available or not, failed or survived. *Figure 9.2* is an example. If the state of the components is properly described by multilevel variables, or even continuous variables, the fault tree methodology can be generalized by using appropriate logical functions, provided only that the system state can be obtained as a function of the component states.

The probability of the top event is a function of the probabilities of the basic events. These probabilities depend on external circumstances, and may vary with time, location, or the state of the other variables. The fault tree is therefore not a complete representation for the reliability of the system. The fault tree must generally be supplemented with other techniques, such as reliability analysis or event tree analysis.

The literature is full of examples of applications of fault trees and event trees: see, for example, Henley and Kumamoto (1981). To provide an introduction to the subject as well as a basis for reflection by those more experienced in risk analysis, two specific but simplified examples of hazards related to materials and transportation are discussed from different viewpoints in the next section. Section 9.2 is somewhat technical in nature and may be skipped without loss of continuity.

9.2. Examples

9.2.1. Pressure vessel test

A reactor vessel has a seal that may fail at pressure X_1 and a relief valve that normally allows the dangerous substances in the vessel to vent into a holding tank if the pressure exceeds a safe low threshold. Seal and valve are drawn independently at random from their respective populations. A fault tree is shown in *Figure 9.3*. The probabilities of failure of the seal and the valve are p_1 and p_2 respectively. The probability of failure of any other component of the system is negligible.

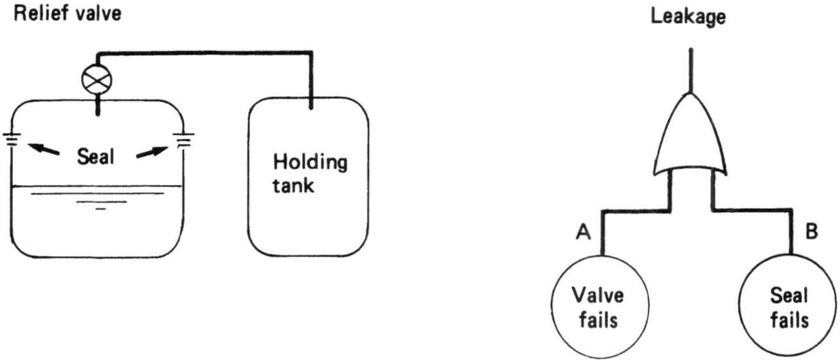

Figure 9.3. Pressure test example: system and fault tree.

To test the system, the valve is adjusted such that the median value m_2 of the relief pressure X_2 is equal to the median of the seal failure pressure m_1. The vessel is then pressurized to the pressure $m = m_1 = m_2$. We wish to predict the probability P_L of leakage of the seal in the test. At the test pressure p_1 and p_2 both equal 1/2. Since the two components are independent, the probability of leakage is calculated formally as $(1/2) \times (1/2) = 1/4$.

Figure 9.4(a) shows an event tree for the same system. If the relief valve fails, it means that the pressure will reach the value m and leakage will occur with probability 1/4. However, if the pressure at which the relief valve operates, X_2, takes a value that is less than m, the seal is tested to a smaller pressure, and leakage will occur with probability $P_L(X_1 < X_2)/2$, giving a total probability of leakage that is greater than was found by the fault tree.

A reliability analysis of the system considers the space of outcomes (X_1, X_2) shown in *Figure 9.4(b)*. There are three compound events: survival of the test, S; pressure relief, R; and leakage, L. The figure also shows level curves of the joint probability density function,

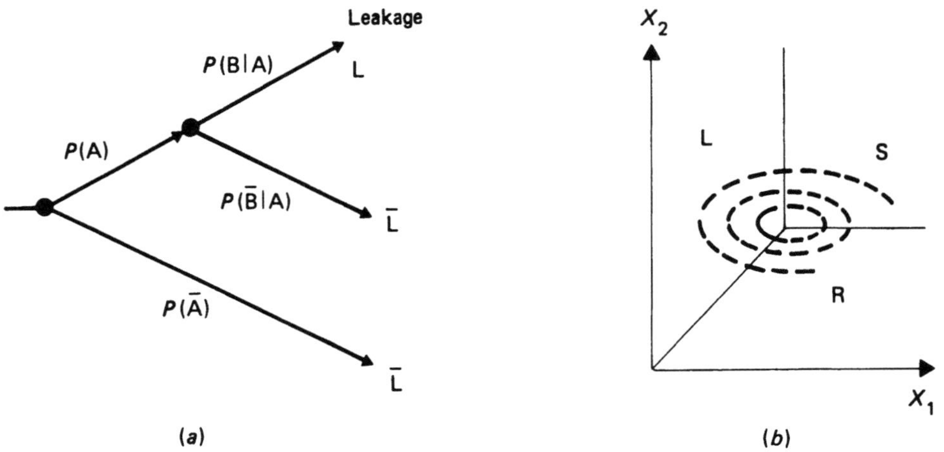

Figure 9.4. **(a) Event tree and (b) event space of reliability analysis for pressure test example.**

$f_1(X_1) f_2(X_2)$, of X_1 and X_2. The probability of leakage is calculated as

$$P_L = \int_L f_1(X_1) f_2(X_2)\, dX_1\, dX_2 \ . \tag{9.2}$$

Equation (9.2) shows that the reliability of the system cannot be determined from the reliability of the components. The probability distributions for the two random variables have to be taken into account to solve the problem. Numerical integration is generally necessary to determine P_L. If we assume that the variables are lognormally distributed with coefficients of variation V_1 and V_2, the result is easily calculated (Madsen *et al.*, 1985) as

$$P_L = 1/4 + [1/(2\pi)] \arctan (V_1/V_2) \tag{9.3}$$

Thus, even for very simple mechanical systems the fault tree must be employed with care, and the conventional information on the failure rates of the components may be insufficient to determine the risk. While the relief valve is an important component of the system, it is observed that leakage occurs if and only if the seal fails.

9.2.2. Collision

It is instructive to consider a model of a simple collision between a vehicle and another object: see *Figure 9.5(a)*. The vehicle may be a hydrofoil boat while the object is a floating log, or the vehicle may be a movable offshore platform and the object an iceberg, and so on. In any case, the probability P_C of a collision depends on the probability of detection (among other parameters), which at any time depends on the distance x between the vehicle and the object [*Figure 9.5(b)*]. P_C also depends on the probability of a successful evasion given that detection occurs, which depends on x [*Figure 9.5(c)*].

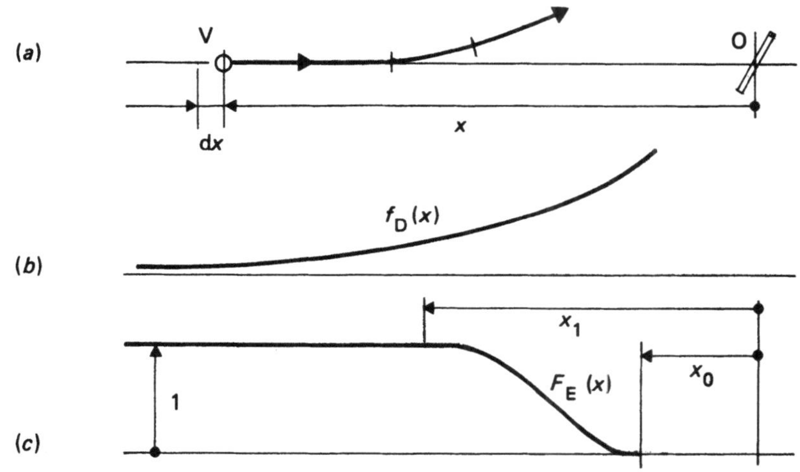

Figure 9.5. (*a*) Collision of a vehicle V with an object O; (*b*) probability of detection per unit of distance; (*c*) probability of success of evasion maneuver initiated at distance x.

A fault tree is shown in *Figure 9.6(a)*. Collision occurs if either the detection system or the evasion system fails. The probability that the detection system fails is denoted by $G_D(x)$, where x is the distance between object and vehicle.

Let $f_D(x)\,dx$ denote the probability of occurrence of detection between x and $x + dx$ given that prior detection has not occurred. The probability of no detection before the distance is reduced to x is

$$G_D(x) = 1 - \int_x^\infty f_D(x)\,dx \quad . \tag{9.4}$$

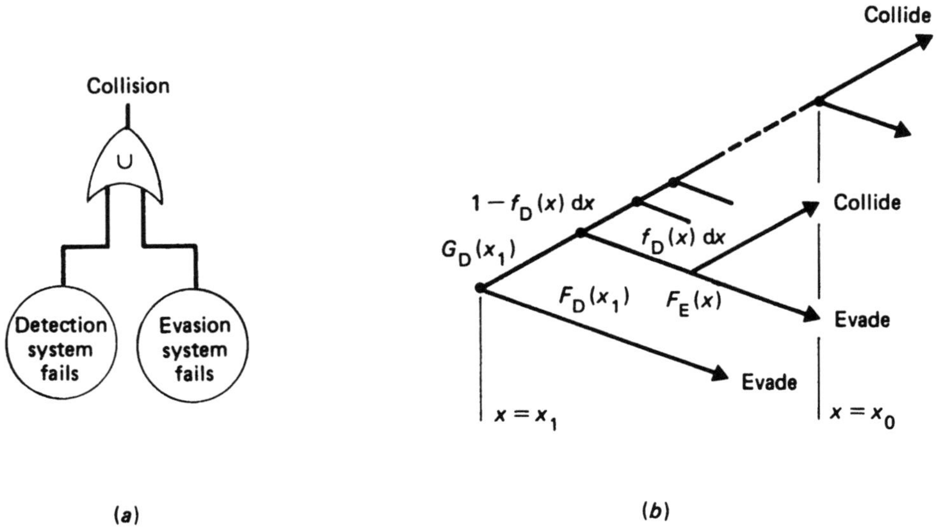

Figure 9.6. **(*a*) Fault tree and (*b*) event tree for the collision example.**

The probability that detection occurs between distances x and $x + dx$ equals $G_D(x) f_D(x)\,dx$, and the probability of collision equals

$$P_C = 1 - \int_0^\infty F_E(x) G_D(x) f_D(x)\,dx \quad . \tag{9.5}$$

An event tree that also leads to equation (9.5) is shown in *Figure 9.6(b)*.

This example illustrates the need to account explicitly for time in dynamical systems. Most of the systems employed in the production and transportation of hazardous substances are dynamical. Fault trees may be an aid in the analysis of such systems, but the problem must be solved by integration of probability functions of time as reflected in the event tree.

The example also gives an appreciation of the degree of complexity that may be encountered in alarm and warning systems [7] and in real collision problems as they occur in the transportation of hazardous materials. Frequently, serious collision situations involve three or more vehicles in relative motion, and they involve the interaction of several detection and evasion systems with incomplete information about the state of relative motion and about each other's future response.

9.3. Data and Uncertainty

If we had ample statistical data on all the modes of failure of a system, we would not need to perform a risk analysis. However, data are scarce, particularly on catastrophic malperformance of very hazardous systems. The reliability must be inferred deductively from the behavior of the components which typically have higher failure rates.

"Probabilities do not exist", as Barlow (1984) has observed. Probabilities cannot be determined, measured, or assessed. Probabilities are assigned, and inserted into models. The outputs of these models are risks, i.e. probabilities associated with elements of sets of undesirable outcomes. Risk analysis thus amounts to the calculation of top event probabilities on the basis of assigned probabilities.

In the public debate over the risks of hazardous systems, discussion often arises over what the "true" (meaning "objective") risks are. Such "true" risks do not exist, however, and the discussion can become fruitful only if it is understood that a believable *consensus* on the probabilities is all that can ever be established. To reach such a consensus among professionals is the central objective of risk analysis, and the search for consensus is prominent when the analysis is used as a tool in design. Consensus among professionals is fundamental in the licensing process for a hazardous facility (e.g. nuclear power plants), which continues until the team of designers have convinced the licensing authority that the risk is acceptable.

Consensus is not easily achieved when data are scarce. It is often necessary to rely on experts' opinions, when data cannot be bought at any price. This happens in siting decisions and design of hazardous plant in earthquake zones in the absence of long-term records. Experts differ not only quantitatively. One may be a specialist on local soil, another on regional geology, a third on earthquakes, and it is necessary for the designer or licensing authority to weigh and compound somewhat conflicting information. One straightforward approach is to treat expert testimony as observations of random variables, and to compound it using Bayesian processing. In a different approach, the weights are generated by the team of experts themselves; in yet another approach the experts are calibrated against independent evidence [8]. One approach formally minimizes the weighted distribution between the experts (whose role is only to provide evidence) and the professionals (who design and certify the plant in the public interest and among whom quantitative risk is established by consensus).

Normally there is no communication between the professionals and the public concerning risk. Civil engineering structures, for example, are formally considered "absolutely safe" by the public, in the sense that the public has normally no fear of failure, no sense of risk. This myth is sustained in part by an excellent performance record, in part by the familiar and unglamorous nature of the facilities, and in part by the profession's reaction to any failure which usually terminates in the assignment of cause to one or several human errors. Individuals, rather than facilities, are seen to cause failures. The absence of some of these attributes in the case of nuclear or chemical plant may contribute to the more skeptical public attitude encountered, and points out the great complexity of the question of communication of risk between the engineering profession and the public.

A paradigm for risk analysis in this role is available in the traditional engineering design process. For example, reinforced concrete highway bridges are produced to perform in a known and accepted way on the basis of a few fast compression tests of concrete samples (among other data). There are many reasons why such tests give an incorrect value of the strength of the concrete. The concrete in the cylinders has cured under conditions that are significantly different from the concrete in situ. The concrete further changes between the date of the test and the time that the most critical loading occurs. Moreover, the prescribed standard statistical procedures to process the test results are quite primitive and may even violate a basic "principle of reliability consistency" (Lind and Chen, 1985). Nevertheless, the traditional procedures to verify the adequacy of the structure are successful, acceptable to the public, and beyond serious dispute with regard to risk. Although these methods are less than rational, they are a sufficient basis for engineering consensus and public trust.

Figure 9.7 illustrates some of the difficulties with conventional statistics when they are applied to the risk analysis context. For a simple system of random capacity R versus demand S the random sample data on capacity are shown as points in a normal probability plot, and a conventional estimate of the capacity is made (see Gumbel, 1954). The result is known to be an optimum estimate within the class of normal distributions. A variety of other distribution functions can be obtained by a homogeneous linear transformation of the normal distribution (as, for example, lognormal distributions), and classical optimal estimation of such distributions can be done in the same way.

Figure 9.7 shows on the left a typical demand density function $f_S(x)$. The probability of failure is

$$P_F = \int_0^\infty F_R(x) f_S(x)\, dx \quad . \tag{9.6}$$

That is, P_F equals the value of the capacity distribution function $F_R(\,)$ weighted by the density function of the demand. Consider now what

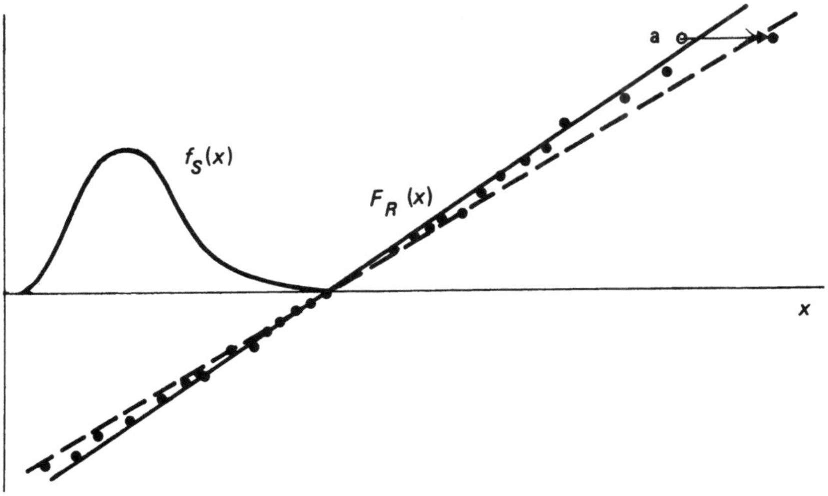

Figure 9.7. Reliability inconsistency of classical mean-and-variance estimation.

happens if one of the high observations of capacity (labeled "a" in *Figure 9.7*) is increased. This causes a change in $F_R(\,)$ as shown by the broken line. For some demands it can therefore happen that the probability of failure inferred from equation (9.6) will increase when observations of high values of resistance increase. This result is absurd, violating the principle of reliability consistency. As Ofverbeck (1980) has suggested, a remedy is to apply differential weighting to the observations, giving greater weight to the low (high) observations in the case of capacity (demand). For further discussion see Lind and Chen (1986).

When data are scarce, legitimate questions can be raised about the accuracy of the calculated probabilities. Top event probabilities are complicated functions of many random variables, namely the estimators of the input probabilities. A reliability analysis is not complete without an estimate of the uncertainty in the result. Monte Carlo methods may be appropriate for this purpose, particularly when ample computing power is available and the number of basic random variables is small. Otherwise, the more powerful point distribution methods can be used (Rosenblueth, 1981; Lind, 1983).

Here is an example. Suppose that V_1 and V_2 in equation (9.3) are random variables with means 0.4 and 0.2 respectively, both with standard deviation 0.1 and correlated negatively with correlation coefficient -0.6. Then the failure probability is given by four point estimates with weights as follows (Rosenblueth, 1981):

$$
\begin{aligned}
p^{--} &= 1/4 + \text{Arctan}\,[(0.4 - 0.1)/(0.2 - 0.1)]/2\pi\ , \\
w^{--} &= (1 - 0.6)/4 \\
p^{-+} &= 1/4 + \text{Arctan}\,[(0.4 - 0.1)/(0.2 + 0.1)]/2\pi\ , \\
w^{-+} &= (1 + 0.6)/4 \\
p^{+-} &= 1/4 + \text{Arctan}\,[(0.4 + 0.1)/(0.2 - 0.1)]/2\pi\ , \\
w^{+-} &= (1 + 0.6)/4 \\
p^{++} &= 1/4 + \text{Arctan}\,[(0.4 + 0.1)/(0.2 + 0.1)]/2\pi\ , \\
w^{++} &= (1 - 0.6)/4\ .
\end{aligned}
\tag{9.7}
$$

From these the mean and standard deviation of the failure probability are calculated to be approximately 0.4237 and 0.0427 respectively. These values compare with p = 0.4133 and s_p = 0.0757 obtained by 500-point Monte Carlo simulation, assuming normally distributed variables.

Uncertainty in data and models eventually translates into risk uncertainty. Uncertainty in the calculated risk is irrelevant if a single decision-maker aims to maximize expected utility. However, it may be important if a collective (public sector) safety decision is to be made, unless the decision-maker is pledged to make rational decisions on behalf of clients or a constituency. In extreme cases of risk uncertainty it is often suggested that risk analysis is inapplicable. Such suggestions carry little weight unless accompanied by a justifiable alternative.

9.4. Concluding Remarks

Quantitative risk calculation can be done effectively for systems that are not subject to major human intervention. Risk can be calculated, on the one hand, for hazards beyond human control (e.g. natural hazards) and, on the other hand, for highly controlled systems (e.g. chemical plant) where human intervention is programmed and controlled. However, even very simple systems may behave in a fashion that defies routine risk analysis except by experts with a knowledge of the hardware as well as of risk analysis techniques. The systems involved in the production, transport, storage, and (to a lesser extent) the disposal of hazardous substances rely on human monitoring and intervention for risk control. The risks depend on factors that are not so easily quantified. This does not imply that risk analysis is impossible or useless. Indeed, risk analysis remains an effective tool in the design of such systems. Risk is then assessed conditional upon a specified human behavior (e.g. "normal workmanship" or "design according to the pressure vessel code").

Such conditional risks reflect the fact that probabilities are not system properties but are assigned quantities that reflect available knowledge and are relative to a conceptual system model. Probabilities, and hence risks, cannot be "true" (in the sense of "objective") but remain a matter of consensus. As such, they may be effective in communication among professionals who share a specialized body of knowledge and beliefs (e.g. in engineering design). They are also suitable in the comparison of policies or projects, when errors and uncertainties in the quantities assigned to the alternatives tend to be compensated.

Risk analysis, then, may be appropriate in selecting policies and in the design of facilities involving hazardous materials. Considerable development is desirable to make risk analysis an effective tool in communication to the public. Also risk analysis must be developed and made more "scientific" if it is to serve as evidence of compliance in the regulatory process. Apart from the development of more effective analytical tools and data bases – processes which are well under way and destined to bear fruit in the long run – there is a need for development of a broad basis for consensus between the public and the professionals. This can take the form both of better professional understanding of the nature and structure of values entertained by the public, and of increased public education in matters of technology, probability, and risks to life and health.

Notes

[1] Findings reported in this chapter were obtained in the course of research projects supported financially by the University of Waterloo Institute for Risk Research, the Natural Sciences and Engineering Research Council of Canada, and the Canadian Committee for the International Institute for Applied Systems Analysis. Thanks are due to Hans Bohnenblust, J.L. Hawksley, Paul Kleindorfer, Howard Kunreuther, Elisabeth Paté-Cornell, and Emilio Rosenblueth for useful discussion and suggestions.

[2] The usefulness of the scenario concept in structural engineering is currently being studied by various international bodies. The ideas expressed here follow unpublished contributions by the Swiss engineers W. Bosshard and J. Schneider. See also Ericksen (1975).

[3] The generalized presentation of threat analysis and a description of the computer program to support the analysis is as yet unpublished Canadian work (Canadian National Railways, and D. Bonyun).

[4] An example is the impact of a meteor of a size sufficient to destroy life over an area the size of Canada, which has been estimated to occur with an annual probability of 10^{-18}. Catastrophes, perhaps of cosmic origin, sufficient to exterminate a large proportion of the species in existence have a recurrence rate of less than 10^8 years (Calder, 1983).

[5] An introduction to event tree analysis and fault tree analysis with a comparison of their salient features has been provided by Paté-Cornell (1984).

[6] See also Barlow and Lambert (1975) or Henley and Kumamoto (1981).

[7] Paté-Cornell (1986) has considered such systems.

[8] Barlow (1984) and Rosenblueth (1981) discuss this and related problems on the processing of experts' opinions from a Bayesian perspective.

References

Barlow, R.E. (1984), *The Bayesian Approach in Risk Analysis*, ASCE Specialty Conference on Probabilistic Mechanics and Structural Reliability, Berkeley, CA, 11–13 January.

Barlow, R.E. and Lambert, H.E. (1975), Introduction to fault tree analysis, in R.E. Barlow, J.B. Fussell, and N. Singhpurwalla (eds), *Reliability and Fault Tree Analysis* (SIAM, Philadelphia).

Calder, N. (1983), *Timescale* (Viking Press, New York, NY).

Ericksen, N.J. (1975), *Scenario Methodology in Natural Hazards Research* (Institute of Behavioral Sciences, University of Colorado, Boulder, CO).

Gumbel, E.J. (1954), *Statistical Theory of Extremes and some Practical Examples*, National Bureau of Standards Applied Mathematics Series, 33 (National Bureau of Standards, Washington, DC).

Henley, E.J. and Kumamoto, H. (1981), *Reliability Engineering and Risk Assessment* (Prentice-Hall, Inc., Englewood Cliffs, NJ).

Lind, N.C. (1983), Modelling of uncertainty in discrete dynamical systems, *Applied Mathematical Modelling*, **7**, 146–152.

Lind, N.C. and Chen, X. (1986), *Consistent Distribution Parameter Estimation for Reliability Analysis*, IRR Paper (University of Waterloo Institute for Risk Research, Waterloo, Ontario); to appear in *Structural Safety*, **3** (in press).

Madsen, H.P., Krenk, S., and Lind, N.C. (1986), *Methods of Structural Reliability* (Prentice-Hall, Inc., Englewood Cliffs, NJ).

Ofverbeck, P. (1980), *Small Sample Control and Structural Safety*, Report TVBK-3009 (Department of Structural Engineering, Lund Institute of Technology, Lund, Sweden).

Paté-Cornell, M.E. (1984), Fault trees vs. event trees in reliability analysis, *Risk Analysis*, **4**(3), 177–186.

Paté-Cornell, M.E. (1986), Warning systems in risk management, *Risk Analysis* (in press).

Rosenblueth, E. (1981), Two-point estimates in probabilities, *Applied Mathematical Modelling*, **5**, 329.

9.D1. Discussion

H. Bohnenblust

The following comments on the chapter by Lind are made by someone who applies risk analysis in his daily work for an engineering firm, where risk analysis is mainly used to help decision-making on safety measures for hazardous systems. The framework of risk analysis, however, can be used to address problems in many different areas. Hence, it is naturally of interest to people from various fields as represented by those attending the conference on Transportation, Storage, and Disposal of Hazardous Materials. When I first read Lind's contribution, I concluded that I basically agree with everything stated there. However, in the area of risk analysis there still remain many points that Lind touches on that need to be discussed in detail and that are of relevance to a broad audience. I would like to draw attention to some essential points which do make difficulties in the practical application of risk analysis methods.

9.D1.1. Risk analysis as a tool in the political decision process?

To start with, let me point out that we encounter a lot of encouragement in our practical work. Actually, we feel quite enthusiastic about the use of risk analysis and believe it to be the best way to solve most safety problems. With respect to the practical performance of risk analysis methods, we are even more optimistic than Lind, who in his chapter considers them to be useful in the design of policies and facilities but not yet applicable either in regulations or in communication with the public. I would maintain that risk analysis – even in its present use – can be an effective tool in aiding management decisions on safety measures as well as in the political decision process on questions of acceptable risks. And, of course, it may also serve as a basis for insurance decisions.

Our firm has been involved in the analysis of risky systems for about 20 years and there have been a lot of successful applications. In this context, "successful" means that risk analysis has been of great importance in the process of deciding about the safety precautions required for hazardous systems (Schneider, 1978). In Switzerland, for example, there do exist regulations which are based on a risk concept. They concern storage, handling, transportation, and manufacture of explosives and ammunition. Also, in many cases, local authorities ground their decisions concerning safety measures on a risk analysis. In this sense, our optimism is founded on the practical performance of risk analysis methods. And, as Lind says, this is the only way risk analysis can prove itself.

But what makes risk analysis an effective tool in the decision process? Lind emphasizes that it needs to be value free. I agree with that. Moreover, I would say that one can hardly stress this enough. But at the same time one needs to underline that risk analysis has to be linked to the appraisal of risk. We like to think of risk analysis as being a plug – a plug which helps to express the risk of various hazardous systems in comparable terms. And, like any plug, this one has to fit into an outlet. In this case, the outlet is supposed to be the process of risk appraisal. To be useful, any risk analysis has to be structured in such a way that it fits into risk appraisal (Bohnenblust and Schneider, 1983). This finding has consequences for how a risk analysis is conducted. Reading Lind's chapter, lay people might get the impression that the task of such an analysis is fulfilled by establishing the probabilities of undesirable events and the extent of the corresponding consequences. We think that there is much more information needed to allow for a serious appraisal of risk. We do not only need a two-pole plug merely supplying probabilities and consequences but a more sophisticated one fitting into the multiple-pole outlet of risk appraisal.

9.D1.2. Risk analysis in a defensive attitude?

Still, even the best risk analysis cannot offer a guarantee of its acceptance in decision processes. There are many examples where it has been forced into a defensive position. Obviously, assuming this attitude it cannot be of much help. Many people accuse risk analysis because it is not part of the exact sciences. However, as Lind says, bridge design and most other engineering disciplines are also pseudosciences. Actually, no analysis can claim to be able to capture reality. It can never be more than a model of reality. This is especially true when we are concerned with events possibly happening in the far future. In that sense, the public, the politicians, and all those involved in safety decisions need to realize the nature of risk analysis. Conducting a risk analysis does not alter a safety problem at all. Also, it will not change anything concerning the knowledge of the physical data and the background information of a problem. The only thing we are able to change by analyzing risks is the way we are investigating the problem. Traditional engineering methods do not explicitly express how safe or how risky a system is. Risk analysis, however, does address this question in a clear-cut way. Performing a risk analysis is an attempt to make one step in the direction of getting a better grasp of a safety problem. It is the effort to handle and examine complex problems in a more transparent and consistent way.

9.D1.3. Risk analysis without confirmed data?

There is another point which misleads many people. They often forget that each risk analysis consists of two parts. The first one is the model, the methodology, and the conceptual framework; the second part is its content – the quantitative knowledge or the data. People often argue that it is not possible to perform a risk analysis when there are not enough data available. In contrast with this, we would suggest reflection on the following thought: the less data, the less knowledge we have, and the more we are in need of a good risk analysis.

In the first place, risk analysis helps to disclose the structure of a problem and its interconnections in a clear and logical way. Even without data we gain insight into a problem. Also, it is much more valuable to capture the basic elements of a problem in a simple but correct and transparent way than to do some fancy mathematical modeling.

The ultimate goal of a risk analysis is to assist in a decision process. To achieve this goal we do not only need a qualitative but also a quantitative result. Of course, it is very nice and makes things easy if we can evaluate statistics, make tests and experiments, etc. in order to obtain confirmed data to calculate risks. It is quite a hard work to perform a risk analysis if we lack data. But again, I believe that it is better to do an analysis mainly based on estimates than to do no analysis at all. Some readers may know the saying "It's better to be roughly right than exactly wrong". Certainly, we will not be able to calculate "true" risks but I agree with Lind that people have to be ready to realize that there are no "true" risks anyway. Even after a system has failed or after an accident we do not know these true risks. In this sense, a risk analysis cannot ever be checked for its correctness. However, that is not the point. The aim of risk analysis is to facilitate a decision to be made. For that purpose we need a broad consensus on the risk estimates and the assumptions leading to them. It is easier to achieve a consensus if you can confront people with confirmed data. But, where such data are lacking it is legitimate and meaningful to base your risk estimates mainly on subjective consensus data. This even has some positive aspects. Applying risk analysis, experts are forced to come to a consensus on the basic assumptions of a safety problem. Without using this method, discrepancies in expert opinions are rarely eliminated, and mostly they are not even recognized. Consider the expression "to the best of one's knowledge and belief". I think this expression applies very aptly to the way one has to perform risk analysis. We have to make use of all information available, be it in terms of "hard" data or "soft" beliefs. Utilizing all information available will ultimately lead to the best decision.

In most of the cases we are fortunate enough to dispose of statistical or experimental data. As Lind shows, the interpretation of these data can be tricky sometimes. He also points out that a risk analysis is not accomplished without an estimate of the uncertainty of the results. In our firm we sometimes have a hard time to make it clear to our clients that there is any uncertainty at all. Many people would prefer to believe that there is only one deterministic and correct answer to each question. But it is indispensable that people accept this uncertainty in the result of a risk analysis. And perhaps it is even more important that they become conscious of the uncertainty in the nature of most safety problems.

9.D1.4. Different types of risk analysis methods

Clearly, Lind's chapter touches on many fundamental questions. I shall make just one more comment. Lind discusses five examples of risk analysis methods. I would like to emphasize that not all of them are of the same level. Rather we can distinguish three different types:

(1) *Reliability analysis, fault tree analysis, and event tree analysis*. These methods can be referred to as techniques of risk analysis. They are more or less mathematical tools.
(2) *Scenarios*. The concept of scenario is somewhat different from the techniques of risk analysis. It is, rather, a conceptual aid for the analyst who investigates the risks of a system as well as for the decision-maker.
(3) *Threat analysis*. Of the five methods discussed, threat analysis is the most comprehensive tool which forms a framework for capturing a safety problem as a whole. I would put threat analysis on the same level as decision analysis and game theory. All these concepts go beyond the analysis of risk and enter the area of risk appraisal. Possibly, threat analysis should be considered not merely a method of risk analysis but rather a method for analyzing risky systems.

References

Bohnenblust, H. and Schneider, T. (1983), From legal principles to engineering decisions – where do we shift to quantification?, *Proceedings of the V International Summer Symposium on Safety Sciences, Leuven, Belgium, 1983*.

Schneider, T. (1978), How much should we be willing to pay for explosives safety?, *Proceedings of the 18th Explosives Safety Seminar of the Department of Defense Explosive Safety Board, San Antonio, TX, USA, 1978*.

9.D2. Discussion

J.L. Hawksley

I do not intend to comment on the mathematical or theoretical rigor of Lind's chapter. Rather I shall make some remarks of support and qualification based upon some experience of risk analysis in the chemical industry.

I would agree that fault tree analysis along with other techniques of risk analysis, can be an effective tool to aid the design of facilities for handling hazardous materials. This is widely accepted, although the extent of use varies. What is more contentious is the extent to which risk analysis can or should be used as an aid to judging the acceptability of the siting of hazardous activities. For that purpose the analysis would concentrate on high-consequence low-probability events and, in the chemical industry, would tend to center on the severe failure of pipework and vessels. Many would argue that risk analysis of that sort is subject to too great an uncertainty to be validly useful.

Risk analysis of some form or other is a necessary part of risk management, which in simple terms requires:

(1) Identification of hazards.
(2) Analysis of the risk of hazards occurring.
(3) Decision on course of action.

The first step is vital and could be said to be the most important. It does not seem to appear in these *Proceedings* and I am tempted to digress slightly from the subject of Lind's chapter to give a reminder that formal techniques of hazard identification are far more widely used in the chemical process industries than are formal techniques of risk analysis. Typical are hazard and operability (HAZOP) studies and failure mode and effect analysis. Both are inductive procedures involving the postulation of failures or deviations and examination of the consequences that might result. Rigorous hazard identification is seen as an essential cornerstone of a system for hazard control. Its use in Imperial Chemical Industries (ICI), for instance, has contributed to a significant improvement in safety performance over the last two decades.

For the safe management of a hazardous activity, hazard identification has to be carried out and needs to be thorough. Risk analysis, however, can be more variably applied. Indeed, it may be bypassed if the decision is to apply an accepted standard or code of practice to avoid the identified hazard. Of course, risk analysis may have contributed towards establishing that standard. I would not agree with Lind that risk analysis is by definition both "formal" and "quantitative"

(Section 9.1, first paragraph). Much informal and qualitative risk analysis takes place in a design process. Also I would submit that not all formal risk analysis need necessarily be quantitative. Indeed, although quantification may give an additional dimension to add perspective to an analysis, I would contend that getting the logic right is the more important. In many cases a qualitative evaluation of, say, a fault tree can be sufficient to identify the main contributors to a risk. Only in selected cases is formal quantitative analysis necessary.

Turning to the methods of risk analysis reviewed by Lind, I read first the abstract of his paper which spoke of comparing fault tree analysis with other methods. I reacted to this as it tended to imply that the various methods might be alternative ways of tackling the same problem. I think this is rarely the case and, as Lind does point out, (Section 9.1) "each method is appropriate for specific purposes". In practice a combination of methods is used depending on the nature of the risk analysis to be done.

In the chemical industry there are two broad areas in which risk analysis is used. Its application to hardware design (e.g. safety shutdown systems) is well established. Its application to choice of site and determination of separation distances is developing. Fault tree analysis is the most widely used technique for the former together with some form of reliability analysis. Event trees and the scenario approach are more suited to the latter. For instance, in an analysis of the off-site risks from two installations handling toxic gases about 60 scenarios involving about 100 failure cases were considered. For only two of the failure cases was fault tree analysis found appropriate. It really is a case of "horses for courses".

I would like to make a few comments on some of the "horses":

(1) *System reliability analysis*. I have no personal experience of this technique. I do not think it is widely used in the chemical industry in relation to the safety problems of hazardous processes. Some of the more complex control and instrumented protective systems may be analyzed in this way but these are the exception rather than the rule (as opposed to the nuclear industry for instance). As a safety man I tend to be wary of techniques that are largely mathematical and highly specialist. In my experience, developing the failure logic of a system in an analytical exercise involving both designer and operator adds more to their understanding of the system and awareness of the hazards than rather esoteric calculations.

(2) *Scenarios*. The scenario approach is now incorporated into legislation in the European Economic Community for the control of major industrial accidents involving chemicals. An operator is required to demonstrate that he appreciates the scale of the

hazard that might arise from his activity. This is, of course, a necessary step towards proper control of the hazard. The scenario approach focuses more on the consequences of a hazard than on the likelihood of that hazard being realized and so has the disadvantage of possibly concentrating attention on, say, the "maximum conceivable accident" or the "worst possible accident" and diverting attention from the more credible accidents. Scenarios should always be qualified by some indication of likelihood of occurrence whether in qualitative or quantitative terms.

(3) *Threat analysis*. This is not a term that I have come across. The various steps in the technique as listed by Lind put me in mind of the technique of hazard analysis that we use in the design of chemical plant: *Figure 9.D2.1* illustrates the general procedure. While in applying this technique we can interpret hazard in its broadest sense – i.e. danger to life, material assets, production, and profits – threat analysis is clearly even broader in its scope in seeking to cover "malicious agents" as well as accidents. This leads to a question that is sometimes asked in risk analysis: should any account be taken of possible terrorist activity and, if

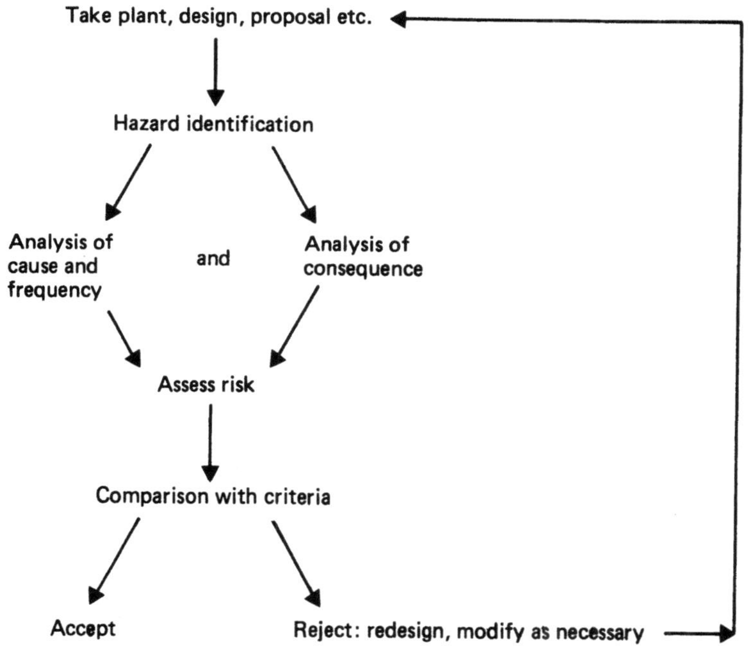

Figure 9.D2.1. Generalized model for hazard analysis.

so, how? I wonder if threat analysis could be useful where that is a concern?

Of more general and perhaps greater interest is the concept of unifying the approach to protection against threats from any cause. It occurs to me that this procedure could perhaps be a useful way of displaying to the public the various risks to which they may be exposed together with protection costs, benefits, etc., so that, for instance, the risks from a chemical hazard could be put into perspective. In other words, could threat analysis be a useful communication tool?

(4) *Event trees and fault trees.* I will comment on these together. Firstly, I had some difficulty with the descriptions! I would not doubt that they are strictly correct but the rather mathematical terminology obscures the appealing basic simplicity of the techniques which makes them so useful. I would recommend two booklets: *Risk Analysis in the Process Industries* (Institution of Chemical Engineers, 1985) and *Methodology for Hazard Analysis and Risk Assessment in the Petroleum Refining and Storage Industry* (CONCAWE, 1982).

One might be left with the impression from Lind's examples that event and fault trees are always alternative ways of analyzing the same problem. That is by no means the case because of the fundamental differences between the techniques that I would like to make clear. Event tree analysis is a "bottom up" procedure that takes an event (cause) and seeks to analyze the various outcomes possible. It does not generate other causes of those outcomes. Fault tree analysis, conversely, is a "top down" procedure that starts with a specific outcome (the top event) and seeks to analyze the various causes of that outcome. It may generate other outcomes but that is not necessarily a prime objective. It is not an analysis of all system failures but only of those that could cause the specified outcome. The techniques are complementary so that the outcome of a fault tree may be further analyzed by an event tree (see *Figure 9.D2.2*).

I will take Lind's pressure test example to illustrate an important point. That is, it is vital to define adequately the starting point of the analysis, the various branch points of the trees, and the events and causes embodied in the trees. I drew up the fault and event trees shown on *Figures 9.D2.3* and *9.D2.4*. The additional definition, I would claim, significantly clarifies the analysis. An attraction of these techniques is that they are vivid methods for communication of information as well as analysis of problems.

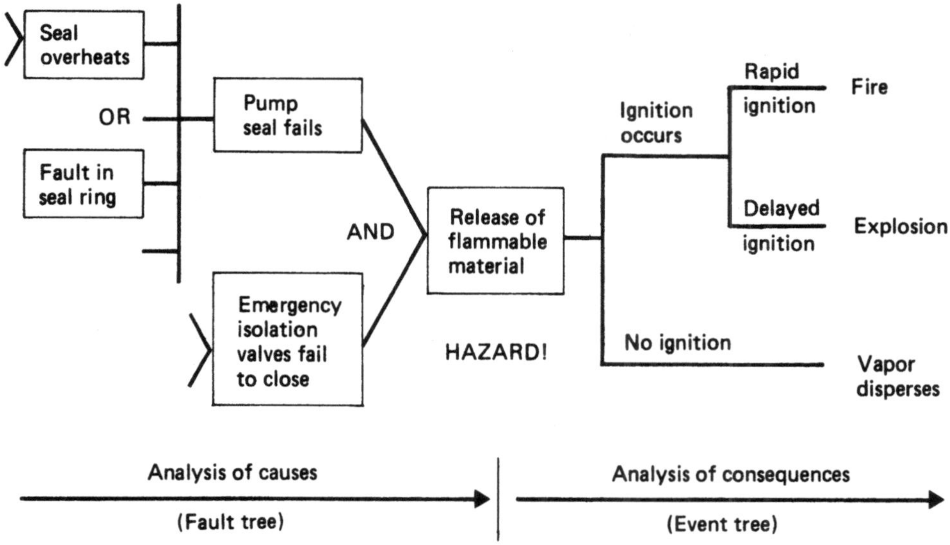

Figure 9.D2.2. Complementary use of fault tree and event tree analysis.

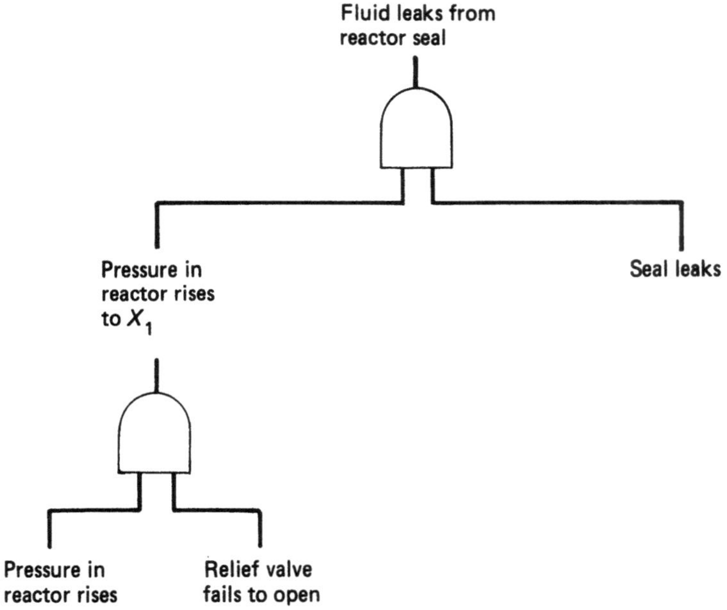

Figure 9.D2.3. Expanded fault tree for pressure test example.

I would now like to make a few general points. In his concluding remarks Lind says that some systems may defy risk analysis except by experts with a knowledge of the hardware as well as risk analysis

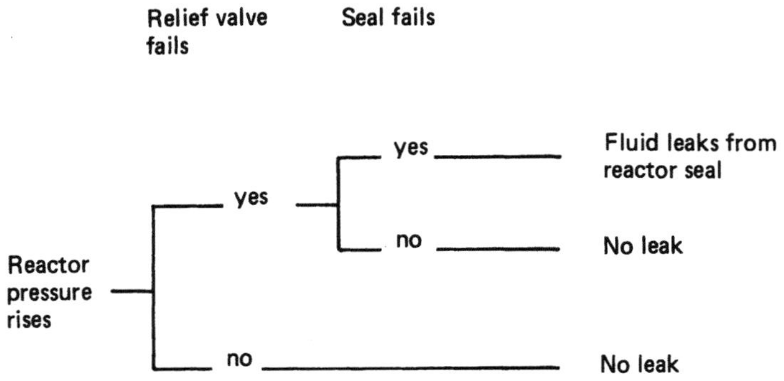

Figure 9.D2.4. **Descriptive event tree for pressure test example.**

techniques. This might be taken to suggest that nonexperts might handle some analyzes. There are, of course, degrees of expertise but I would suggest that for *any* risk analysis to have credibility it has to be done by persons with adequate training and technical experience of the systems, both hardware and software, that are to be analyzed.

He describes threat analysis as not being a panacea. I would extend that to all methods of risk analysis. The various procedures can be a powerful aid to the safe operation of hazardous activities but none can be a substitute for experienced engineering and management judgment.

A major limitation of risk analysis is the difficulty of allowing for the human influence. Lind clearly recognizes this, and it is worth highlighting some statements: e.g. in Section 9.1.1, "calculated reliability ... is conditional upon the absence of human influences"; the opening words of Section 9.4, "Quantitative risk calculation can be done effectively for systems not subject to major human intervention" (we do not have many of those!); and later in Section 9.4, "Risk is then assessed conditional upon a specified human behavior (e.g. ...'design according to the pressure vessel code')". To summarize this in general terms, risk analysis might be able to show that a hazardous activity *can* be safe but it cannot show that it *will* be safe. That is dependent on satisfactory management to ensure, among other things, that the assumptions implicit in the risk analysis are valid and remain so. Perhaps the Bhopal disaster provides the most recent example. On paper the plant would seem to have had adequate safety systems, but in the event they did not work. Some would use this to argue that risk analysis has little value. I would not agree. Risk analysis, involving those who will manage an activity, can be an effective means of highlighting those aspects that are vital for safe operation and of increasing the awareness of where stringent management control is required.

Further problems with risk analysis that Lind rightly acknowledges are the inadequate availability of failure data and uncertainties in calculation procedures – both for event frequencies and, more particularly, for the consequences of accidents. This problem becomes more acute for risk analysis of low-probability/high-consequence events, i.e. the consideration of risks to the public. An estimate put the uncertainty in a recent risk analysis of that type at about ±30 times. Such levels of uncertainty seem to severely limit the usefulness of risk analysis in the public domain although this is an application being pursued by a number of state regulatory authorities primarily for planning control purposes – notably in the Netherlands and the UK. The acceptability of this application of risk analysis depends on the approach adopted. The approach in the UK of seeking to use risk analysis to establish broad guidelines for planning control with flexibility to judge specific cases on the merits of the local situation is to be preferred to the setting of fixed numerical criteria against which to judge the results of a risk analysis as seems to be being pursued in the Netherlands.

This is an appropriate aspect on which to finish as it is perhaps the most contentious and where there are problems to be resolved.

A consensus needs to be reached on the extent to which risk analysis can be validly applied to risks in the public domain. This needs to recognize that the uncertainty in the technique is such that only relatively large differences in risk can reliably be identified, so that analysis might, at best, be only justified in selected cases. Also techniques are in almost a continuous state of development and many aspects require judgment so that any statutory, say, requirement for risk analysis has to be flexible. The public remains to be convinced of the validity of risk analysis and is more inclined to believe Murphy's Law, i.e. if it can happen it will. A recent example is the rejection of Union Carbide's plan to open a factory at Livingstone in Scotland that would handle small quantities of toxic gases (e.g. arsine) in cylinders. A risk analysis by an independent consultant that estimated very low risks to local people was rejected; the worst likely accident that was assumed in the analysis was the escape of the whole contents of one cylinder only, whereas local opinion was that it should be assumed that the total inventory of gases might escape despite being distributed in many separate cylinders. Admittedly there were emotive issues with Union Carbide's involvement and had the matter been dealt with prior to December 1984 there might well have been a different outcome. It does raise a final question, though: How can the validity of a risk analysis be tested?

References

CONCAWE (1982), ***Methodologies for Hazard Analysis and Risk Assessment in the Petroleum Refining and Storage Industry*** [CONCAWE (The Oil Companies' International Study Group for Conservation of Clean Air and Water – Europe), The Hague].

Institution of Chemical Engineers (1985), ***Risk Analysis in the Process Industries***, Report of the International Study Group on Risk Analysis, European Federation of Chemical Engineering Publications Series No. 45 (Institution of Chemical Engineers, Rugby, England).

CHAPTER 10

Informing and Educating the Public About Risk

P. Slovic [1]

> To effectively manage ... risk, we must seek new ways to involve the public in the decision-making process.... They [the public] need to become involved early, and they need to be informed if their participation is to be meaningful. (William Ruckelshaus, 1983, p 1028.)

10.1. Introduction

In a bold and insightful speech before the National Academy of Sciences at the beginning of his second term as administrator of the Environmental Protection Agency (EPA), William Ruckelshaus called for a government-wide process for managing risks that thoroughly involved the public. Arguing that government must accommodate the will of the people, he quoted Thomas Jefferson's famous dictum to the effect that "If we think (the people) not enlightened enough to exercise their control with a wholesome discretion, the remedy is not to take it from them, but to inform their discretion".

Midway into his tenure as EPA administrator, Ruckelshaus' experiences in attempting to implement Jefferson's philosophy led him to a more sober evaluation: "Easy for *him* to say. As we have seen, informing discretion about risk has itself a high risk of failure" (Ruckelshaus, 1984, p 160).

This chapter attempts to illustrate why the goal of informing the public about risk issues, which seems easy to attain in principle, is surprisingly difficult to accomplish. To be effective, risk communicators must recognize and overcome a number of obstacles that have their roots in the limitations of scientific risk assessment and the idiosyncrasies of the human mind. Doing an adequate job of communicating means finding comprehensible ways of presenting complex technical material that is clouded by uncertainty and inherently difficult to understand. Awareness of the difficulties should enhance the chances of designing successful informational programs.

10.2. Limitations of Risk Assessment

Risk assessment is a complex discipline, not fully understood by its practitioners, much less the lay public. At the technical level, there is still much debate over terminology and techniques. Technical limitations and disagreements among experts inevitably affect communication in the adversarial climate that surrounds many risk issues. Risk communicators must be fully aware of the strengths and limits of the methods used to generate the information they are attempting to convey to the public. In particular, communicators need to understand that risk assessments are constructed from theoretical models which are based on assumptions and subjective judgments. If these assumptions and judgments are deficient, the resulting assessments may be quite inaccurate.

Nowhere are these problems more evident than in the assessment of chronic health effects due to low-level exposures to toxic chemicals and radiation. The typical assessment uses studies of animals exposed (relatively briefly) to extremely high doses of the substance to draw inferences about the risks to humans exposed to very low doses (sometimes over long periods of time). The models designed to extrapolate the results from animals to humans and from high doses to low doses are controversial. For example, some critics have argued that mice may be from 3×10^4 to 10^9 times more cancer prone than humans (Gori, 1980). Different models for extrapolating from high-dose exposures to low doses produce estimated cancer rates that can differ by factors of 1000 or more at the expected levels of human exposures (which themselves are often subject to a great deal of uncertainty). Difficulties in estimating synergistic effects (interactions between two or more substances, such as occur between cigarette smoking and exposure to asbestos) and effects on particularly sensitive people (e.g. children, pregnant women, and the elderly) further compound the problems of risk assessment. In light of these various uncertainties, one expert concluded that "Discouraging as it may seem, it is not plausible that animal carcinogenesis experiments can be improved to the point where quantitative generalizations about human risk can be drawn from them" (Gori, 1980, p 259).

In the adversarial climate of risk discussions, these limitations of assessment are brought forth to discredit quantitative risk estimates. To be credible and trustworthy, a communicator must know enough to acknowledge valid criticisms and to discern whether the available risk estimates are valid enough to have value for helping the public gain perspective on the dangers they face and the decisions that must be made. On the positive side, there are some hazards (e.g. radiation, asbestos) whose risks are relatively well understood. Moreover, for many other hazards, risk estimates are based on a chain of conservative decisions at each choice point in the analysis (e.g. studying the most sensitive species, using the extrapolation model that produces the highest risk estimate, giving benign tumors the same weight as malignant ones, etc.). Despite the uncertainties, one may have great confidence that the "true risk" is unlikely to exceed the estimate resulting from such a conservative process. In other words, uncertainty and subjectivity do not imply chaos. Communicators must know when this point is relevant and how to make it when it applies.

Parallel problems exist in engineering risk assessments designed to estimate the probability and severity of rare high-consequence accidents in complex systems such as nuclear reactors or liquefied natural gas (LNG) plants. The risk estimates are devised from theoretical models (in this case fault trees or event trees) that attempt to depict all possible accident sequences and their (judged) probabilities. Limitations in the quality or comprehensiveness of the analysis, the quality of the judged risks for individual sequences, or improper rules for combining estimates can seriously compromise the validity of the assessment.

10.3. Limitations of Public Understanding

Just as they must understand the strengths and limitations of risk assessment, communicators must appreciate the wisdom and folly in public attitudes and perceptions. Among the important research findings and conclusions are the following.

10.3.1. People's perceptions of risk are often inaccurate

Risk judgments are influenced by the memorability of past events and the imaginability of future events. As a result, any factor that makes a hazard unusually memorable or imaginable, such as a recent disaster, heavy media coverage, or a vivid film, could seriously distort perceptions of risk. In particular, studies by Lichtenstein *et al.* (1978), Morgan *et al.* (1985), and others have found that risks from dramatic or sensational causes of death, such as accidents, homicides, cancer, and natural disasters, tend to be greatly overestimated. Risks from undramatic causes such as asthma, emphysema, and diabetes, which take one life at a time and are common in nonfatal form, tend to be underestimated. News media coverage of hazards has been found to be biased in much the same direction, thus contributing to the difficulties of obtaining a proper perspective on risks (Combs and Slovic, 1979).

10.3.2. Risk information may frighten and frustrate the public

The fact that perceptions of risk are often inaccurate points to the need for warnings and educational programs. However, to the extent that misperceptions are due to reliance on imaginability as a cue for riskiness, such programs may run into trouble. Merely mentioning possible adverse consequences (no matter how rare) of some product or activity could enhance their perceived likelihood and make them appear more frightening. Anecdotal observation of attempts to inform people about recombinant DNA hazards supports this hypothesis (Rosenburg, 1978) as does a controlled study by Morgan *et al.* (1985). In the latter study people's judgments of the risks from high voltage transmission lines were assessed before and after they read a brief and rather neutral description of findings from studies of possible health effects due to such lines. The results, shown in *Figure 10.1*, clearly indicated a shift toward greater concern in three separate groups of subjects. Whereas mere mention and refutation of potential risks raises concerns, the use of conservative assumptions and "worst case scenarios" in risk assessment creates extreme negative reactions in people because of the difficulty of appreciating the improbability of such extreme but imaginable consequences. The possibility that imaginability may blur the distinction between what is (remotely) possible and what is probable obviously poses a serious obstacle to risk information programs.

Other psychological research shows that people may have great difficulty making decisions about gambles when they are forced to resolve conflicts generated by the possibility of experiencing both gains and losses, and uncertain ones at that (Slovic, 1982; Slovic and Lichtenstein, 1983). As a result, wherever possible, people attempt to reduce the anxiety generated in the face of uncertainty by denying that uncertainty, thus making the risk seem either so small that it can safely be ignored or so large that it clearly should be avoided. They rebel against being given statements of probability, rather than fact; they want to know exactly what will happen.

Given a choice, people would rather not have to confront the gambles inherent in life's dangerous activities. They would prefer being

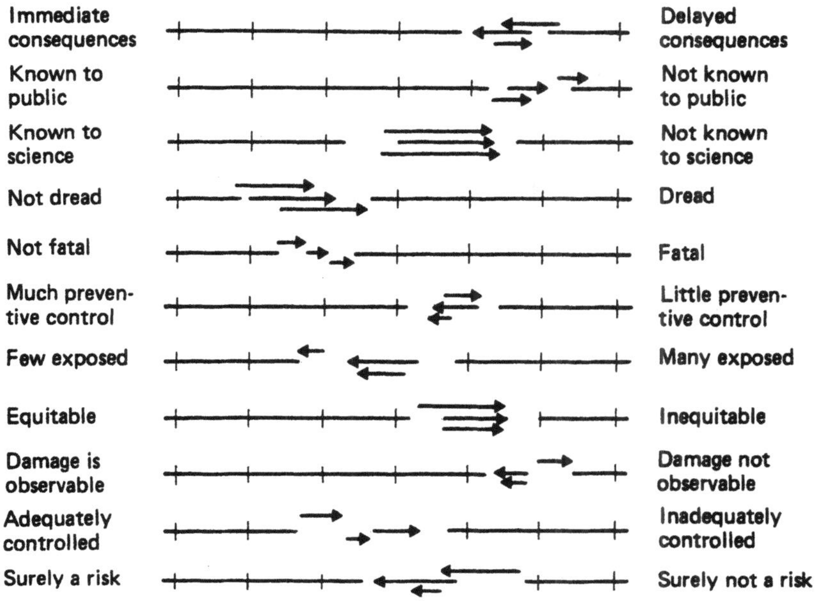

Figure 10.1. Comparison of the direction and magnitude of shifts in mean scores on nine risk characteristics and two evaluative measures before (tail of vector) and after (point of vector) receiving specific information on possible health effects of transmission lines. Results from three separate groups of respondents are shown for each scale. (*Source*: Morgan *et al.*, 1985.)

told that risks are managed by competent professionals and are thus so small that one need not worry about them. However, if such assurances cannot be given, they will want to be informed of the risks, even though doing so might make them feel anxious and conflicted (Alfidi, 1971; Fischhoff, 1983; Weinstein, 1979).

10.3.3. Strong beliefs are hard to modify

It would be comforting to believe that polarized positions would respond to informational and educational programs. Unfortunately, psychological research demonstrates that people's beliefs change slowly and are extraordinarily persistent in the face of contrary evidence (Nisbett and Ross, 1980). Once formed, initial impressions tend to structure the way that subsequent evidence is interpreted. New evidence appears reliable and informative if it is consistent with one's initial belief; contrary evidence is dismissed as unreliable, erroneous, or unrepresentative.

10.3.4. Naive views are easily manipulated by presentation format

When people lack strong prior opinions, the opposite situation exists: they are at the mercy of the way that the information is presented. Subtle changes in the way that risks are expressed can have a major impact on perceptions and decisions. One dramatic recent example of this comes from a study by McNeil *et al.* (1982), who asked people to imagine that they had lung cancer and had to choose between two therapies, surgery or radiation. The two therapies were described in some detail. Then, some subjects were presented with the cumulative probabilities of surviving for varying lengths of time after the treatment. Other subjects received the same cumulative probabilities framed in terms of dying rather than surviving (e.g. instead of being told that 68% of those having surgery will have survived after one year, they were told that 32% will have died). Framing the statistics in terms of dying dropped the percentage of subjects choosing radiation therapy over surgery from 44% to 18%. The effect was as strong for physicians as for lay persons.

Numerous other examples of "framing effects" have been demonstrated by Tversky and Kahneman (1981) and Slovic *et al.* (1982). The fact that subtle differences in how risks are presented can have such marked effects suggests that those responsible for information programs have considerable ability to manipulate perceptions and behavior. This possibility raises ethical problems that must be addressed by any responsible risk information program.

10.4. Placing Risks in Perspective

10.4.1. Choosing risk measures

When we know enough to be able to describe risks quantitatively, we face a wide choice of options regarding the specific measures and statistics used to describe the magnitude of risk. Fischhoff *et al.* (1981) point out that choosing a risk measure involves several steps:

(1) Defining the hazard category.
(2) Deciding what consequences to measure (or report).
(3) Determining the unit of observation.

The way the hazard category is defined can have a major effect on risk statistics. For example, statistics on the risks from coal mining would look very different depending on whether or not they combined underground and strip mining. Deaths, injuries, and illnesses are the most commonly measured consequences; these can be assessed as a function of age, cause, type, or severity. Sometimes aggregate measures are created such as person-days lost from work or loss of life expectancy. Units of measurement can range from population values to risks indexed according to some unit of exposure or unit of work produced.

Crouch and Wilson (1982) provide some specific examples of how different measures of the same risk can sometimes give quite different impressions. For example, they show that accidental deaths per million tons of coal mined in the USA have decreased steadily over time. In this respect, the industry is getting safer. However, they also show that the rate of accidental deaths per 1000 coal mine employees has increased. Neither measure is the "right" measure of mining risk. They each tell part of the same story.

The problem of selecting measures is made even more complicated by the framing effects described earlier. Thus, not only do different measures of the same hazard give different impressions, the *same* measures, differing only in (presumably) inconsequential ways, can lead to vastly different perceptions.

Sharlin's case study of the communication of information about the risks of the pesticide ethylene dibromide (EDB) points to an important distinction between macro and micro measures of risk (Sharlin, 1986). The EPA, which was responsible for regulating EDB, broadcast information about the aggregate risk of this pesticide to the exposed population. While the media accurately transmitted this macro analysis, newspaper editorials and public reaction clearly indicated an inability to translate this into a micro perspective on the risk to the exposed individual. In other words, the newspaper reader or television (TV) viewer had trouble inferring an answer to the question, "Should I eat the bread?", from the aggregate risk analysis.

10.4.2. Basic statistical presentations

In this section, we shall describe a few of the statistical displays most often used to educate people about general and specific risks. We do not mean to endorse these presentations as optimal. They simply represent the favored formats of statisticians and risk assessors [2]. To date, there has been little systematic effort to develop and test methods for maximizing clarity and understanding of quantitative risk estimates. As a result, we know of no "magic displays" that guarantee understanding and appreciation of the described risks at the "micro level".

Among the few "principles" in this field that seem to be useful is the assertion that comparisons are more meaningful than absolute numbers or probabilities, especially when these absolute values are quite small. Sowby (1965) argued that to decide whether or not we are responding adequately to radiation risks we need to compare them to "some of the other risks of life" and Rothschild (1979) observed, "There is no point in getting into a panic about the risks of life until you have compared the risks which worry you with those that don't, but perhaps should."

Familiarity with annual mortality risks for the population as a whole or as a function of age may provide one standard for evaluating specific risks. Sowby (1965) took advantage of such data to observe that one hour riding a motorcycle was as risky as one hour of being 75 years old. *Table 10.1* provides annual mortality rates from a wide variety of causes.

Table 10.1. Annual fatality rates per 100000 persons at risk.

Risk	*Rate*
Motorcycling	2000
Aerial acrobatics (planes)	500
Smoking (all causes)	300
Sport parachuting	200
Smoking (cancer)	120
Fire fighting	80
Hang gliding	80
Coal mining	63
Farming	36
Motor vehicles	24
Police work (nonclerical)	22
Boating	5
Rodeo performer	3
Hunting	3
Fires	2.8
1 diet drink/day (saccharin)	1.0
4 tbs. peanut butter/day (aflatoxin)	0.8
Floods	0.06
Lightning	0.05
Meteorite	0.000006

Source: Adapted from Crouch and Wilson (1982).

Mortality rates fail to capture the fact that some hazards (e.g. pregnancy, motorcycle accidents) cause death at a much earlier age than others (e.g. lung cancer due to smoking). One way to provide perspective on this consideration is to calculate the average loss of life expectancy due to the exposure to the hazard, based on the distribution of deaths as a function of age. Some estimates of loss of life expectancy from various causes are shown in *Table 10.2*.

Yet another innovative way to gain perspective was devised by Wilson (1979), who displayed a set of activities (*Table 10.3*), each of which

Table 10.2. Estimated loss of life expectancy due to various causes.

Cause	*Days*
Cigarette smoking (male)	2250
Heart disease	2100
Being 30% overweight	1300
Being a coal miner	1100
Cancer	980
Stroke	520
Army in Vietnam	400
Dangerous jobs, accidents	300
Motor vehicle accidents	207
Pneumonia, influenza	141
Accidents in home	95
Suicide	95
Diabetes	95
Being murdered (homicide)	90
Drowning	41
Job with radiation exposure	40
Falls	39
Natural radiation (Beir)	8
Medical X rays	6
Coffee	6
All catastrophes combined	3.5
Reactor accidents (UCS)	2[a]
Radiation from nuclear industry	0.02[a]

[a]These items assume that all US power is nuclear. UCS is the Union of Concerned Scientists, the most prominent group of critics of nuclear energy.
Source: Cohen and Lee (1979).

was estimated to increase one's chance of death (during any year) by one in a million.

Comparisons within lists of risks such as those in *Tables 10.1*, *10.2*, and *10.3* have been advocated not just to gain some perspective on risks but as guides to decision-making. Thus Cohen and Lee (1979) argued that "to some approximation, the ordering [in *Table 10.2*] should be society's order of priorities" and Wilson (1979) claimed that the comparisons in *Table 10.3* "...help me evaluate risk and I imagine that they may help others to do so, as well. But the most important use of these comparisons must be to help the decisions we make, as a nation, to improve our health and reduce our accident rate." However, Slovic *et al.* (1980a) argued that such claims could not be logically defended. Although carefully prepared lists of risk statistics can provide some degree of insight, they provide only a small part of the information needed for decision-making. As a minimum, inputs to decision-making should include a detailed account of the costs and benefits of the available options, as well as an indication of the uncertainty in these

Table 10.3. Risks estimated to increase chance of death in any year by 0.000001 (1 chance in 10^6).

Activity	*Cause of death*
Smoking 1.4 cigarettes	Cancer, heart disease
Spending 1 hour in a coal mine	Black lung disease
Living 2 days in New York or Boston	Air pollution
Traveling 10 miles by bicycle	Accident
Flying 1000 miles by jet	Accident
Living 2 months in Denver on vacation from New York	Cancer caused by cosmic radiation
One chest X ray taken in a good hospital	Cancer caused by radiation
Eating 40 tablespoons of peanut butter	Liver cancer caused by aflatoxin B
Drinking 30 12-oz cans of diet soda	Cancer caused by saccharin
Drinking 1000 24-oz soft drinks from recently banned plastic bottles	Cancer from acrylonitrile monomer
Living 150 years within 20 miles of a nuclear power plant	Cancer caused by radiation
Risk of accident by living within 5 miles of a nuclear reactor for 50 years	Cancer caused by radiation

Source: Wilson (1979).

assessments. As we have seen, uncertainties in risk estimates are often quite large. Failure to indicate uncertainty not only deprives the recipient of information needed for decision-making, but it spawns distrust and rejection of the analysis.

Some hazards, such as radiation, are present in nature and in many commonplace activities. For these hazards, comparisons of "non-natural" exposures (e.g. medical X rays) with the natural or "everyday" exposures may prove instructive.

10.5. Beyond Numbers: A Broader Perspective on Risk Perception and Communication

A stranger in a foreign land would hardly expect to communicate effectively with the natives without knowing something about their language and culture. Yet risk assessors and risk managers have often tried to communicate with the public under the assumption that they and the public share a common conceptual and cultural heritage in the domain of risk. That assumption is false and has led to failures of communication and rancorous conflicts.

10.5.1. The psychometric paradigm

Evidence against the "commonality assumption" comes from sociological, psychological, and anthropological studies directed at understanding the determinants of people's risk perceptions and behaviors. In psychology, research within what has been called the "psychometric paradigm" has explored the ability of psychophysical scaling methods and multivariate analysis to produce meaningful representations of risk attitudes and perceptions (see, for example, Brown and Green, 1980; Gardner *et al.*, 1982; Green, 1980; Green and Brown, 1980; Johnson and Tversky, in press; Lindell and Earle, 1983; Macgill, 1983; Renn, 1981; Slovic *et al.* 1980b, 1984; Vlek and Stallen, 1981; von Winterfeldt *et al.*, 1981).

Researchers employing the psychometric paradigm have typically asked people to judge the current riskiness (or safety) of diverse sets of hazardous activities, substances, and technologies, and to indicate their desires for risk reduction and regulation of these hazards. These global judgments have then been related to judgments about the hazard's status on various qualitative characteristics of risk, some of which are shown in *Table 10.4*.

Table 10.4. Characteristics examined in psychometric studies of perceived risk.

Voluntary – Involuntary
Chronic – Catastrophic
Common – Dread
Injurious – Fatal
Known to those exposed – Not known to those exposed
Known to science – Not known to science
Controllable – Not controllable
Old – New

Among the generalizations that have been drawn from the results of the early studies in this area are the following:

(1) Perceived risk is quantifiable and predictable. Psychometric techniques seem well suited for identifying similarities and differences among groups with regard to risk perceptions and attitudes.

(2) "Risk" means different things to different people. When experts judge risk, their responses correlate highly with technical estimates of annual fatalities. Lay people can assess annual fatalities if they are asked to (and produce estimates somewhat like the technical estimates). However, their judgments of risk are sensitive to other characteristics as well and, as a result, often differ markedly from experts' assessments of risk. In particular, perception of risk is greater for hazards whose adverse effects are uncontrollable, dread, catastrophic, fatal rather than injurious,

not offset by compensating benefits, and delayed in time so the risks are borne by future generations.

(3) Many of the risk characteristics are highly correlated with each other, across a wide domain of hazards. For example, hazards rated as "voluntary" tend also to be rated as "controllable" and "well known"; hazards that threaten future generations tend also to be seen as having catastrophic potential, etc. Investigation of these interrelationships by means of factor analysis has shown that the broader domain of characteristics can be condensed to two or three higher-order characteristics or factors, as shown in *Figure 10.2*.

The factor space presented in *Figure 10.2* has been consistently replicated across groups of lay persons and experts judging large and diverse sets of hazards. The factors in this space reflect the degree to which a risk is understood, the degree to which it evokes a feeling of dread, and the number of people exposed to the risk. Most important is the factor "Dread Risk". The higher a hazard's score on this factor (i.e. the farther toward the right a hazard lies in the space), the higher its perceived risk, the more people want to see its current risks reduced, and the more they want to see strict regulation employed to achieve the desired reduction in risk.

Another useful concept that has emerged from this research is the notion that the societal cost of an accident or mishap is determined to an important degree by what it signifies or portends (Slovic *et al.*, 1984). An accident that takes many lives may produce relatively little social disturbance (beyond that caused the victims' families and friends) if it occurs as part of a familiar and well understood system (e.g. a train wreck). However, a small accident in an unfamiliar system (or one perceived as poorly understood), such as a nuclear reactor or a recombinant DNA laboratory, may have immense social consequences if it is perceived as a harbinger of further and possibly catastrophic mishaps [3]. The informativeness or "signal potential" of a mishap, and thus its potential social impact, appears to be systematically related to both Dread Risk and Unknown Risk factors (see *Figure 10.3*).

10.5.2. Other paradigms

Other important contributions to our current understanding of risk perception have come from geographers, sociologists, and anthropologists. The geographical research focused originally on understanding human behavior in the face of natural hazards, but it has since broadened to include technological hazards as well (Burton *et al.*, 1978).

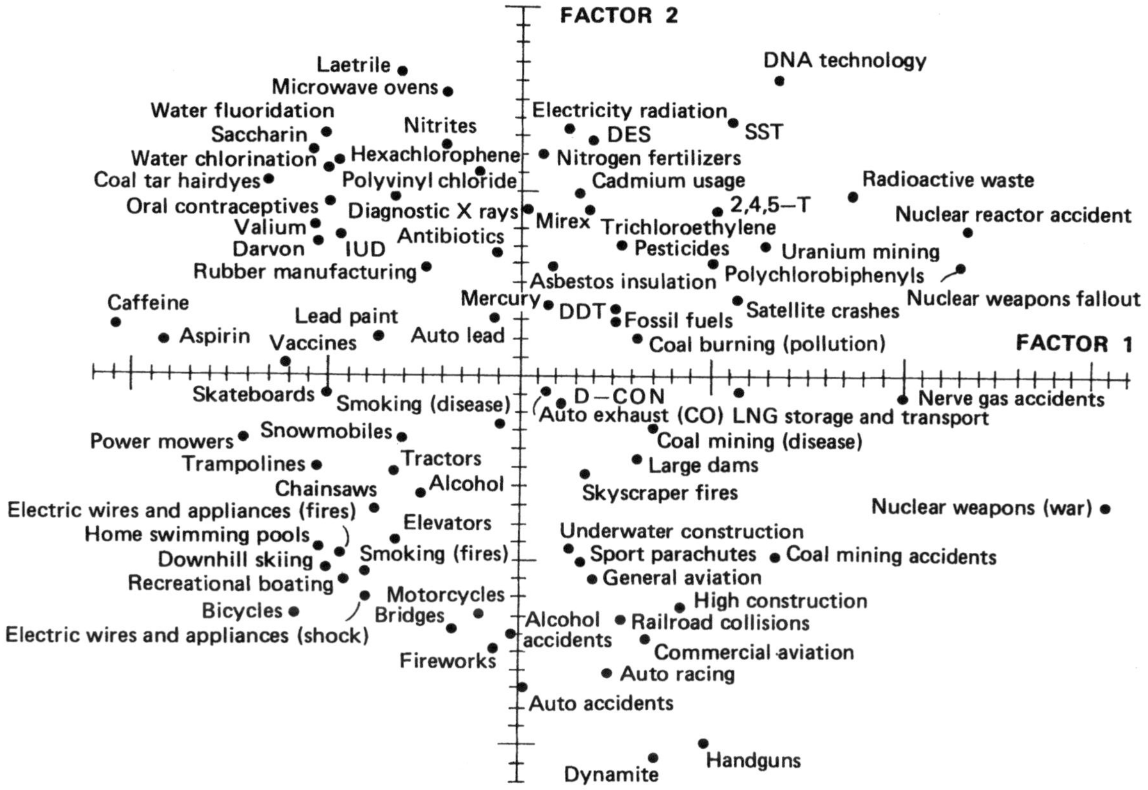

***Figure 10.2(a)*. Hazard locations on factors 1 and 2 derived from interrelationships among 16 risk characteristics. Each factor is made up of a combination of characteristics, as indicated by *Figure 10.2(b)*. (*Source*: Slovic *et al.*, 1985.)**

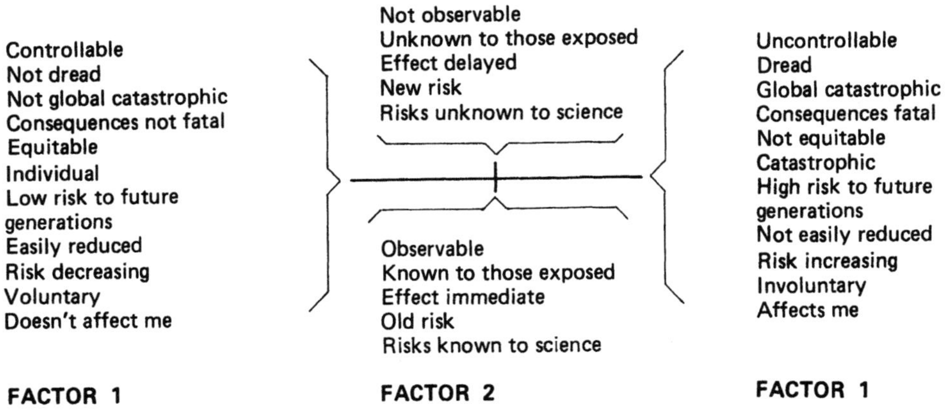

Figure 10.2.(b) Each factor in *Figure 10.2.(a)* is made up of these combinations of characteristics. (*Source*: Slovic *et al.*, 1985).

The sociological work (Moatti *et al.*, 1984; Mazur, 1984) and anthropological studies (Douglas and Wildavsky, 1982) have shown that the perceptions of risk that have been identified within the psychometric paradigm may have their roots in social and cultural factors. Mazur argues that, in some instances, response to hazards is caused by social influences transmitted by friends, family, fellow workers, and respected public officials. In these cases, risk perception may form afterwards, as part of the ex post facto rationale for one's own behavior. In a similar vein, Douglas and Wildavsky assert that people, acting within social organizations, downplay certain risks and emphasize others as a means of maintaining the viability of the organization.

10.5.3. Implications for risk communication

Risk perception research has a number of direct implications for communication efforts. Psychometric studies imply that comparative examination of risk statistics, such as those in *Tables 10.1–10.3*, will not, by themselves, be adequate guides to personal or public decision policies. Risk perceptions and risk-taking behaviors appear to be determined not only by accident probabilities, annual mortality rates or mean losses of life expectancy, but also by numerous other characteristics of hazards such as uncertainty, controllability, catastrophic potential, equity, and threat to future generations. Within the perceptual space defined by these and other characteristics, each hazard is unique. To many persons, statements such as "the annual risk from living near a nuclear power plant is equivalent to the risk of riding an extra three miles in an automobile" appear ludicrous because they fail to give adequate consideration to the important differences in the nature of the risks from these two technologies.

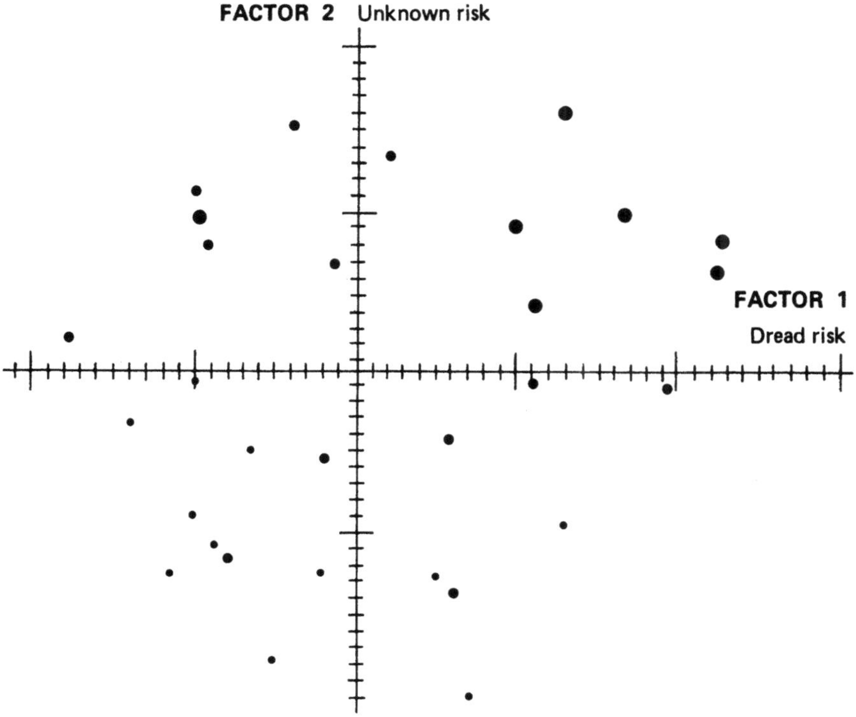

Figure 10.3. Relation between signal potential and risk characterization for 30 hazards in *Figure 10.2.* The larger the point, the greater the degree to which an accident involving that hazard was judged to "serve as a warning signal for society, providing new information about the probability that similar or even more destructive mishaps might occur within this type of activity." (*Source*: Slovic *et al.*, 1984.)

Psychometric research indicates that attempts to characterize, compare, and regulate risks must be sensitive to the broader conception of risk that underlies people's concerns. Fischhoff *et al.* (1984) have made a start in this direction by demonstrating how one might go about constructing a more adequate definition of risk. They advocated characterizing risk by a vector of measures (mortality, morbidity, concern due to perceived uncertainty, concern due to dread, etc.).

The concept of accidents as signals indicates that, when informed about a particular hazard, people's concerns will generalize beyond the immediate problem to other related hazards. For example, with regard to the EDB scare, one newspaper editor wrote: "The cumulative effect – the 'body burden count' as scientists call it – is especially worrisome considering the number of other pesticides and carcinogens humans are exposed to" (*The Sunday Star – Bulletin and Advertiser*, Honolulu, 5

February 1984). On the same topic, another editor wrote: "Let's hope there are no cousins of EDB waiting to ambush us in the months ahead" (*San Francisco Examiner*, 10 February 1984).

As a result of this broad (and legitimate) perspective, communications from risk managers pertaining to the risk and control of a single hazard, no matter how carefully presented, may fail to alleviate people's fears, frustrations, and anger. If people trust the ability of the risk manager to handle the broader risk problems, these general concerns will probably not surface.

Whereas the psychometric research implies that risk debates are not merely about risk statistics, the sociological and anthropological work implies that some of these debates may not even be about risk. Risk may be a rationale for actions taken on other grounds or it may be a surrogate for social or ideological concerns. When this is the case, communication about risk is simply irrelevant to the discussion. Hidden agendas need to be brought to the surface for open discussion, if possible (Edwards and von Winterfeldt, 1984).

Perhaps the most important message from the research done to date, is that there is wisdom as well as error in public attitudes and perceptions. Lay people sometimes lack certain basic information about hazards. However, their basic conceptualization of risk is much richer than that of the experts and reflects legitimate concerns that are typically omitted from expert risk assessments. As a result, risk communication efforts are destined to fail unless they are structured as a two-way process (Renn, 1984). Each side, expert and public, has something valid to contribute. Each side must respect the insights and intelligence of the other.

10.6. Role of News Media in Informing People About Risk

10.6.1. Critics of the media

The mass media exert a powerful influence on people's perceptions of the world, the world of risk being no exception. Each morning's paper and each evening's TV newscast seems to include a report on some new danger to food, water, air, or physical safety. It is not surprising, given the actual and perceived influence of the media and the stakes involved in risk issues, that media coverage of risk has been subjected to intense scrutiny and harsh criticism. Content analysis of media reporting for specific hazards (DNA research, nuclear power, cancer) and the domain of hazards in general (e.g. diseases, causes of death) has documented a great deal of misinformation and distortion (Burger, 1984; Freimuth *et al.*, 1984; Combs and Slovic, 1979; Kristiansen, 1983); this distortion has caused critics such as Cirino (1971) to assert: "No one can be free from the effects of bias that exist in the mass media...

Decisions based on distorted views of the world resulting from [such] ... bias have resulted in tragically mistaken priorities, death and suffering" (p 31).

More than a few observers have blamed the media for what they see as public overreaction to risk. Among the most vehement is physicist Bernard Cohen who argued that:

> Journalists have grossly misinformed the American public about the dangers of radiation and of nuclear power with their highly unbalanced treatments and their incorrect or misleading interpretations of scientific information. This misinformation is costing our nation thousands of unnecessary deaths and wasting billions of dollars each year. (Cohen, 1983, p 73.)

10.6.2. In defense of the media

A balanced examination of media performance needs to consider the difficulties faced by the media in reporting risk stories. Journalists operate under many constraints, including tight deadlines, the pressure of competition to be first with a story, and limitations on space or time (for TV reports). But the major difficulty stems from the inherent complexity of risk stories as outlined in Section 10.2. Because of the technical complexity of the subject matter, journalists must depend on expert sources. But a risk story may involve such diverse problems that the journalist might need to interview specialists in toxicology, epidemiology, economics, hydrology, meteorology, emergency evacuation, etc., not to mention a wide variety of local, state, and federal officials. Even then, there is no assurance of completeness. No one may know what all the pieces are or recognize the limits of their own understanding (Fischhoff, 1985a). Few journalists have the scientific background to sort through and make sense of the welter of complex and often contradictory material that results from such a search.

10.6.3. Improving media performance

Despite the difficulties, there seem to be a number of actions that might help the media to improve their performance in communicating risk information. Some of these actions are professional, others involve research. At the professional level, the following steps may be useful.

Acknowledge the problem

The first step in addressing any deficiency is to recognize it as an important problem. We now know that an understanding of risk is central to decisions that are of great consequence to individuals and to society, that risk and uncertainty are inherently difficult to communicate, and that the media are a dominant source of risk information. The combination of these factors highlights the role of the media as a problem worthy of explicit, sustained attention, in high level meetings between journalists, scientists, and risk managers.

Enhance science writing

Reporters obviously need to be educated in the importance and subtleties of risk stories. Fischhoff (1985a) suggests a number of checklists and protocols that a reporter might use as a guide to understanding and clarifying risk issues. One of these, titled "Questions to Ask of Risk Analysis", is shown in *Table 10.5*. There should be scholarships to induce students and young journalists to pursue science writing as a profession, accompanied by awards and prizes to recognize and reward good science journalism when it occurs.

Table 10.5. Questions to ask of risk analyses.

Reporters should consider the following questions whenever a risk analysis is produced for use in policy decisions:

(1) Does the risk analysis state the *probability* of the potential harm as well as the amount of harm expected?
(2) Does the risk analysis disclose forthrightly the points at which it is based on assumptions and guesswork?
(3) Are various risk factors allowed to assume a variety of values depending on uncertainties in the data and/or various interpretations of the data?
(4) Does the risk analysis multiply its probabilities by the number of people exposed to produce the number of people predicted to suffer damage?
(5) Does the risk analysis disclose the confidence limits for its projections and the method of arriving at those confidence limits?
(6) Are considerations of individual sensitivities, exposure to multiple hazards, and cumulative effects included in the risk analysis?
(7) Are all data and processes of the risk analysis open to public scrutiny?
(8) Are questions of (i) involuntary exposure, (ii) who bears the risks and who reaps the benefits, and (iii) alternatives to the hazardous activity considered in the risk analysis?
(9) Are the processes of risk analysis and risk policy separate?

If the answer to any of these questions is "no", then the use of that risk analysis should be questioned.

Source: Adapted from Fischhoff (1985a).

Develop science news clearinghouses

Science journalists need access to knowledgeable and cooperative scientists. A few organizations, such as the Scientists' Institute for Public Information, have performed an important service along this line, and some professional societies, such as the American Psychological Association, maintain offices that provide journalists with the names of scientists knowledgeable about specific topics. More needs to be done to help journalists obtain reliable information about risk topics.

10.7. Research Directions

Although much progress has been made toward understanding risk attitudes, perceptions, and behaviors, we still lack definitive understanding of many important issues relevant to risk communication. Some recommended research directions are described in this section.

10.7.1. Informed consent

The right of citizens, patients, and workers to be informed about the hazards to which they are exposed from their daily activities, their medical treatments, and their jobs provides the motivation behind much of the efforts to communicate information about risks. Within the context of any information program, research is needed to determine what people know and what they want to know about the risks they face and how best to convey that information. Moreover, there is need for a deeper understanding of the concept of consent (MacLean, 1982) as well as for a theory of informed consent that sets out criteria for evaluating the adequacy of information presentations. Fischhoff (1983, 1985b) has made a start in the latter direction by characterizing the problem of informed consent as a decision problem. In this view, the goal of informed consent is to enable the individual to make decisions that are in his or her best interests. Fischhoff points out that there are both cognitive and institutional barriers to achieving informed consent. Research is needed to understand these barriers and overcome them.

To facilitate the process of informed consent, we need better ways to convey quantitative risk information. There is widespread agreement that casting individual risks in terms such as 10^{-x} per year is not helpful to people. We need creative new indices and analogies to help individuals translate risk estimates varying over many orders of magnitude into simple, intuitively meaningful terms. The task will not be easy. Ideas that appear, at first glance, to be useful, often turn out, upon testing, to make the problem worse. For example, an attempt to convey the smallness of one part of toxic substance per billion by drawing an analogy with a crouton in a five ton salad seems likely to enhance one's misperception of the contamination by making it more easily imaginable. The proposal to express very low probabilities in terms of the conjunction of two or more unlikely events (e.g. simultaneously being hit by lightning and struck by a meteorite) also seems unwise in light of experimental data showing that people greatly overestimate the likelihood of conjunctive events. Perhaps we can learn, by studying people's understanding of commonly used measures such as distance, time and speed, whether and how their understanding of quantitative risk can be improved.

The sensitivity of risk communications to framing effects points to another avenue for research. We need a better understanding of the magnitude and generality of these effects. Are people's perceptions really as malleable as early results suggest? If so, how should the communicator cope with this problem? One suggestion is to present information in multiple formats – but does this help or confuse the recipient? Finally, the possibility that there is no neutral way to present information, coupled with the possibility that people's preferences are very easily manipulated, has important ethical and political implications that need to be examined.

Because of the complexity of risk communications and the subtlety of human response to them, it is extremely difficult, a priori, to know whether a particular message will adequately inform its recipients. Testing of the message provides needed insight into its impacts. In the light of the known difficulties of communicating risk information, it could be argued that an informer who puts forth a message without testing its comprehensibility is guilty of negligence. This assertion raises a host of research questions. How does one test a message? How does the communicator judge when a message is good enough in the light of the possibility that not all test subjects will interpret it correctly? Can testing be used against the communicator by providing evidence that not everyone understood the message?

Risk is brewed from an equal dose of two ingredients – probabilities and consequences. But most of the attention pertaining to informed consent seems to focus on the probabilities. It is assumed that once the potential consequence is named – lung cancer, leukemia, pneumoconiosis – one need say little else about it. We believe that neglecting to educate people about consequences is a serious shortcoming in risk information programs. For example, an adequate discussion of risk cannot assume that people have good knowledge of what it is like to experience a certain disease – the pains, the discomforts, the treatments and their effects, etc. This sort of information might best come from those who are the victims of such diseases. Research is needed to determine how best to deepen perspectives about the novel, unfamiliar consequences associated with the outcomes of illnesses, accidents, and their treatments.

10.7.2. Information relevance

What lessons do people draw about their own vulnerability to a hazard on the basis of risk information? For example:

(1) What do residents living near the Union Carbide pesticide plant at Institute, West Virginia, infer about their personal risk as a result of the Bhopal accident?
(2) What does a heterosexual individual infer about personal vulnerability to AIDS from statistics based on homosexuals?
(3) What does a resident of the West Coast infer about his or her risk from cancer due to polluted groundwater upon receiving risk estimates for residents of the East Coast?

Obviously, the personal message one draws from risk information will depend upon the perceived relevance of that message – but the determinants of relevance are by no means understood. There are always differences between the time and place and population (or species) from which risk information is derived and the time, place, and population with which the recipient identifies. When are these differences magnified into barriers justifying denial of relevance ("those statistics don't really pertain to me") and when are the barriers made permeable and the message assimilated? Such questions are fundamental to the process of risk communication, yet we know virtually nothing about them.

10.7.3. Cognitive representations of perceived risk

People's cognitive representations of risk dictate the sorts of information they will find necessary for participating in risk management decisions. Thus, if we believe in the two-factor representation shown in *Figure 10.2*, we will need to provide people with information about how well a hazard is known to science, the extent of its catastrophic potential, etc. If people examine accident reports for their signal value, then methods are needed to assess this factor and communications techniques are needed to express it meaningfully.

However, we still lack a full understanding of the ways in which people characterize risk. There is evidence that the representation arising out of factor analytic studies is not unique. Johnson and

Tversky (in press), for example, have shown that applications of multidimensional scaling or tree analysis to judgments of similarity between risks produce very different representations from those obtained by factor analysis. Research is needed to provide a clearer picture of the multiple ways to represent perceptions and the variations of these representations across different individuals and groups (Harding and Eiser, 1984; Kuyper and Vlek, 1984; Kraus, 1985).

The multivariate characterizations that have emerged from psychometric studies demonstrate that there are many things to be considered when thinking about risk and many (possibly incommensurable) factors to bear in mind when assessing the riskiness of different hazards. The need for some convenient general summary measure of risk seems apparent. Reliance on multiattribute utility theory to construct such an index (Fischhoff *et al.*, 1984) provides one approach, but research is needed to determine if people can provide the explicit judgments needed to create such an index. Given an index, can people absorb the information it summarizes in a way that is meaningful and will they make or accept decisions based on it? Would they feel more comfortable being shown, in matrix or vector form, the component information it summarizes?

10.7.4. Risk and the media

We need a theoretical framework to understand and improve the media's role in communicating risk. Some theorists, such as Gans (1980) have proposed that one major role of journalism is to report events that threaten or violate important values – such as preserving a stable social order. In this light, things that "go awry", and thereby threaten natural, technological, social or moral disorder, become prime news topics. The relation between hazard characteristics and news coverage should be examined to discern more precisely how the media interpret their responsibility to warn society.

One possibility is that coverage of risk incidents is systematically related to threat potential or signal value. If so, such coverage (as measured by frequency, size, and prominence of reports) should be related to the same risk factors that predict other risk perceptions and attitudes. Thus, incidents involving hazards perceived as unknown and dread (potentially catastrophic) would be expected to receive much greater coverage than incidents involving hazards with other characteristics. Data reported by Kristiansen (1983) provide some support for these notions. Her study of seven British daily newspapers found that threats with high signal value, such as infectious diseases, food poisoning, and rabies, were disproportionately reported relative to their frequency of occurrence.

Content analyses of media reports need to be supplemented by more controlled studies. An intriguing example of a controlled study was done by Johnson and Tversky (1983) who asked subjects to judge the perceived frequency of death from various causes after reading a single newspaper-style story about a tragic incident involving the death of a young man. The cause of death was either leukemia, homicide or fire, depending on the story. They expected to find that a story would increase perceived frequency most for the specific hazard involved in the story, with somewhat smaller increases for similar hazards. Instead, the results indicated large increases in perceived frequencies for all hazards, with size of increase being unrelated to similarity. They hypothesized that the stories aroused negative effect which had a general influence on perception. This hypothesis is an important one, in need of further study, because it implies that media coverage might influence our perceptions of threat in subtle and pervasive ways.

Other topics that could be studied by means of controlled news simulations are the reporting (or deletion) of uncertainties in risk estimates and the treatment given to expert disagreements. How, for example, would journalists report a story in which 20 experts argued one way and one argued another? Would it matter if the ratio were higher or lower or if the dissenter had more or less prestigious credentials? Would experienced journalists or their editors treat the story differently from inexperienced reporters? Would the type of medium (TV, radio, print) make a difference? In sum, studies like these could point out biases or inadequacies in reporting about which journalists need to be informed.

10.8. Conclusions

Some observers, cognizant of the communication difficulties described above, have concluded that they are insurmountable. This seems an unreasonably pessimistic view. Upon closer examination, it appears that people understand some things quite well, although their path to knowledge may be quite different from that of the technical experts. In situations where misunderstanding is rampant, people's errors can often be traced to biased experiences, which education may be able to counter. In some cases, people's strong fears and resistance to experts' reassurances can be traced to their sensitivity to the potential for catastrophic accidents, to their perception of expert disagreement about the probability and magnitude of such accidents, to their knowledge of serious mistakes made by experts in the past, and to their sensitivity to many qualitative concerns not included in technical risk analyses. Even here, given an atmosphere of trust in which both experts and lay persons recognize that each group has something to contribute to the discussion, exchange of information and deepening of perspectives may well be possible.

Notes

[1] The text of this chapter draws heavily on the author's joint work with his colleagues, Baruch Fishhoff and Sarah Lichtenstein. Support for the writing of this chapter was provided by the National Science Foundation under contract No. PRA-8419168 to the University of Southern California.

[2] We make no attempt to defend the validity of the statistics presented in Section 10.4.2. We take them directly from published studies. In Section 10.4.1 we printed out the problems that one must be aware of when using and interpreting risk data.

[3] The concept of accidents as signals was eloquently expressed in an editorial addressing the tragic accident at Bhopal, India:

> What truly grips us in these accounts [of disaster] is not so much the numbers as the spectacle of suddenly vanishing competence, of men utterly routed by technology, of fail-safe systems failing with a logic as inexorable as it was once – indeed, right up until that very moment – unforeseeable. And the spectacle haunts us because it seems to carry allegorical import, like the whispery omen of a hovering future. (*The New Yorker*, 18 February 1985.)

References

Alfidi, R.J. (1971), Informed consent: A study of patient reaction, *Journal of American Medical Association*, **216**, 1325.

Brown, R.A. and Green, C.H. (1980), Precepts of safety assessments, *Journal of the Operational Research Society*, **11**, 563–571.

Burger, E.J., Jr. (1984), *Health Risks: The Challenge of Informing the Public* (The Media Institute, Washington, DC).

Burton, I., Kates, R.W., and White, G.F. (1978), *The Environment as Hazard* (Oxford University Press, New York, NY).

Cirino, R. (1971), *Don't Blame the People: How the News Media use Bias, Distortion and Censorship to manipulate Public Opinion* (Random House, New York, NY).

Cohen, B. (1983). Lies, damned lies, and news reports, *Policy Review*, February, 70–74.

Cohen, B., and Lee, I. (1979), A catalog of risks, *Health Physics*, **36**, 707–722.

Combs, B. and Slovic, P. (1979), Newspaper coverage of causes of death, *Journalism Quarterly*, **56** (4), 837–843.

Crouch, E.A.C. and Wilson, R. (1982), *Risk/benefit analysis* (Ballinger, Cambridge, MA).

Douglas, M. and Wildavsky, A. (1982), ***Risk and Culture*** (University of California Press, Berkeley and Los Angeles, CA).

Edwards, W. and von Winterfeldt, D. (1984), ***Public values in risk communication***, Unpublished manuscript (Institute for Systems and Safety Management, University of Southern California, Los Angeles, CA).

Fischhoff, B. (1983), Informed consent for transient nuclear workers, in R. Kasperson and R.W. Kates (eds), ***Equity Issues in Radioactive Waste Disposal*** (Oelgeschlager, Gunn, & Hain, Cambridge, MA).

Fischhoff, B. (1985a), Environmental reporting: What to ask the experts, ***The Journalist***, Winter, 11–15.

Fischhoff, B. (1985b), Cognitive and institutional barriers to "informed consent", in M. Gibson (ed), ***Risk, Consent, and Air*** (Rowman & Allenheld, Totowa, NJ).

Fischhoff, B., Lichtenstein, S., Slovic, P., Derby, S.L., and Keeney, R.L. (1981), ***Acceptable Risk*** (Cambridge University Press, New York, NY).

Fischhoff, B., Watson, S., and Hope, C. (1984), Defining risk, ***Policy Sciences***, **17**, 123–139.

Freimuth, V.S., Greenberg, R.H., De Witt, J., and Romano, R. (1984), Covering cancer: Newspapers and the public interest, ***Journal of Communications***, **34** (1), 62–73.

Gans, H.J. (1980), ***Deciding What's News*** (Vintage, New York, NY).

Gardner, G.T., Tiemann, A.R., Gould, L.C., DeLuca, D.R., Doob, L.W., and Stolwijk, J.A.J. (1982), Risk and benefit perceptions, acceptability judgments, and self-reported actions toward nuclear power, ***Journal of Social Psychology***, **116**, 179–197.

Gori, G.B. (1980), The regulation of carcinogenic hazards, ***Science***, **208**, 256–261.

Green, C.H. (1980), Risk: Attitudes and beliefs, in D.V. Canter (ed), ***Behaviour in Fires*** (Wiley, Chichester).

Green, C.H. and Brown, R.A. (1980), ***Through a Glass Darkly: Perceiving Perceived Risks to Health and Safety***, Research paper (School of Architecture, Duncan of Jordanstone College of Art/University of Dundee, Scotland).

Harding, C.M. and Eiser, J.R. (1984), Characterizing the perceived risk of some health issues, ***Risk Analysis***, **4**, 131–141.

Johnson, E.J. and Tversky, A. (1983), Affect, generalization, and the perception of risk, ***Journal of Personality and Social Psychology***, **45**, 20–31.

Johnson, E.J. and Tversky, A. (in press), Representations of perceptions of risks, ***Journal of Experimental Psychology: General***.

Kraus, N.N. (1985), ***Taxonomic Analysis of Perceived Risk: Modeling the Perceptions of Individuals and Representing Local Hazard Sets***, Doctoral dissertation (University of Pittsburgh, Pittsburgh, PA).

Kristiansen, C.M. (1983), Newspaper coverage of diseases and actual mortality statistics, ***European Journal of Social Psychology***, **13**, 193–194.

Kuyper, H. and Vlek, C. (1984), Contrasting risk judgments among interest groups, ***Acta Psychologica***, **56**, 205–218.

Lichtenstein, S., Slovic, P., Fischhoff, B., Layman, M., and Combs, B. (1978), Judged frequency of lethal events, ***Journal of Experimental Psychology: Human Learning and Memory***, **4**, 551–578.

Lindell, M.K. and Earle, T.C. (1983), How close is close enough: Public perceptions of the risks of industrial facilities, ***Risk Analysis***, **3**, 245–254.

Macgill, S.M. (1983), Exploring the similarities of different risks, *Environment and Planning B: Planning and Design*, **10**, 303–329.

MacLean, D. (1982), Risk and consent: Philosophical issues for centralized decisions, *Risk Analysis*, **2**, 59–67.

Mazur, A. (1984), *Perception and Communication of Risk: Sociological Issues*, Paper presented at a meeting entitled Risk Perception and Risk Communication, Long Beach, CA.

McNeil, B.J., Pauker, S.G., Sox, H.C., Jr., and Tversky, A. (1982), On the elicitation of preferences for alternative therapies, *The New England Journal of Medicine*, **306**, 1259–1262.

Moatti, J.P., Stemmelen, E., and Fagnani, F. (1984), *Risk Perception, Social Conflicts and Acceptability of Technologies (An Overview of French Studies)*, Paper presented at the Annual Meeting of the Society for Risk Analysis, Knoxville, TN, October.

Morgan, M.G., Slovic, P., Nair, I., Geisler, D., MacGregor, D., Fischhoff, B., Lincoln, D., and Florig, K. (1985), Powerline frequency and magnetic fields: A pilot study of risk perception, *Risk Analysis*, **5**, 139-149.

Nisbett, R. and Ross, L. (1980), *Human Inference: Strategies and Shortcomings of Social Judgment* (Prentice-Hall, Englewood Cliffs, NJ).

Renn, O. (1981), *Man, Technology, and Risk: A Study on Intuitive Risk Assessment and Attitudes towards Nuclear Power*, Report Jul-Spez 115, Jülich, June (Nuclear Research Center, FRG).

Renn, O. (1984), *Premises of Risk Communication: Results from the West-German Planning Cell Experiments*, Unpublished manuscript.

Rosenburg, J. (1978), A question of ethics: The DNA controversy, *American Educator*, **2** (27).

Rothschild, N. (1979), *Coming to Grips with Risk*, Address presented on BBC television, November 1978; reprinted in *The Wall Street Journal*, 13 May 1979.

Ruckelshaus, W.D. (1983), Science, risk, and public policy, *Science*, **221**, 1026–1028.

Ruckelshaus, W.D. (1984), Risk in a free society, *Risk Analysis*, **4**, 157–162.

Sharlin, H.I. (1986), EDB: A case study in communicating health risk, *Risk Analysis*, **6**, 61–68.

Slovic, P. (1982), Toward understanding and improving decisions, in W.C. Howell and E.A. Fleishman (eds), *Human Performance and Productivity: Vol. 2, Information Processing and Decision Making* (Erlbaum, Hillsdale, NJ).

Slovic, P. and Lichtenstein, S. (1983), Preference reversals: A broader perspective, *American Economic Review*, **73**, 596–605.

Slovic, P., Fischhoff, B., and Lichtenstein, S. (1980a), Informing people about risk, in L. Morris, M. Mazis, and I. Barofsky (eds), *Product Labeling and Health Risks*, Banbury Report 6 (The Banbury Center, Cold Spring Harbor, NY).

Slovic, P., Fischhoff, B., and Lichtenstein, S. (1980b), Facts and fears: Understanding perceived risk, in R. Schwing and W.A. Albers Jr. (eds), *Societal Risk Assessment: How Safe is Safe Enough?* (Plenum, New York, NY).

Slovic, P., Fischhoff, B., and Lichtenstein, S. (1982), Response mode, framing, and information processing effects in risk assessment, in R.M.

Hogarth (ed), *New Directions for Methodology of Social and Behavioral Science: The Framing of Questions and the Consistency of Response* (Jossey-Bass, San Francisco, CA).
Slovic, P., Lichtenstein, S., and Fischhoff, B. (1984), Modeling the societal impact of fatal accidents, *Management Science*, **30**, 464–474.
Slovic, P., Fischhoff, B., and Lichtenstein, S. (1985), Characterizing perceived risk, in R.W. Kates, C. Hohenemser, and J.X. Kasperson (eds), *Perilous Progress: Managing the Hazards of Technology*, pp 91–125 (Westview, Boulder, CO).
Sowby, F.D. (1965), Radiation and other risks, *Health Physics*, **11**, 879–887.
Tversky, A. and Kahneman, D. (1981), The framing of decisions and the psychology of choice, *Science*, **211**, 453–458.
Vlek, C.A.J. and Stallen, P.J. (1981), Judging risk and benefits in the small and in the large, *Organizational Behavior and Human Performance*, **28**, 235–271.
von Winterfeldt, D., John, R.S., and Borcherding, K. (1981), Cognitive components of risk ratings, *Risk Analysis*, **1**, 277–288.
Weinstein, N.D. (1979), Seeking reassuring or threatening information about environmental cancer, *Journal of Behavioral Medicine*, **2**, 125–139.
Wilson, R. (1979), Analyzing the daily risks of life, *Technology Review*, **81** (4), 40–46.

10.D1. Discussion

T.R. Lee

In discussing the chapter by Slovic, I begin by drawing attention to its *stance* – an orientation so pervasive and so thoroughly compatible with the scientific culture that it is very easily overlooked.

In spite of his even-handedness and characteristic modesty in discussing what he refers to as the "wisdom and the folly" of the public, both his explicit title and his implicit stance suggest that someone, or some institution with expertise, should "inform and educate" the public. With a little diligent searching, one can find acknowledgment that each side, expert and public, has "something valid to contribute" and "must respect the insights and intelligence of the other". But these are not up front. In this position, we find reference to the "idiosyncrasies" of the human mind, implying the existence elsewhere of a different kind of mind, one which is not prey to these peculiar irrationalities – like, for example, the mind of the scientist. We find a one-way communication of scientific risk information from scientist to public.

Risk assessment, though properly described as an approximate science, is presented as if the "true risk" actually has an existence out there, where its appearance as reality is veiled only by the limitations of our present methods of analysis. If the risk could only be quantified more accurately, then one major facet of our problem would, it seems, be resolved. The other, of course, is how to educate and inform the public about what we have ascertained.

My first thesis, then, is that this superordinate stance, from which even social and behavioral scientists find it hard to deliver themselves, is at the root of the many social misunderstandings and confrontations over hazards. I plead personally guilty in full measure but it is easier to see the mote in my brother's eye. There is every indication that the physical and biological scientists and engineers who play the leading role in risk assessment and management also have their vision occluded by some pretty sizable beams.

Communication about hazards, I would suggest, is intractable for two reasons: firstly, because while modestly acknowledging that as scientists we can only make *estimates* of reality, we believe that this reality actually exists; secondly, because we believe that our task is to communicate this reality from the scientist to the public.

I repeat that there are signs that Slovic is not wholly fettered by this way of thinking, for he refers approvingly to the "other paradigms" that are beginning to emerge from sociology and anthropology. These emphasize that our attitudes towards hazards may be even further from "reality" than the psychometrician would suppose. It is not merely that people's perception of the "true risk" is biased by faulty estimates or imperfect heuristics – instead, perceptions of risk are governed by and transmitted through the total culture in which we are embedded. They are subject to selective exposure and then to group pressures to conform to risk norms; they are dependent on the view of reality we are obliged to adopt as part of our roles within society.

Moreover, I would submit that we do not actually need to turn to these "other paradigms" to confirm such an approach, for it can be found within the later developments of the psychometric approach, of which Slovic and his group are preeminent exponents.

I stress "later developments" because the earliest scalings of perceived risks were made on the assumption of unidimensionality. They were based on a strong principle propagated by the probabilistic risk assessors, i.e. that riskiness is a unidimensional function of *severity*, measurable mainly by mortality per unit of exposure. Thereafter, however, perceived risks began to be "unpacked" by investigators to show that each one had many attributes and that some of these attributes (e.g. perceived voluntariness, catastrophic potential) were correlated with the severity dimension while others (e.g. familiarity, delayed versus immediate effect, etc.) were relatively independent. The fact that principal components analysis can be used to discern the underlying vectors of our cognitive structures should not blind us to the fact that the ordinary antagonist or protagonist has a cognitive profile composed of a particular set of these attributes.

Some of the attributes already mentioned have a flavor of social if not cultural determination, but they are based on comparisons between many risks and they are restricted to the potentially adverse qualities. What about the beneficial aspects of the hazardous objects or activities? These most assuredly have some influence on people's evaluations of hazards and the behavioral decisions they make, even if it is possible to abstract comparative severity perceptions as a cognitive phenomenon.

This aspect becomes strongly evident if we move from comparative scaling to place a single hazard under the microscope, to examine the mental representation which is organized around such concepts as nuclear power or pesticide spraying.

People are found to harbor much more than an anticipated likelihood that things could go *wrong*, with associated degrees of anxiety; they anticipate also that things could go *right* and have ideas about these more benign consequences. It must be acknowledged that Slovic recognizes that communicators should take benefits into account, but we must surely go further to say firmly that researchers must not stop at providing the public's counterpoint to scientific risk assessment and assuming that the debate hinges on the gap between them. They need at least to elicit the attitude profiles of the main groups of actors, the pros, antis, and uncommitted. Only then can an "attitude change" strategy be appropriately designed and implemented.

10.D1.1. Attitudes to hazards

An attitude may be defined as a constellation of knowledge, beliefs, feelings, and potential actions in relation to some aspect of the environment. It is only in respect of nuclear power that much progress has been made in decomposing attitudes into these elements with the aim of improving understanding and deciding effective communications.

The joint International Institute for Applied Systems Analysis/ International Atomic Energy Agency (IIASA–IAEA) Group, under the leadership of Harry Otway, blazed the trail. They adopted the attitude model constructed by Fishbein and Ajzen (1975), who aver that the main constituent elements of an attitude are *beliefs* of differing degrees of salience, which combine in characteristic packages. The most detailed study to emerge from IIASA was an application of this approach to a comparison of a small range of energy systems (Thomas *et al.*, 1980) and this is reproduced in *Figure 10.D1.1*.

My own group (Lee *et al.*, 1983, 1984) has attempted to extend this approach by including perceived risks and benefits as well as beliefs, and by adding people's factual knowledge and their predisposition to act in various ways (e.g. whether they know the main fuel used in nuclear reactors and whether they would be prepared to sign petitions or join marches).

We found that our samples' beliefs were more likely to discriminate between pros and antis than their perceived risks and benefits, that knowledge levels hardly differed, and that the pros strongly endorsed the more passive forms of involvement in the nuclear debate, while the antis would, of course, go much further.

A smallest space analysis using the Guttman–Lingoes multiple discriminant analysis (MDA) program is illustrated in *Figure 10.D1.2*. This computes a correlation matrix for all the items in the analysis and then seeks an optimal resolution for the distribution of these items in a two-dimensional space so that similarity (correlation) is represented by spatial proximity. Two items close together are likely to be endorsed in the same way by the respondents, while those placed wide apart bear no relation to each other.

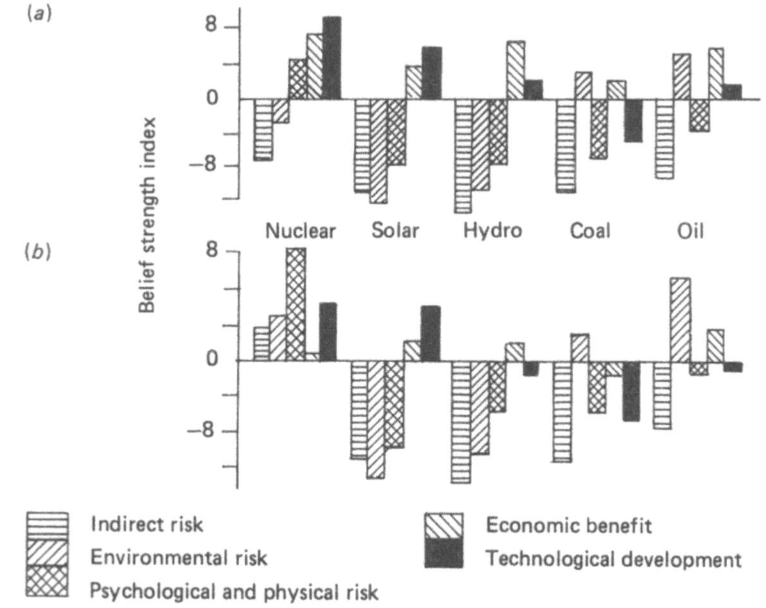

Figure 10.D1.1. Beliefs about five energy systems held by those pro (*a*) and con (*b*) the use of nuclear energy. Subgroups of Austrian public sample: pro nuclear group, $n = 48$; con nuclear group, $n = 47$. (*Source*: Thomas *et al.*, 1980.)

10.D1.2. Can attitudes to hazards be changed?

These and other studies have led us to several conclusions which, although they apply to the limited area of nuclear power, can probably be generalized. They suggest that although it is necessary and commendable to provide information which places technological risks in comparative order, we could go further than this.

For example, for both the establishment and the opposition communicators, it is the "uncommitted" group that is most amenable to change. Slovic points out that established attitudes are extremely intractable; the reason is that they do not live (I use the verb intentionally to convey a dynamic existence) in isolation, but in a complex of related attitudes with which they must retain some kind of harmony. Brenot *et al.* (1984) have shown that for the French people, support for nuclear power is part of a right-wing "sacred mission of France" cluster of attitudes, against strikes and pornography, and in favor of God and of censorship, while the opposition is located at the *profane* end of this dimension where people are also in favor of abortion and the abolition of marriage, and against military service and the death penalty, etc.

The whole complex of such attitudes, as mentioned earlier, depends for its existence on continuous sustenance from the community of people who share the same set of attitudes. Change means estrangement at best and betrayal at worst.

Another argument that underlies the whole approach to "public relations" in the nuclear industry is that if only people understood the technology they would accept it. Hence, many information pamphlets and films give clear, accurate, and often multicolored accounts of particle physics or the main engineering alternatives for nuclear power stations. But, although we found the public's factual knowledge of the technology to be sparse indeed, with much room for education, it did not differ between the pros and the antis, and it seems likely that a knowledge of particle physics can be used to support either a pro or an anti posture.

The differences between the three positions on the perceived risks and benefits associated with nuclear power was also less than might be expected. The pros are not indifferent to the risks, by any means. As others have also found, it is their expectation of economic benefit that mainly distinguishes them. These perceived risks and benefits are not included in the analysis shown in *Figure 10.D1.2* but they overlap with the sets of beliefs which are shown there.

For these beliefs, we asked our respondents to endorse one from each of a number of pairs of opposing belief statements. It will be seen that these were highly discriminating. These beliefs were about the likelihood of more jobs, the potential of "alternative sources" of

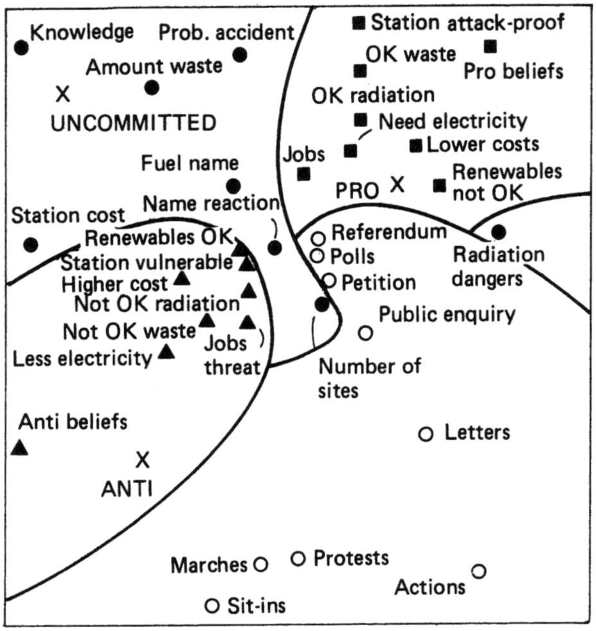

Figure 10.D1.2. Guttman–Lingoes smallest space analysis of knowledge, beliefs, action disposition, and attitudes toward nuclear power. (*Source*: Lee *et al.*, 1984.)

energy, and the likelihood of safely managing terrorism and radioactive waste. They may be called "nonfactual" or "nonverifiable" beliefs, though it has to be pointed out that beliefs are, in a sense, nonverifiable by definition. Anyway, it is these beliefs that distinguish people and it is about these rather than numerical probabilities of mortality that people wish to receive communications. In making this point, I am underlining rather than supplementing Slovic's argument, for he points out that, although Cohen and Lee (1979) advocate that clear and comprehensible communication tables should be the basis for risk decisions, he himself accepts this only as a beginning and asserts that the communication should include a detailed audit of costs and benefits. Sadly, if the costs in risk terms are hard to quantify, the benefits are even more elusive. It has to be reiterated that benefits are also multidimensional, personal, and intangible.

Unfortunately, matters which may be classified as moral questions about the future of society, or even as political issues, are not ones with which scientists feel comfortable. Even for political administrators, there are sensitive thresholds between the advocacy of policy, mass media persuasion, and propaganda – dreaded word! These intangibles are by no means restricted to nuclear power: they apply just as forcefully to major developments in pharmaceuticals, food additives, chemical plants, and LNG containers. They often have to do with the kind of society that people envisage for their children.

One further point about the "unpacking" of attitudes before we leave this theme. Very little progress has been made in identifying the emotional or "feeling" aspects of attitudes to hazards. It is acknowledged that some people may experience anxiety or fear from a raised level of autonomic nervous system activity, but such effects are extremely difficult to measure in their milder forms, either with physiological transducers or by means of "state anxiety" questionnaires. It can certainly be speculated from psychotherapy that communications *can* change cognitions and hence emotions, but no evidence of this exists so far in the risk field, although anxiety is postulated as an important element. It seems likely from our own researches on nuclear power that the large majority of people have attitudes which are almost exclusively cognitive and motivational, concerned with the moral, "future of civilization" aspects of the technology and with the means of expressing views on these issues. This does not exclude the possibility, of which our only evidence at present is derived from the qualitative observation of active protesters, that some people experience a sense of personal danger and associated anxiety.

What is clear is that one of the areas where this is most likely to occur is in respect of the central theme of this Conference, i.e. in relation to the storage of hazardous materials. In a national survey in the United Kingdom (UK) which we have recently completed, it was found that no fewer than 50% of our sample perceived it as "possible" that something would "go seriously wrong" within five years. Just under half of these thought it "likely" or "very likely".

10.D1.3. Some reflections on the social psychology of communicating

There is an extensive literature in social psychology on the effects of communication (Lee, 1986). Changes in some attitudes in response to some communications undoubtedly take place and are measurable. This is reflected in Slovic's chapter where, although pointing out that entrenched attitudes are resistant, he refers to newly formed or "naive" attitudes that are "easily manipulated". For example, imaginability, as he puts it, "may blur the distinction between the possible and the probable".

There is a tendency, here and elsewhere, to believe that attitudes towards risks cannot easily be changed positively by the establishment, but that they are highly susceptible to negative change by irresponsible journalists. I would question this, but more on this below.

What is mainly needed, in my submission, is communication research which uses an experimental mode and is specifically in the area of risk. The paper Slovic quotes, by Morgan *et al.*, is an excellent example (Morgan *et al.*, 1985).

Researchers have found it useful to distinguish between the source, the message, and the medium, and to manipulate each of these as independent variables while measuring attitude change as the dependent variable.

The structure of the message is well covered by Slovic, but there are some findings from research that are worthy of additional mention. For example, messages may be presented "logically" or "emotionally". McGuire (1969) usefully distinguished as logical "...those appeals which argue for the truth of a given belief by presenting evidence in favor of the probable truth and premises (antecedents) from which the given belief follows" and as emotional "those appeals arguing for a given belief by pointing out the desirability of consequences that would follow from the given belief". The success of these alternatives cannot be judged in isolation from the issue, but it is worth noting that where they have been compared, they are either equally effective or the emotional message is more effective. Although emotional communications and especially their extreme form, i.e. "fear appeals", may be better suited to activist groups, it is not inconceivable that messages arguing in the opposite direction, i.e. advocating the adoption of technology, could have more emotional flavor than is currently the case, if our earlier analysis is valid.

One message factor where results are unequivocal relates to the use of implicit versus explicit conclusions. It appears that the conclusions of an argument should be made absolutely explicit and that there is no special virtue in allowing the public to arrive at them by the "discovery" route. It is easy to overestimate the grasp that others may have of matters that are strongly evident to ourselves. A risk table, however clear, may not be enough to convey a message about a particular risk, and the inference needs to be brought out and put on the table in simplistic terms, even if this is an uncomfortable process for the scientist who wishes to draw attention to a whole set of reservations and caveats.

Again, there is an option to mention only one or both sides of the argument and, on balance, research comes out in favor of the latter. Of course the message is strongest if the opposing arguments are specifically refuted and, other things being equal, when the positive arguments are presented first in order. This approach will evoke or set up a schema or frame of reference in the receiver which is then consolidated by successive evidence. When the opposition arguments are mentioned, their refutation strengthens rather than weakens this emergent structure.

This generalization may have to be modified in circumstances where there are pro and anti speakers presenting their cases in succession, as in legal advocacy. In such circumstances, there is a special advantage in criticizing the exposed arguments of the previous speaker and this may favor speaking last. The complexities are, of course, great. For example, the balance between presenting one's arguments first or last will depend upon when the audience has to make a decision. There is some advantage in presenting arguments close in time to the decision point.

10.D1.4. The credibility of the communicator

Turning to the communicator him/herself, the credibility of the *source* of a message strongly influences its acceptance. For example, the perceived *expertise* of a communicator is a highly relevant aspect of information dissemination about risks. The recipient has to ask him/herself two questions. Firstly, does the source possess relevant expertise? Secondly, is the expertise being honestly purveyed? The scientist generally receives a high rating from the public on both scores. However, the lay person has a healthy skepticism and is familiar enough with examples of fallibility, bias, or ineptitude to be able to reject evidence, even from this source. If the scientist is seen as wholly independent, then perceived bias is reduced, but otherwise the lay person is well aware that people select themselves for occupations and are then molded to the accepted belief system of their organization and depend on it for much of their reward. Hence, a university scientist has more chance of credibility than one employed by a hazardous industry.

Another important aspect of a communicator's credibility is his/her *attractiveness*. People want to share the same attitudes as those they like and admire, a finding duly exploited by advertisers. This is not to suggest that an industry whose activities are perceived as hazardous should employ TV stars to improve their image but they should certainly consider carefully who are their most charismatic and personable representatives.

Similarity to the target group is another crucial aspect of credibility. If a communicator appears to share the same human needs and goals, then his propositions are more likely to be perceived as "right for us", "right for our sort of people". Conversely, if he is clearly a member of an "out group" of remote, blue-suited, well-off technocrats from the capital city, then it will be assumed that he has "out group" motives and his arguments are less likely to be trusted.

This identification with the target audience can be promoted by employing a style and method of communication which is a dialogue between equals. Those who are equal are more likely to see each other as similar and hence to share the same goals. The method is called *co-orientation*, employing an interactive exchange, a consultative model as distinct from a pedagogical one.

In our South West Study, we asked our respondents which communicators were perceived by them as credible: the results are shown in *Table 10.D1.1*.

Table 10.D1.1. Perceived credibility of providers of information.

Rank order	*Provider of information*[a]	*All* (%)	*In favor* (%)	*Against* (%)	*Uncommitted* (%)
1	UKAEA	45	52	30	37
2	Secretary of State for Energy	40	44	29	37
3	NRPB	31	32	31	26
4	Physicist	27	30	25	15
5	Local MP	21	21	19	30
6	Conservationist	21	19	30	17
7	CEGB PR Department	21	24	14	15
8	Chairman of CEGB	20	24	12	18
9	National anti group	18	15	30	16
10	CEGB engineer	18	22	10	13
11	Local anti group	15	13	22	13
12	Pro nuclear group	14	15	12	13
13	Local GP	9	9	10	13
14	Local Councillor	9	7	12	15
15	Reporter	9	7	16	7

[a]UKAEA, United Kingdom Atomic Energy Authority; NRPB, National Radiation Protection Board; MP, Member of Parliament; CEGB, Central Electricity Generating Board; PR, Public Relations; GP, general practitioner (medical).

10.D1.5. The medium for risk communication

The most effective medium for communications about risk raises questions for which we have few answers at present. It is known that direct face-to-face interaction is relatively highly effective but, of course, extremely limited in scope. The nuclear industry makes much use of pamphlets, but our evidence suggests that this medium is least preferred by the public. Newspapers and TV are most preferred, but we have seen that the newspaper or TV reporter is given very low credibility. However, distinction should be drawn between newspaper information attributed to journalists and that which reports high-credibility scientific sources.

While the vivid and dramatic report often appears to be influential, there is a real dearth of evidence that journalism has more than a very temporary effect on attitudes. People may overvalue certain kinds of hazards and this may be the cause of newspaper overreporting and not vice versa. Also, people buy the newspapers that conform to their existing attitudes; they are therefore rarely exposed to contrary views through this medium and are usually able to deploy good reasons for rejecting them when they occur. The TV is more intrusive, but easily switched to another channel. Local newspapers are bought non-selectively and are correspondingly important, but this applies mainly to local siting issues.

The area where mass media may be most influential is when attitudes are newly formed and in a highly labile state, e.g. in relation to a new hazardous technology such as the irradiation of food (currently not allowed in the UK) or the spraying with a new pesticide of an unexpected infestation.

Also, we know from our South West Study that the uncommitted read certain tabloid newspapers and watch TV channels which at present contain zero communications on environmental hazards; also, that the forms of pamphlets produced to persuade them contain too little information on their particular concerns and are couched in a language and form that is hard for them to understand, even if it were interesting enough to capture their attention. (The respondents' preferred information sources are presented in *Table 10.D1.2*.) People are not passive and labile targets for information campaigns: they are

Table 10.D1.2. Preferred information sources.

Rank order	*Source*	*All (%)*
1	TV program	48
2	Newspapers	36
3	Radio program	29
4	Books	26
5	Government White Papers	24
6	CEGB[a] pamphlet	24
7	Technical journal	22
8	Public meeting	22
9	Exhibition	21
10	Casual discussion	16
11	Antinuclear literature	16
12	Publicity poster	13
13	Film	8
14	Phone-in	7

[a]CEGB, Central Electricity Generating Board.

Table 10.D1.3. Most frequently chosen information units.

Rank order	*Providers*[a]	*Issues*	*Medium*
1	Conservationist	Effects on wildlife	TV
2	Secretary of State	Security of station	TV
3	UKAEA	Security of station	TV
4	NRPB	Health and safety of workforce	TV
5	Secretary of State	Health and safety of workforce	TV
6	UKAEA	Health and safety of workforce	TV
7	Secretary of State	Security of station	Government White Papers
8	Secretary of State	Cost	Government White Papers
9	Secretary of State	Cost	TV
10	UKAEA	Security of station	Government White Papers
11	Secretary of State	Health and safety of workforce	Government White Papers
12	Physicist	Health and safety of workforce	TV
13	UKAEA	Health and safety of workforce	Technical journal
14	UKAEA	Security of power station	Newspapers
15	Conservationist	Effects of wildlife	Public meeting

[a]UKAEA, United Kingdom Atomic Energy Authority; NRPB, National Radiological Protection Board.

active and selective seekers after what concerns them and selective filterers against what does not. Respondents were asked to combine their preferred providers of information, issues of concern, and source of information, i.e. to say what they wanted to hear, from whom, and in what form. These are given in *Table 10.D1.3*. In all, 1470 combinations were possible, but in the event 15 accounted for 53% of choices.

10.D1.6. The need for evaluation

Although there is a useful stock of generalizations available from communication researchers, the permutation of variables of source, message, medium, and target means that there is no substitute for evaluating the effectiveness of particular communications or campaigns. This kind of research can be ex post facto or "formative". We have, for example, explored the impact for five United Kingdom Atomic Energy Authority (UKAEA) pamphlets, using questionnaires and group discussions with ordinary members of the public. Also, the information gain has been measured of a small sample of the public who watched a film entitled *Using Radioactivity*. Another study was conducted of a UKAEA exhibition stand on nuclear power, where comparisons were made between a sample of the public who were approaching the stand, one which was leaving, and a third who were not attracted at all. This research showed that people with more than average concern over nuclear power were inclined to enter the exhibit, but, on leaving, their concern was measurably greater.

Finally, we have evaluated a video tape entitled *Uranium the Magic Metal* at the formative stage. A draft script addressed to 11-year old children had been prepared and the intended film sequences and animations were "mocked up" with hundreds of artists' drawings which were made into a tentative prototype tape with the script dubbed over. This early version was then taken into the schools and shown to children and science teachers, leading to many recommendations for a change in content, terminology and level. For example, the original version was found to be pitched at a level suitable for 15-year olds instead of 11-year olds. This is a common fault in all communications which are designed by sophisticated highly educated science information staff. This video is now available from the UKAEA Information Services Branch.

References

Brenot, J., Fabre, J., Morlat, G., Pages, J.P., and Steminella, E. (1984), *Structure of public opinion and nuclear debate in France*, unpublished paper (Commisariat a l'Energie Atomique, IPSN, Fowtenay-aux-Roses).

Cohen, B. and Lee, I. (1979), A catalog of risks, *Health Physics*, **36**, 707–722.

Fishbein, M. and Ajzen, I. (1975), *Belief, Attitude, Intention and Behaviour: An Introduction to Theory and Research* (Addison-Wesley, Reading, MA).

Lee, T.R. (1986), The effective communication of information about chemical risks, *Science of the Total Environment*, **51**, 149–183.

Lee, T.R., Brown, J., and Henderson, J. (1983), *Psychological Perspectives on Nuclear Power*, Report No. 2, Public Attitude Surveys Towards Nuclear Power Stations on South West England (University of Surrey, Guildford, Surrey, UK).

Lee, T.R., Brown, J., and Henderson, J. (1984), The public's attitudes towards nuclear power in the South West, *Atom*, **336**, 8–11.

McGuire, W.J. (1969), The nature of attitudes and attitude change, in G. Lindsey and E. Anouson (eds), (Addison-Wesley, Reading, MA), Ch 3, 136–314.

Morgan, M.G., Slovic, P., Nair, I., Geisler, D., MacGregor, D., Fischhoff, B., Lincoln, D., and Florig, K. (1985), Power-line frequency and magnetic fields: A pilot study of risk perception, *Risk Analysis*, **5**, 139–149.

Thomas, K., Maurer, D., Fishbein, M., Otway, H.J., Hinkle, R., and Simpson, D. (1980), *A Comparative Study of Public Beliefs about Five Energy*

Systems, Research Report RR-80-15 (International Institute for Applied Systems Analysis, Laxenburg, Austria).

10.D2. Discussion

O. Renn

10.D2.1. Results of and conclusions from risk perception studies

In general people are doing a good job in assessing the magnitude of a risk that is familiar to them. They underestimate high risks and overestimate low risks, but otherwise they are quite aware of the threats and dangers which they are exposed to.

The perceived degree of severity of risks is almost independent of the perceived number of expected losses. The disaster potential (dread) and the degree of uncertainty (individual familiarity and scientific knowledge) are two of the key factors in determining the perceived level of risk. In addition, the perceived justice in the risk–benefit distribution, the potentials for individual and societal control of the risk, and voluntariness of the risky activity influence personal judgment on how society should deal with risks and what regulations are required.

Social psychological and sociological studies show that judgments on risky technologies or activities depend not only on psychological factors like the ones mentioned above but also on reference group judgments, salient beliefs about the risk source, perception of the proponents and opponents of the risk source, degree of loyalty toward official policymakers, and commitment to social values and cultural goals. Since all these factors, including the psychological ones, are interrelated and sometimes reflect mere post-rationalizations of unconscious feelings and social constraints, it is very difficult to set up a reliable model of how people actually perceive risks and evaluate them. What we know – and to what degree we know it – is what matters; but analysts are still searching for a theory that can explain the process of people's judgment on risks.

We certainly do know that people make judgments on technologies, events, or activities, and not on an abstract notion like risk. There are indications that people assess the potential for threat in a similar way and that similar mechanisms of processing and evaluating information about risk are operating under the psychological premises of common sense. But the meaning of risk differs not only among individuals: it differs also among technologies and activities perceived by one and the same individual.

Hence, there is no universal risk threshold which allows distinction between accepted and nonaccepted risks. What kind of risks are acceptable or not differs among people and risk sources considerably and is almost independent of the actual probabilities of being affected. A compound model of psychological, group-related, and value-oriented factors can best predict the individual judgment on the acceptance of technologies or activities.

10.D2.2. Lessons for risk communications

Risk comparisons usually fail to convince anyone except the professional risk analyst. Since risk has a different meaning in different contexts, comparisons make little sense to the public. Only in cases where risk sources have a very similar structure and serve the same purpose should comparisons be used in communication.

Risk communication should not concentrate on conveying probabilities and their meaning to the public. Probabilities will be intuitively learnt by experience (although biased by personal performance) or can be indirectly communicated by describing the safety measures taken to protect individuals and society.

Risk communication should focus on the key aspects of risk perception: on the disaster potential, on the management of uncertainty, on the means of societal control and monitoring, and, of course, on the benefits which are given to the society and/or the individual.

Acceptance of risk sources is highly influenced by the trust in the fairness and rationality of the decision-making procedure and the credibility of the actors involved. Psychological barriers of perception can easily be overcome if those two conditions are met. Unfortunately or fortunately – depending on which side one stands – all tricks and recipes for retaining or gaining trust and credibility usually fail in an open society with a pluralist value structure and a free press. Trust and credibility are dependent on honesty, transparency, competence, and a good past record.

Risk communication can only be effective if the communication process is structured as a two-way information exchange. Regulators can learn something from the public and vice versa. It is also essential that the communication process allows for alterations of the final decision. Nobody is motivated to communicate if he cannot change parts of the issue.

10.D2.3. Lessons for compensation

Since there is no universal risk acceptance threshold and people's judgments vary over the severity of risk posed by a technology or activity, risk perception studies provide no recipes for determining how much compensation is needed to level off the exposure to a specific risk.

Because of the different marginal utilities of income and variations of perceived danger among individuals, compensation on the basis of perceived personal loss will inevitably lead to different compensational sums being handed out to individuals with respect to the same risk taken. Since people demand that risk spreading as well as compensation should be just, variations in payment will not be accepted.

Taking the mean value of all revealed demands for compensation would probably be rejected by those persons whose compensation was lower than originally demanded and would enhance the distrust of those who were overpaid ("if they pay me more than I demand, there has to be something wrong"). Choosing the upper limit of all revealed demands would exceed any financial limits.

Compensation might be a viable and acceptable means of risk management if it were comprised of better or cheaper access to services (e.g. electricity) – disregarding marginal utility – or collective goods (e.g. improvement of the infrastructure). Although people use the community infrastructure in varying degrees, the axiom of perceived justice is met, since anyone has the same chance of using the facility.

Since the exact amount of compensation cannot be calculated on the basis of perceived risk and will run into acceptance problems if the sum is calculated only on the basis of "objective" probabilities, either an open bargaining process with representatives of the community (which will work best if alternative sites are available and compete for this specific risk source) should be initiated or a panel group of selected citizens (in particular ones not directly affected by the risk source to avoid strategic responses) should determine the amount of compensation which they feel just and fair (after being informed about the risks involved).

The above points concerning compensation do not apply if risks can be reduced by compensating measures (e.g. rebuilding houses to increase protection).

CHAPTER 11

Value Tree Analysis: An Introduction and an Application to Offshore Oil Drilling

D. von Winterfeldt [1]

11.1 Introduction

11.1.1. Background and motivation for value tree analysis

One of the most common difficulties when solving complex public policy problems is the existence of multiple interest groups that are affected by the policies and who therefore have a stake in their outcomes. During siting of a hazardous waste facility, for example, nearby residents will express their concerns with pollution and property values; environmentalists are likely to question the impacts of the facility on the flora and fauna; government agencies may raise issues of health and safety; and members of the local chamber of commerce will probably point to the facility's employment and business benefits.

These interest groups or *stakeholders* typically meet in hearings of local governmental agencies, at council meetings, at the negotiation table, and ultimately, in the courts to resolve their differences. The discussions in these *arenas* vary, but they are usually about facts: issues center on whether a particular environmental impact is likely to occur, what types of laws and regulations are applicable to a proposed project, whether dangers and risks have been overestimated or underestimated, and whether someone's rights have been violated.

These factual debates are, however, often merely the surface of more deeply rooted differences in the values of the stakeholders. In the example of siting a hazardous waste facility, value differences appear in a variety of shades: in disputes about the relative importance of providing employment versus protecting the environment; in the "not in my backyard" phenomenon; in concerns with inequities; and in fundamental moral or ethical issues about the value of life.

Unfortunately, while our technical and institutional tools for addressing and resolving factual disputes are adequate, our tools for resolving value conflicts are feeble. On the institutional side, arbitration and mediation have been explored as alternatives to the often costly and unsatisfactory legal procedures. On the technical side, a variety of new models and techniques are being developed to assist stakeholders in understanding each others' positions and in resolving their disputes. In this chapter, I will describe and illustrate one such technique, called "value tree analysis".

11.1.2. A brief history of value tree analysis

Value tree analysis originated in several applications of multiattribute utility techniques (MAUTs) to controversial public policy problems. MAUT, as originally conceived, was designed to aid an individual decision-maker in making a choice among alternatives that vary on several value-relevant dimensions (attributes). The MAUT process usually consists of carefully laying out the decision alternatives, defining objectives and attributes, assessing the impact of each alternative for each objective, and weighting the objectives. In its most common form alternatives are evaluated by a weighted additive model (see, for example, Keeney and Raiffa, 1976; von Winterfeldt and Edwards, 1986).

When decision analysts began to apply MAUT to controversial public policy problems like the siting of nuclear power plants (Keeney, 1980) or the evaluation of desegragation plans for the Los Angeles Unified School District (Edwards, 1980), they soon recognized that the existence of multiple stakeholders required modification of the MAUT process in at least two ways: at the front end, a much more detailed process for structuring the opposing stakeholders' values was needed; and, at the tail end, new procedures for using the MAUT results for conflict analysis and resolution were required.

Independently of each other, several researchers began to experiment with approaches to the "front end" problem. Keeney and Raiffa (1976) – building on Manheim and Hall (1968), Miller (1970), and Raiffa (1969) – developed the useful concept of objectives hierarchies. These are tree structures that logically lay out the ends that a decision-maker would like to achieve when choosing among the decision alternatives (means). At the top of the tree are general value categories like "environmental quality" or "cost". Below, at increasing levels of detail, are the definitions of what these value categories mean in terms of objectives like "reduce SO_2 pollution", "minimize visual impact", or "minimize investment cost". At the bottom are so-called attributes which specify in detail how the upper level objectives and values should be measured.

Edwards (1980) found that objectives hierarchies could be constructed together with several stakeholders, even if they disagreed strongly on the issues. In his study of Los Angeles School Desegregation plans, a very large common objectives hierarchy, which he called a "value tree", was a central part of the evaluation. Saaty (1983) found objectives hierarchies, which he calls "analytic hierarchies", useful in thinking about such conflicts as the Falkland Islands crisis or apartheid politics.

Today, value trees have become a common tool for the initial stage of structuring a conflict among multiple stakeholders. An example of a value tree for evaluating offshore oil development plans is shown in *Figure 11.1*. This tree reflects the concern of fishermen with oil development. After building several individual trees, the analyst combines them into a joint, or common, value tree.

Having built a common value structure, it is possible to construct MAUT models separately for each stakeholder group, and subsequently analyze the conflicts as expressed in the MAUT models. Because of the extensive use of value trees in this type of application of MAUT, the term *value tree analysis* is often used. At its present stage it is merely a diagnostic tool, i.e. it allows the pinpointing of the nature of the factual and value conflicts in the value trees, impact assessments, and weights assigned by stakeholders.

Recent research on bargaining and negotiation processes (e.g. Raiffa, 1982) as well as studies of compensation mechanisms (Kunreuther *et al.*, 1984) could make value tree analysis useful for conflict resolution too. In particular, the process of generating and inventing new compromise solutions, guided by the value structures of the opposing stakeholders, seems to be a promising direction. Applications of MAUT to assist in negotiations (e.g. Ulvila and Snider, 1980) as well as several, as yet unpublished, value tree analyses involving direct interaction between stakeholders clearly show the potential for conflict resolution inherent in the process.

11.1.3. Value tree analysis: an overview of the technique

Value tree analysis goes through the following generic steps:

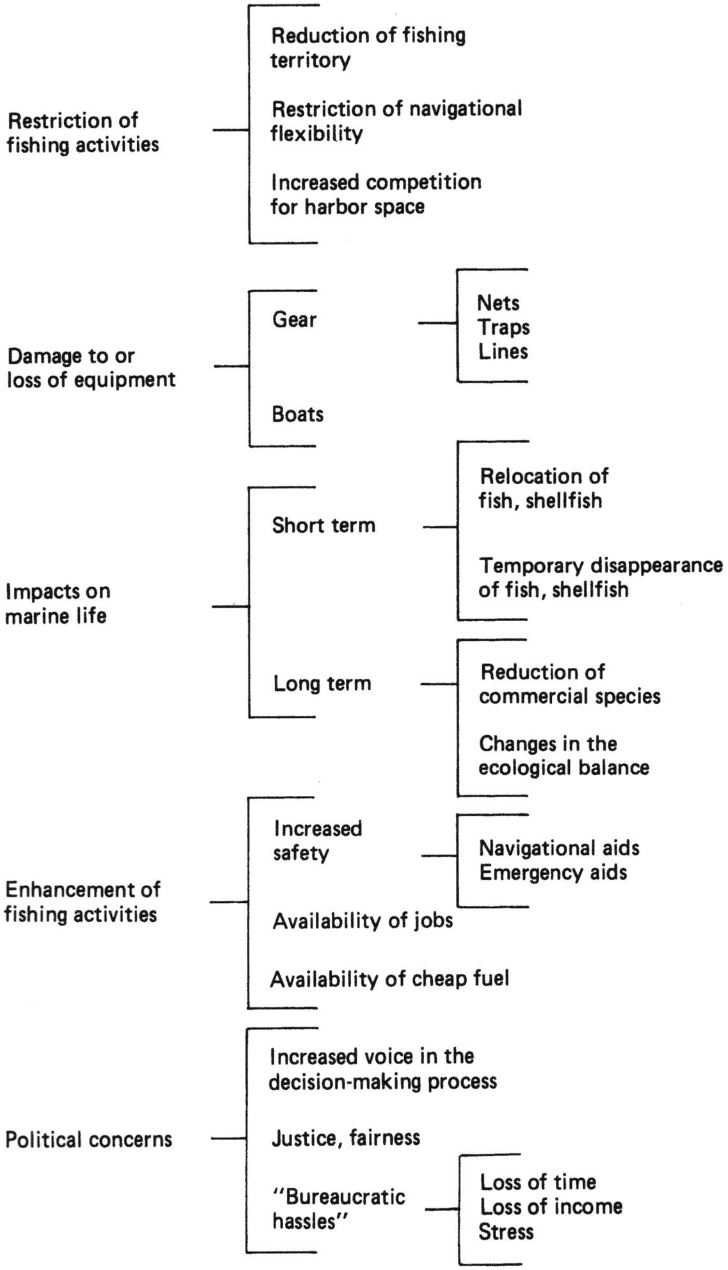

Figure 11.1. A value tree reflecting the concern of fishermen with offshore oil development.

(1) Structure the problem as a decision one.
(2) Identify stakeholders.
(3) Build separate value trees in interviews with leading representatives of each stakeholder group.
(4) Construct a combined value tree.
(5) Construct a multiattribute utility model for each stakeholder group.
(6) Carry out sensitivity analyses for each stakeholder model.
(7) Analyze stakeholder agreements and disagreements and generate compromise options.

While these seven steps appear quite linear and sequenced, they are, in fact, recursive, with frequent feedback, restructuring, and possibilities for change in repeated consultations with the stakeholder representatives. Anonymity and stakeholder control of the outputs of the analysis should be assured at each stage.

The following sections will briefly discuss each of the seven steps. The first step is to identify what decisions are to be made and what alternatives exist for that decision. Typical classes of decisions problems are choosing a location for a production facility, selecting the best transportation route or mode of transportation, or selecting an acceptable waste disposal site. Although sometimes phrased in terms of accepting or rejecting a particular decision or proposal, the problem is usually much more complex, and different stakeholders often have quite different problem formulations. In the problem of siting a hazardous facility, residents may prefer a simple yes–no formulation: either the facility is located in "our backyard" or it is not. Public officials, having responsibility for resource allocation with a wider perspective, tend to formulate the problem as a choice among alternative specified sites each having its own pros and cons. Environmentalists, concerned with the environmental constraints, often impose such severe restrictions that the "choice" simply boils down to a search for the one site that can meet these restrictions. And, to complicate matters, some political groups see the problem, not in the siting issue at all, but rather in the design of equitable and fair decision-making processes such as having public participation in the siting debate.

Creating a common problem formulation can therefore be quite an accomplishment by and of itself. It may be useful to include innovative options as well as some process options in order to accommodate the different problem perspectives of the stakeholders. For example, when siting a hazardous waste facility, it might be useful to explore participation and compensation options. It is recommended to consult with some of the stakeholder representatives early in the process to generate a common understanding of the problem.

Identification of stakeholders (step 2) is usually a straightforward task. In many disputes involving hazardous materials or facilities, stakeholders emerge who voice their opinions loudly, and they frequently meet in public hearings or comment on official proposals. Environmental impact statements are a good source for the identification of stakeholders. While stakeholders are usually numerous (a list of 200 would not be excessive), they usually cluster around common issues or common values. This clustering allows reduction of most stakeholder lists to between five and ten typical classes.

Building separate value trees (step 3) is still an art. The process of constructing these trees is highly interactive and should be iterative. It begins with an initial interview with one to three leading representatives of the organizations whose values are to be structured. One analyst should pursue the questioning, while a second analyst should take notes and interject on occasions to pursue missing values or to review the discussion. The analysts ask questions like the following: What are your general concerns with this or that particular site or several sites? Why are these concerns important? What does this or that concern or value category mean? How does this or that value differ from others? What makes this or that particular option better than another one? By pursuing the explication of general values of the respondents the analyst typically assembles a long list of values and concerns, roughly ordered by broad categories. At the end of the interview the analyst probes for missing values, and asks respondents to stretch their imagination and think of good aspects of the least preferred option and of bad aspects of the most preferred option.

After the interview, the analyst constructs a value tree and sends it back to the stakeholder group for comments and possible revisions. Usually, the revisions consist of additional values or of wording changes. In some instances the complete tree is revised. Since the interview partners are likely to be skeptical about the uses of the interview material and their trees, it is critical that they be assured complete confidentiality until they accept a revised and finalized version of the initial tree.

Building a combined value tree (step 4) is again done in the analysts' office. Stress should be on comprehensiveness while maintaining the internal logic and structure of each individual tree. All stakeholders should be able to identify all their values and concerns in the tree. Ideally, each stakeholder tree should be identifiable as an overlay over the combined tree.

The separate and combined trees are products in themselves and often can be quite useful in structuring the debate and pinpointing areas of agreement and disagreement. Often they help stakeholders to understand each other's concerns better.

The fifth step, building a multiattribute utility model for each stakeholder, is carried out using standard MAUT methods. It begins with building of an impact matrix in which the alternatives that are to be evaluated are crossed with the attributes derived from the value tree. The analyst summarizes the relevant data and knowledge in each cell of the impact matrix. Wherever he discovers expert disagreement, it should be noted in the respective cell by outlining ranges of impacts.

The impact matrix, complete with expert assessments and data, is discussed in a meeting with stakeholder representatives who are encouraged to change any of the data based on their own expertise. Subsequently they are asked to rate the impacts in each attribute and to weight attributes. Combining ratings and weights, the analyst can then calculate an aggregate utility for each of the alternative sites. Formally, the result of this analysis is expressed by

$$v_k(x_j) = \sum_{i=1}^{n} w_{ik} v_{ik}(x_{ijk}) , \tag{11.1}$$

where x_{ijk} is the impact of the jth alternative on the ith attribute assessed by (or agreed upon by) the kth stakeholder, v_{ik} is the rating of that impact in the ith attribute by the kth stakeholder, w_{ik} is the weight of attribute i assigned by the kth stakeholder, and v_k is the aggregate utility function that evaluates the jth alternative x_j for the kth stakeholder.

After completing step 5 of the analysis, the analyst typically discovers some disagreement between the model-based evaluation (the "v_k"s) and the direct rankings of the alternatives supplied by the stakeholders. Thus, the first part of the sixth step consists of helping stakeholders to straighten out their own inconsistencies. The goal is to gain insights into the reasons for inconsistencies and to produce what ultimately amounts to a satisfying logical formulation of the stakeholders' evaluations of the alternatives.

The model thus should capture the essence of the stakeholders' arguments and consequently can be used to perform "dialogues" with the stakeholders in the form of sensitivity analyses. Sensitivity analysis programs are very useful for this purpose.

A coherent model for each stakeholder having been produced, the seventh and final step of the value tree analysis is to compare stakeholder models, diagnose the sources of the conflicts, and, if possible, invent options or versions of existing alternatives that help reduce the conflict. In most applications of this technique the conflicts have been predominantly in values expressed as disagreements in value structures and weights. In contrast, surprising agreements have existed on the "factual" side of the model, namely the impact estimates x_{ijk} and the ratings v_{ik} associated with them. Overall, value tree analysis discovers more agreement than disagreement, since the procedure, unlike adversarial processes and techniques, encourages discussion of both, and since most reasonable people tend to agree on a good number of issues.

The conflict having been diagnosed, the remaining task is conflict resolution. At this point it is useful to bring the stakeholders together, present the model results, and discuss the outstanding issues. Resolution of the conflict probably lies in the joint exploration of novel alternatives like compensation schemes or option packages.

11.2. An Application of Value Tree Analysis to Offshore Oil Drilling

Soon after his appointment as Secretary of Interior in 1981, James Watt developed a long-term strategy to reduce oil imports into the USA. Part of this strategy was an ambitious and aggressive offshore oil leasing plan that was to speed up oil and gas exploration and development in virtually all federal waters off the coast of the USA (see US Geological Service, 1982). Not surprisingly, this expansive offshore oil leasing plan soon met with opposition from many groups, including local counties and cities, which were concerned with overcrowding and unbalanced growth, and fishermen, who objected to the likely interference with their fishing activities.

A particularly interesting and heated case of opposition to offshore oil development occurred in Southern California, which is relatively rich in offshore oil resources but also has a very beautiful shoreline and many environmentally valuable marine resources. The following study focused on the debate surrounding offshore oil lease no. 80 which was scheduled for February 1984, and which considered leasing almost the complete area outside of a three-mile zone off the shore of Southern California, between Point Conception and the Mexican border (see *Figure 11.2*).

As originally proposed by the Minerals Management Service of the Department of the Interior, lease sale 80 was to exclude only a few tracts in the ecological preserves surrounding the Channel Islands, Santa Barbara, and San Nicholas Islands (Minerals Management Service, 1983).

If the sale were held as planned, oil companies could place bids on any of the 4-square-mile tracts, and, upon winning a bid, could begin exploration, followed by development, and then later production of oil and gas. Thus, the lease sale is the starting point of a 20–30 year process that is likely to result in construction of oil platforms, pipelines, and processing facilities.

Southern California had its share of offshore oil developments prior to lease sale 80. Previous lease sales were held in the 1960s, the 1970s, and in 1980, 1981, and 1982; these led to exploration and production mainly in the Santa Barbara Channel. In addition, several tracts had been leased and developed in the State Waters within the three-mile zone, in the Santa Barbara Channel and off the Coast of Long

Figure 11.2. The leasing area for the proposed Southern California lease sale 80 (February 1984). The figure shows the Southern California Planning Area. The area outlined is the proposed lease offering area. Tracts marked by a square grid were to be deleted as part of Option 2 (see text): Excluded areas are the Channel Islands National Marine Sanctuary and the Santa Barbara Ecological Preserve and Buffer Zone. (*Source*: Minerals Management Service, 1983.)

Beach and Newport Beach. Although offshore oil development was not new to Southern California, the scope of the planned lease and development was quite dramatic. If the whole area were offered for sale, an expected 270 million barrels of oil was likely to be produced over a 25-year life of the oil development. Between five and ten drilling platforms would be erected temporarily, and also between five and ten production platforms would likely be constructed and operate for some 20 years. Many of these production platforms would be seen from the shore.

Environmentalists, county and city officials, the Coastal Commission of California, and the fishermen started to fight lease sale 80 as soon as it appeared on the drawing board. This opposition was voiced in the courts by arguing that this lease sale was inconsistent with the Coastal Management Act, in public hearings about environmental impacts, and in Washington, where Southern California representatives fought Watt's plan with a series of moratoria.

Our study team entered this controversy in June 1982, after lease sale 68, the most recent Southern California lease sale, had been completed. Lease sale 68 was somewhat less controversial, since it only offered some 200 tracts as opposed to the more than 2000 tracts that were offered in lease sale 80. In addition, lease sale 68 had carefully excluded tracts in the Santa Monica Bay and off the coast of San Diego, both highly disputed areas. Nevertheless, lease sale 68 had encountered stiff opposition. It appeared that the battle about lease sale 80 would be even hotter.

11.2.1. Identifying the decision

The alternatives that define the decision problem are listed in *Table 11.1*. The "complete sale" alternative (1) and the "cancellation" alternative (5) were analyzed by the Minerals Management Service in its *Environmental Impact Statement* (Minerals Management Service, 1984). Alternative (2) deletes the "hottest" tracts which had encountered stiff opposition. Strong and powerful opposition existed in Los Angeles and San Diego, with support from the mayors of these cities and backing in Washington by Senators Cranston and Wilson. Deletion of tracts in the Santa Monica Bay and off the shore of San Diego would have gone a long way to address that opposition. Minerals Management Service considered these deletions as separate alternatives in the *Environmental Impact Statement* and, in addition, considered deletion of environmentally sensitive tracts and a few tracts that could interfere with shipping off the Los Angeles Harbor. All these "hot" tracts are deleted in alternative (2).

Table 11.1. Lease sale alternatives.

No.	*Description of alternatives*
(1)	Hold complete lease sale 80 as planned (area to be offered is marked by the outer heavy line in *Figure 11.2*)
(2)	Delete hotly disputed tracts: off the shore of Santa Monica and San Diego, and around San Nicolas Island and in front of Los Angeles Harbor (area marked by a square grid in *Figure 11.2*)
(3)	Delete all tracts in the Santa Barbara Channel (area described by heavy line around Santa Barbara Channel in *Figure 11.2*)
(4)	Delete all tracts in Santa Barbara Channel *and* in the Inner Basins (area marked by heavy lines around the Santa Barbara Channel and Inner Basins in *Figure 11.2*)
(5)	Cancel lease sale 80 or delay indefinitely

Alternatives (3) and (4) were created somewhat artificially in order to tap the underlying conflict. Alternative (3) was meant to be equitable to the counties of Santa Barbara and Ventura which had already shared a large part of the burden of offshore oil development. Alternative (4) was meant to be the "beach lovers'" choice, so to speak. It put the possible development out of sight and out of mind.

11.2.2. Identifying stakeholders

A survey of several environmental impact statements, the responses provided to them, and about 20 interviews identified some 100 stakeholders' groups involved in and/or affected by offshore oil development in Southern California. Most stakeholders had made public statements and expressed their overall attitude towards the development (pro, anti, or neutral), and had formulated their main arguments and the values that drove these arguments.

These lists were used to identify a generic structure of opponents, proponents, and regulatory stakeholders, as shown in *Table 11.2*. Opponents are grouped into five classes: environmentalists, local communities and governments, beach residents and "beach lovers", fishermen, and "special issue" opponents. Environmentalists were mainly concerned with oil spills, water and air pollution, preservation of endangered species, and protection of environmentally valuable coastal areas. Many local communities and governments opposed oil development because of the impacts it would have on land use and infrastructure and because it might lead to pollution and unbalanced growth. Beach residents and those who like to spend time at the beaches feared that oil development would alter the recreational and beach life-styles, especially in those large stretches of the California

Table 11.2. Stakeholder groups and a selection of their members.

Opponents	*Proponents*	*Regulators*[a]
Environmentalists:	Oil companies	Federal agencies:
Environmental Defense Center	Exxon	Bureau of Land Management (DoI)
Environmental Coalition	Union	US Dept. of Fish and Wildlife (DoI)
Sierra Club	Shell	National Oceanographic and Atmospheric
Greenpeace	Western Oil and Gas Association	Administration (DoI)
Center for Environmental	Chambers of commerce:	Marine Mammal Commission
Education	California Chamber of commerce:	Department of the Army
Friends of the Sea Otter	Greater Bakersfield Chamber	Corps of Engineers
Whale Center	of Commerce	Bureau of Mines (DoI)
Cities and counties:	Special interest proponents:	US Coast Guard
City of Avalon	International Union of Operating	Environmental Protection Agency
City of Huntington Beach	Engineers	Federal Energy Regulatory Commission
City of Los Angeles		US Geological Survey
City of Santa Monica		National Park Service
County of Santa Barbara		State agencies:
County of Ventura		South Coast Air Quality District
San Diego County		Department of Water Resources
Fishermen		Air Resources Board
Special interest opponents:		California Coastal Commission
Pacific Palisades		Public Utilities Commission
Homeowners' Association		Department of Fisheries and Games
Port of Los Angeles Authority		Department of Lands and Recreation
Dept. of Navy		State Lands Commission
Beach residents and		
beach lovers:		
Ocean Park		
Community Association		
(Mainly comments from		
individuals)		

[a] DoI, Department of Interior.

Coast that have not yet experienced any development. Fishermen were concerned about interference with fishing activities, loss of equipment, and economic losses. Finally, a series of special issue opponents were opposed for diverse reasons like interference with port entry and departures (Los Angeles port authorities), reduction of property values (property owners), or interference with military training activities (Department of the Navy).

Supporters of offshore oil development fall into three main groups: oil companies, business communities, and special issue supporters. The interest of the oil companies in offshore oil development is self-evident. Business communities and some chambers of commerce expressed favorable attitudes towards oil development because of the business stimulation and growth it provides, and also because of the general idea of providing for more national energy independence. Special issue supporters range from those who stand to profit indirectly from the oil development (e.g. subcontractors) to those who have professional interests in oil exploration and production (e.g. the Society of Petroleum Engineers).

Regulators mix support and opposition typically in a way that is highly predictable from their statutory mandate. The California Coastal Commission was opposed to lease sale 80 unless many tracts that their members perceived as threatening the coastline were deleted. The Minerals Management Service of the Department of Interior plays an interesting role as joint regulator and promoter of offshore oil development. Since it is the federal lead agency in the leasing process, it should be viewed, in principle, as the final balancer and decision-maker. The general impression, however, was that Minerals Management Service was clearly in the support group.

Two additional stakeholder groups played interesting roles. During Governor Brown's tenure the state government tended to be opposed to offshore oil development, but this attitude changed during the Deukmejian administration. The new administration favors offshore oil development but is eager to stress that it would like to see pollution and other detrimental effects reduced. In Washington, several congressmen representing coastal regions combined to create a federal opposition to offshore oil development. For all practical purposes the reasons for their opposition coincide with the concerns of their constituencies.

11.2.3. Building value trees for selected stakeholders

Interviews were conducted with one or two individual representatives of five stakeholder groups that seemed to span the range of conflicting values and opinions: the Environmental Defense Center, the County of Santa Barbara, the City of Los Angeles, the Santa Barbara fishermen, and oil companies. From the interview materials and from related materials and comments on the environmental impact statements, five separate value trees were constructed. These trees reflect the biases and inclinations of the interview partners as well as the analysts and therefore should not necessarily be interpreted as "the tree" of the organization or group that our respondents belonged to.

The first drafts of these trees were returned to the stakeholders for comments and revisions. The revised trees for the environmentalists and the oil companies are presented in *Figures 11.3* and *11.4*. The revised tree of the fishermen is shown in *Figure 11.1*. The only tree that points clearly in the direction of favoring oil exploration and production is the tree constructed from interviews with oil company representatives. All other trees include a majority of concerns that tend to oppose oil development. In fact, in most discussions the positive aspects of oil development were only brought out after some probing by the analyst.

The trees emphasize the particular concern of the different stakeholders, but they do not seem incompatible with each other. All trees include environmental concerns, although they appear at different levels of detail, ranging from very detailed for the environmentalists to a "mentioned only" for the oil companies. Air quality impacts emerged as a major concern in many interviews, and it takes some prominence in the trees for Los Angeles City, Santa Barbara County, and the environmentalist group. Santa Barbara County officials had main concerns with onshore socioeconomic impacts (housing, water supply, etc.) and the possible regional inequities resulting from offshore oil development. Aesthetic impacts were not mentioned very often, and sometimes were explicitly called "unimportant".

The oil company's tree reflects concern with amounts of oil that are recoverable. By meeting legal and regulatory requirements the oil company representatives felt that they addressed environmental concerns. Our team did not interview any organizational units representing a combination or association of oil companies like the Western Oil and Gas Association (WOGA). The values of WOGA are likely to be somewhat broader than the tangible exploration and production concerns of the individual managers of oil companies, and probably include national economic concerns as well as environmental and other values.

11.2.4. Building the combined value tree

It proved not to be difficult to combine the four "anti" oil development trees; because of some difficulty in incorporating the oil company's values in that combined tree, oil company values were simply added. The combined tree is shown in *Figure 11.5*.

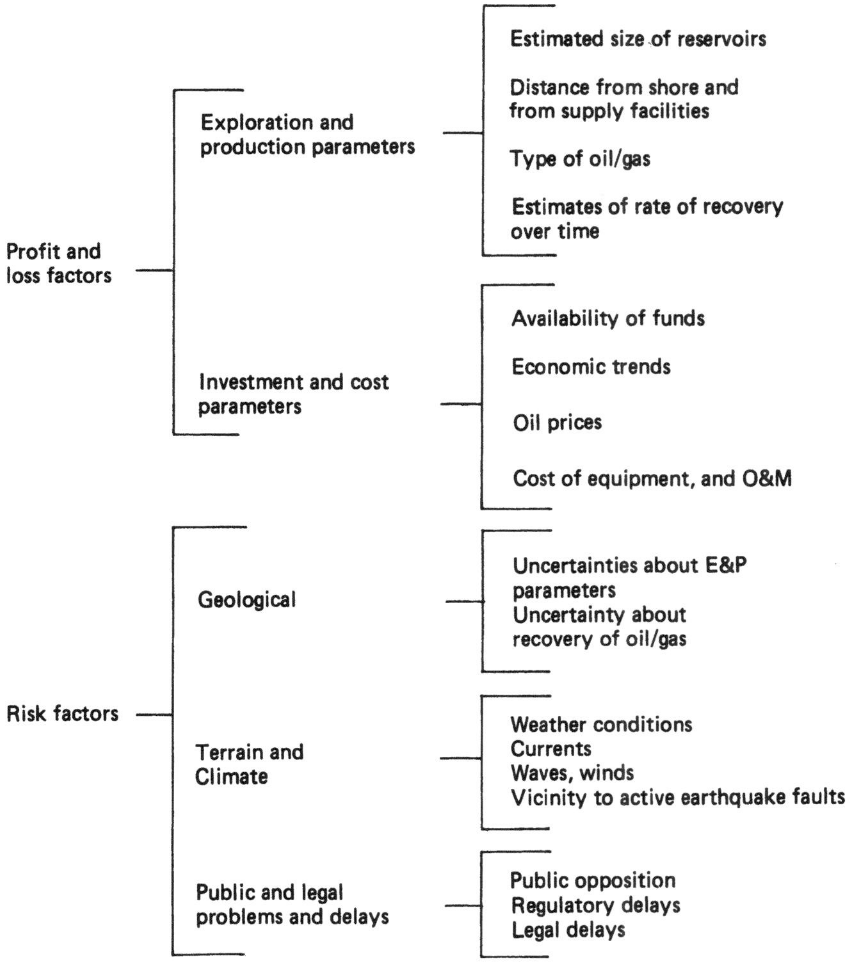

Figure 11.3. Value tree reflecting the concerns of oil companies in offshore oil development.

Perhaps owing to the selection of the stakeholder groups (four out of five were opposed to lease sale 80) the larger number of values in this combined tree reflects anti-oil sentiments. However, the combined tree appears to represent a broad spectrum of values and concerns.

The next task was to determine an appropriate level of values in the combined tree which could be used for an evaluation of the alternatives described in *Table 11.1.* The task should be simple, and, in particular, it should not overburden the stakeholder representatives with large numbers of detailed elicitation questions.

For these reasons the evaluation was constructed at the level of the eleven main value categories of the combined value tree. First an

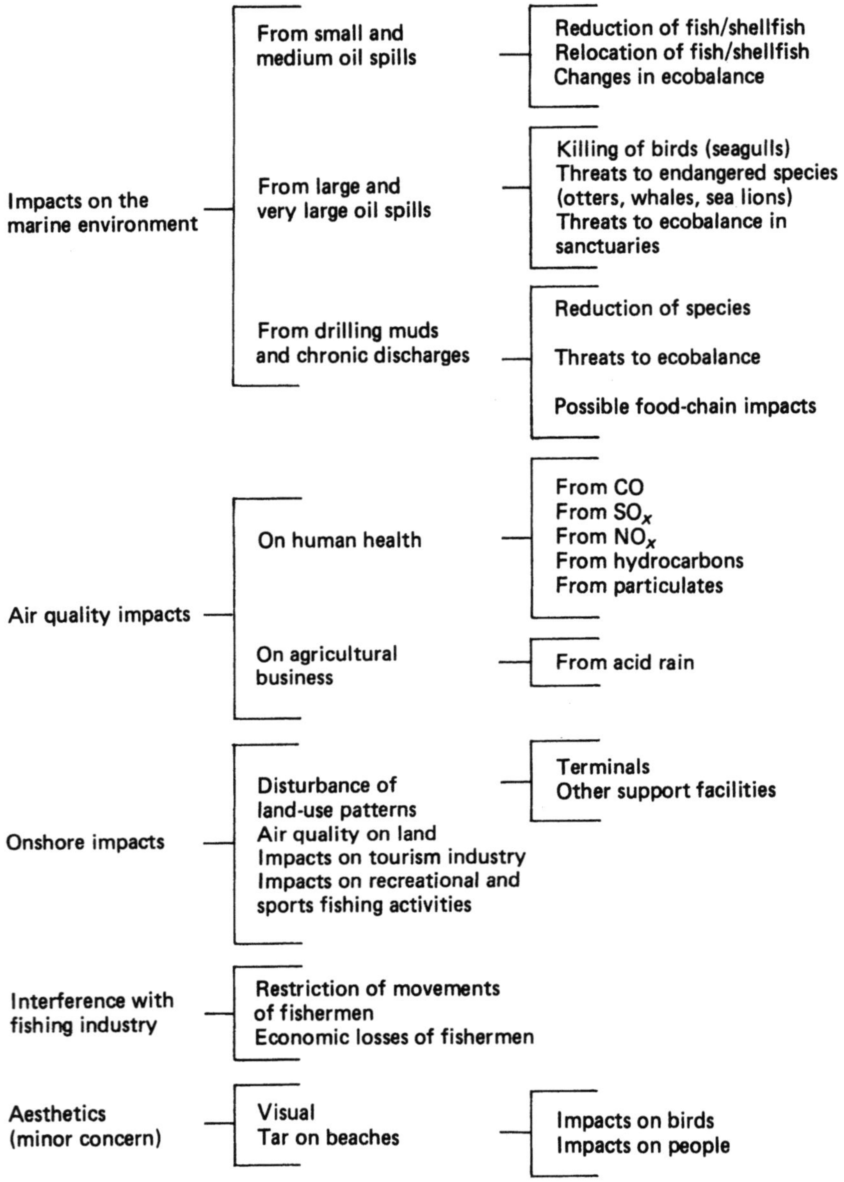

Figure 11.4. Value tree reflecting the concerns of environmentalists about offshore oil development.

impact matrix of five alternatives by eleven criteria was constructed. Subsequently, the analysts began to collect the materials relevant for describing the impacts in each cell of the matrix, using mainly published materials and relying strongly on the environmental impact

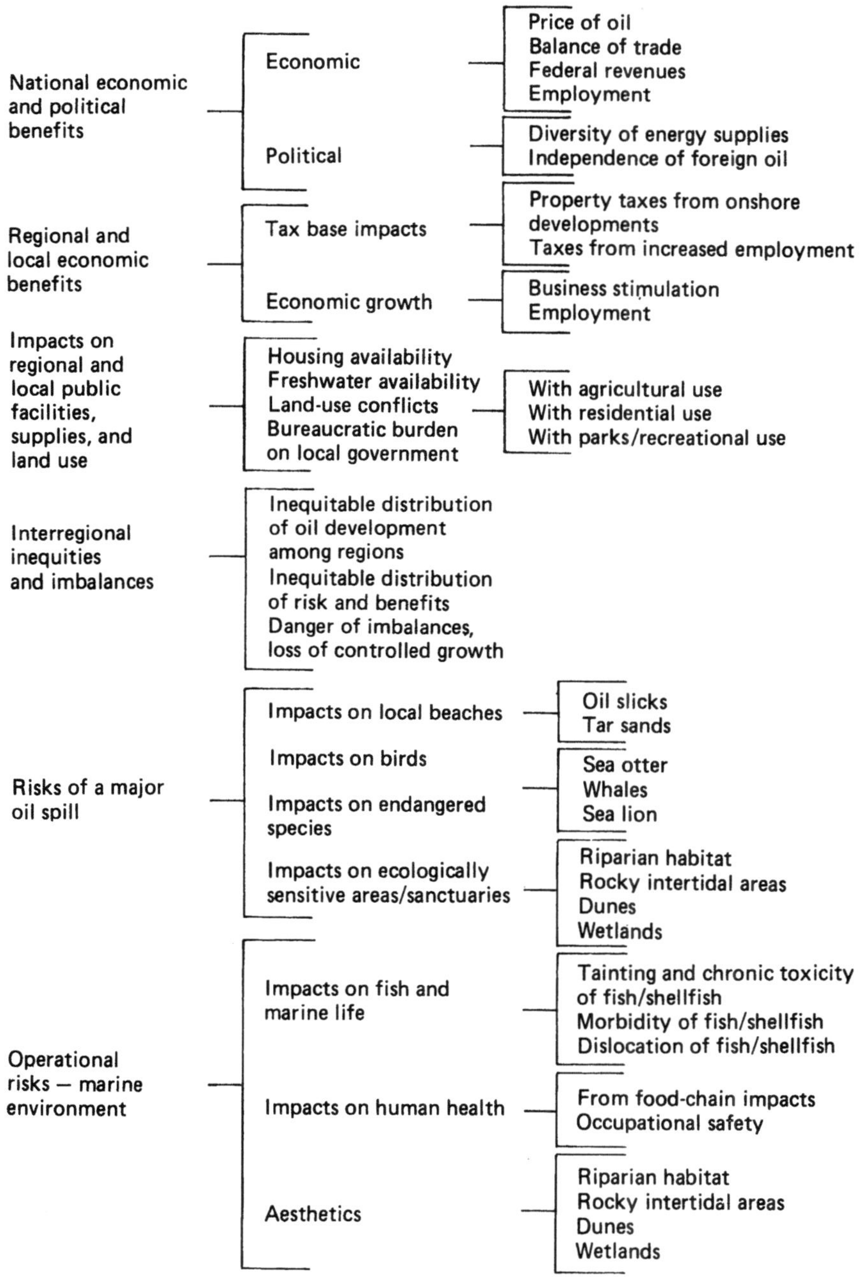

Figure 11.5(a). **Joint value tree reflecting concerns of various parties with offshore oil development.**

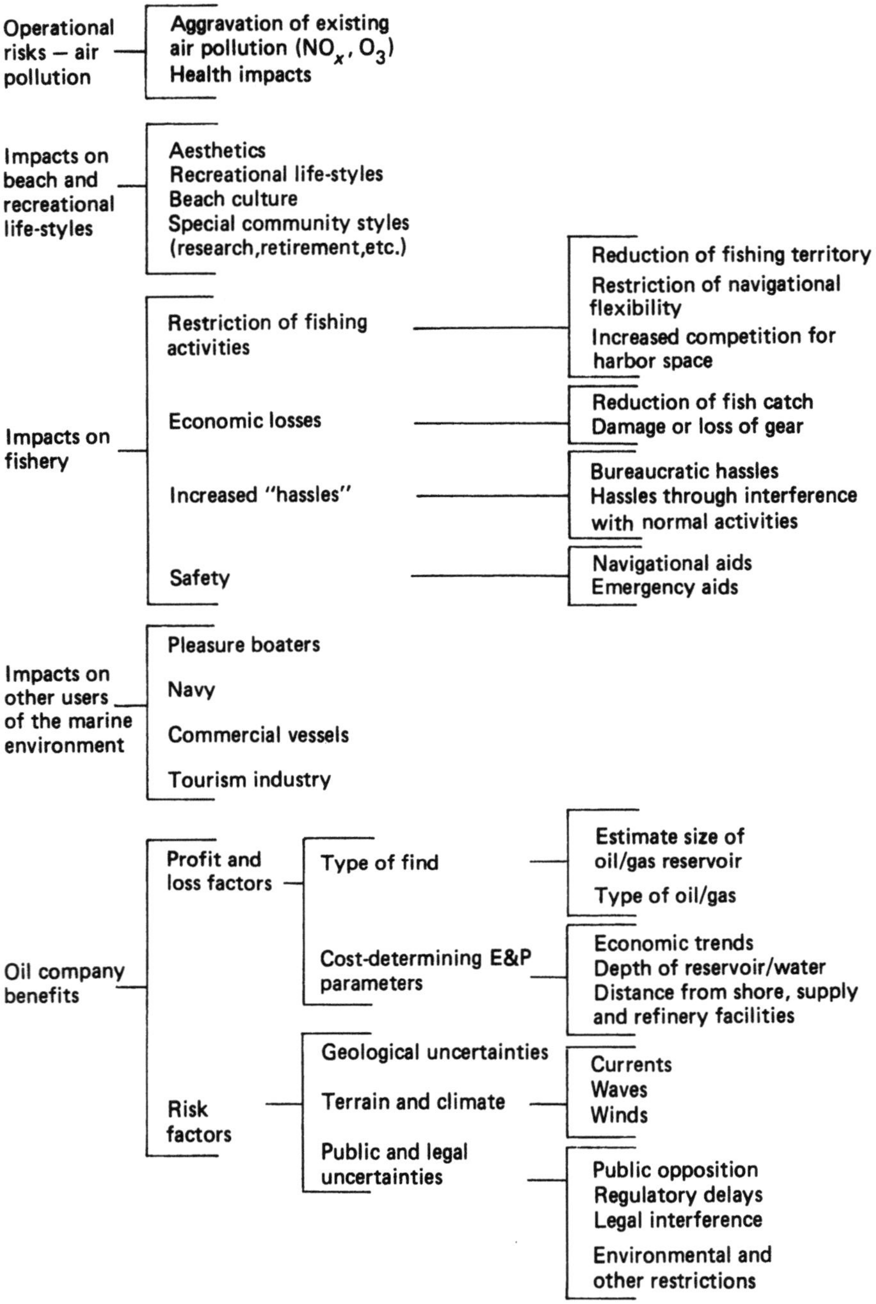

Figure 11.5(b). **Joint value tree reflecting concerns of various parties with offshore oil development.**

Table 11.3. Segment of the impact matrix.

	National political economic benefits	*Interregional inequities and imbalances*[a]	*Risks of a major oil spill*[b]
Option 1: Hold lease sale 80 as planned in February 1984	270 M bbl of oil = 2–3% and 510 B ft³ of gas = 7% of import reduction at peak production, resp. Very minor (positive) impacts on oil prices, federal revenues	Relatively inequitable: large share of burden in SB and Vent Possibilities of imbalanced growth in SB and Vent	$P = 30\%$ # = 0.36 SB + VENT = 2% SD + LA = 4%
Option 2: Delete tracts in Santa Monica Bay, off San Diego, off San Pedro, and around San Nicolas Island	170 M bbl of oil = 1% and 470 B ft³ of gas = 6% import reduction at peak production, resp. Extremely minor (positive) impacts on oil prices, federal revenues	Inequitable: no share of burden in SD and LA Possibility of imbalanced growth in SB and Vent	$P = 15\%$ # = 0.18 SB + VENT = 2% SD + LA = 2%
Option 3: Delete all tracts in Santa Barbara Channel	180 M bbl of oil = 1% and 250 B ft³ of gas = 3–4% of import reduction at peak production, resp. Extremely minor (positive) impacts on oil prices, federal revenues	More equitable: puts burden on counties that have little share now (SD, LA)	$P = 2\%$ # = 0.24 SB + VENT = 0% SD + LA = 2%
Option 4: Delete all tracts in Santa Barbara Channel and inner basins	110 M bbl of oil = < 1% and 280 B ft³ of gas = 4% import reduction at peak production, resp. No discernible impact on oil prices, federal revenues	Relatively inequitable: gives no special relief to SB and Ventura	$P = 11\%$ # = 0.12 SB + VENT = 0% SD + LA = 0%
Option 5: Cancel lease sale or delay indefinitely	No reduction of oil and gas imports below baseline No positive impact on oil prices No additional federal revenues	Relatively inequitable: leaves inequity status quo	No additional risk of oil spill but 92% existing risk from tankers, 96% existing risk from leases # = 6

[a]SB, Santa Barbara; Vent, Ventura; SD, San Diego; LA, Los Angeles.
[b]P, probability of 1 or more very large oil spills (>10 000 bbl); #, expected number of very large oil spills over 25-year life; SB (VENT, SD, LA), probability of a very large oil spill that hits Santa Barbara (Ventura, San Diego, Los Angeles).

statement of the Minerals Management Service. In some cases the analysts added their own judgments, when "hard" data were missing.

Three columns of the impact matrix are shown in *Table 11.3*. They represent a heroic attempt to compress more than 1000 pages of environmental impact assessments onto a few pages of text, and thus presented a condensed factual side of the offshore oil problem. In addition to impacts, relative ratings of the impacts on a 0–100 scale were provided by the analyst team. This turned out to be relatively easy, since all criteria but one are either monotone increasing or monotone decreasing with the amount of oil resources that are likely to be recovered under the given option. The relative spacing reflected judgments about how the numerical differences in impacts would translate into value-relevant differences.

11.2.5. Multiattribute utility models for five stakeholders

In a second round of interviews stakeholders were presented with the impact matrix including the analysts' ratings and they were asked to criticize impacts and ratings. Very few changes were requested by the stakeholders and the necessary modifications were relatively minor. In some instances the stakeholder representatives asked how the analysts had obtained a particular datum. The matrix was left with them for further comments.

Thus the analysis produced a fair amount of agreement about the "factual" shades of possible conflicts. The remaining conflicts were likely to be on the value side.

To operationalize the value conflicts in the weights of the multiattribute utility models a two-stage procedure was used. First, respondents considered only the risk attributes. They were presented with a profile of all the "best" (smallest-risk) and all the "worst" (highest-risk) impacts and were asked to consider a situation in which they were "stuck" with all the worst impacts. For practical purposes this is the situation created by the complete lease sale option 1. Subsequently, respondents picked the one impact that they would most like to eliminate. Presumably that impact that was considered the most severe and thus the value difference (improvement) by stepping from the "worst" to the "best" impact was largest. They then picked the second most severe impact, and so on, thereby rank ordering the value differences in the eight impact categories. Having rank ordered the impact differences, respondents then assigned a value of 100% to the impact category that they had ranked the highest and expressed as a percentage less than 100% the relative difference in the impacts for the remaining impact categories. The 0% point was meant to characterize a hypothetical "no difference" impact category. The resulting "raw" weights for the risk categories were then normalized to add to 1.

Similarly, respondents were asked to rank order the benefit categories in terms of the difference between the "smallest" and "largest" benefits in the impact matrix. After rank ordering they again assigned 100 points to the benefit category that they felt represented the largest improvement in impacts, and assigned relative weights were normalized to sum to 1. Respondents then assigned weights to the combined risks versus the benefits. They were asked to assume that all the "worst" negative impacts were stacked up on one side and all the "best" positive impacts were stacked up on the other side. They were reminded that, for practical purposes, this is like considering the complete sale option 1 versus the cancellation of the sale, option 5. They were then asked whether they preferred all the benefits *and* all the risks (option 1) or neither (option 5). The answer to this question indicated whether the benefits outweighed the risks or vice versa. Finally, they were asked to assign a factor by which risks outweighed the benefits or by which benefits outweighed the risks. The risk–benefit weight ratio was then converted into weights for risk and benefits that sum to 1. Final weights for each of the eleven impact categories were obtained by multiplying the risk or benefit weight with the weight on the particular impact category.

The results of the weighting analysis are shown in *Table 11.4*. They clearly show that by far the main difference is the relative

Table 11.4. Weights for five stakeholder groups.[a]

Impacts	*Environmentalists lists*	*Santa Barbara County*	*Los Angeles City*	*Fishermen*	*Oil company*
Benefits:	17%	17%	25%	50%	75%
National economic	11%	9%	15%	36%	21%
Local economic	6%	7%	7%	14%	25%
Oil company	0%	2%	3%	0%	28%
Risks and negative impacts:	83%	83%	75%	50%	25%
Local infrastructure	3%	14%	1%	12%	1%
Regional inequities	13%	14%	2%	4%	1%
Oil spill risks	10%	14%	16%	7%	7%
Marine pollution	13%	7%	6%	7%	5%
Air pollution	16%	14%	20%	4%	3%
Beach and recreation	2%	8%	10%	1%	5%
Fishery impacts	13%	9%	8%	13%	2%
Impacts on other users	0%	4%	12%	1%	3%
(Conservation)[b]	13%	0%	0%	0%	0%

[a]All weights are percentages (rounding effects cause the apparent discrepancies).
[b]Concern added by the environmentalists.

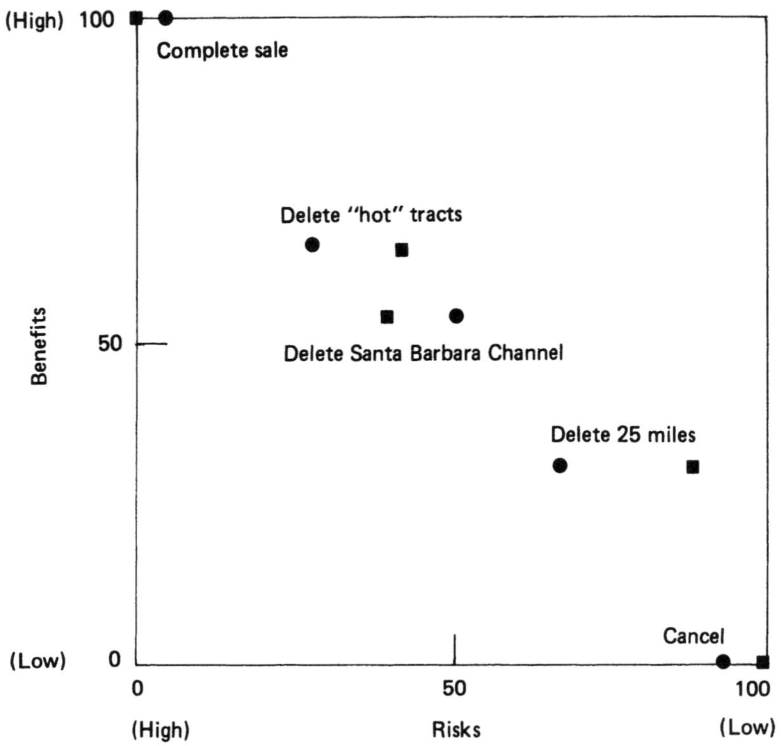

Figure 11.6. Negative correlation between risks and benefits: circle, environmentalists; square, oil company.

weight put on overall risks versus benefits. This weight also strongly determines the rank ordering of the options derived from the weighted additive model. This sensitivity can be demonstrated as follows. *Figure 11.6* shows the two dimensional plot of the five options on aggregated risks and aggregated benefits. The two aggregated variables are almost perfectly correlated. This is understandable, since the main variable that mediates both risks and benefits is the amount of oil resources that can be developed as a result of the lease.

As a consequence, whenever the overall risk weight is significantly larger than that of the benefit weight (e.g. 75% versus 25%) the model prefers the cancellation option followed, in the order of the amount of oil resources, by the other options. The least preferred option would be the complete sale option 1. Whenever benefits strongly outweigh the risks the rank ordering is reversed.

Several sensitivity analyses determined the effects, for example, of deleting air pollution or of reducing concerns about fishing impacts. In general, such deletions had little effect whenever the overall weights on risk versus benefits were highly skewed but could have a substantial effect when they were between 60–40 and 40–60. For example, eliminating some concerns of the fishermen would swing the fishermen's model to prefer the full lease sale 80, while elimination of air pollution would not swing any of the other stakeholders.

11.2.6. Multiattribute utility models for residents of Santa Monica and Ontario, California

Similar results were obtained in a small written survey of residents of Santa Monica and Ontario, an inland city of California, using an abbreviated version of the multiattribute utility models that was developed with the stakeholders (Marks and von Winterfeldt, 1984). It included only three risk attributes (oil spills, water pollution, and aesthetic impacts) and three benefit attributes (independence of foreign oil, employment, and government revenues). Sixty-three respondents in Santa Monica and Ontario carried out the steps of a multiattribute utility analysis, similar to that described in the previous section, to evaluate a single proposal for developing offshore oil resources in the Santa Monica Bay. That proposal envisioned a development of about 50 million barrels of oil in total and construction of 2–3 drilling and production platforms. These platforms would be constructed within viewing distance of the Santa Monica shoreline and would operate for about 25 years. Separate samples in Santa Monica and in Ontario were asked to evaluate a similar proposal that was to occur in Mobile, Alabama, to test how judgments would change when the oil development occurred in "somebody else's backyard".

The results were quite striking, and are interpretable in spite of the fact that the samples were small. The percentage of respondents opposing the development are shown in *Table 11.5*. Residents of Santa Monica generally judged the development as undesirable when it was to occur in their own backyard, the Santa Monica Bay. They were evenly divided between pro and anti development attitudes when the development was to occur in Mobile.

Ontario residents favored the development and did not distinguish between locations. When analyzed in terms of the parameters of the multiattribute utility models, the most striking result was a difference in the relative weights of the risk versus the benefits, as shown in *Table 11.5*. Residents of Santa Monica placed a weight of 79% on the risk attributes when the development was to occur in their backyard, but only 45% when it was planned for Mobile, Alabama. Ontario residents placed about equal weights on risks versus benefits,

Table 11.5. Results of a questionnaire survey of California residents.[a]

Location of development	*Residents of Santa Monica* (n = 30)	*Residents of Ontario, CA* (n = 33)
Santa Monica Bay	81 (0.79)	33 (0.43)
Mobile, Alabama	43 (0.45)	21 (0.49)

[a]The table gives the percentage of residents opposing development and, in parentheses, the weight they attach to the risks.

independently of the location of the development. Thus the "backyard effect" manifests itself mainly in an exaggeration of the weight put on risks.

11.2.7. Discussion

The analysis shows that different value structures and different weights for the risks versus the benefits are the most important reason for the conflict about offshore oil drilling. It also indicates a strong "not in my backyard" phenomenon that manifests itself in accentuating the weights for the risk attributes over the benefit attributes. On the whole, there was substantial agreement among stakeholders about the factual side of the problem.

Because of the large weight differences and because of the correlation between risks and benefits, small manipulations of the options (e.g. reducing air pollution or eliminating a particular environmentally valuable tract) will not generate any compromises. The option to delete all tracts within a 25-mile zone seemed attractive to most opponents, and it may be possible to develop a compromise based on it.

The fishermen are the only opponents who could possibly be convinced to accept an altered version of the complete lease sale alternative. Elimination or reduction of some of the more severe fishing impacts coupled with generous compensation mechanisms for loss of catch or equipment will, according to the models, go a long way towards a compromise between the oil industry and the fishermen.

A Final Note

After Watt's resignation, a temporary moratorium was imposed on offshore oil lease sales. Currently, a modified and much smaller lease sale has been proposed by Interior Secretary Hodel. The plan still encounters stiff opposition, and the arguments have changed little since 1983. The stakeholder models described in the previous sections could provide a blueprint for designing compromise solutions for a new lease sale.

11.3. Lessons, Pitfalls, and Recommendations

In this concluding section we will discuss some of the insights generated by the application of value tree analysis to the offshore oil controversy and by other recent applications. First, some of the possible uses of value tree analysis in diagnosis and resolution of conflicts are described. Subsequently, possible pitfalls of value tree analysis are discussed together with some recommendations for overcoming them.

11.3.1. Does value tree analysis help in diagnosis and resolution of conflicts?

In most applications, the main contribution of value tree analysis has been in the improvement of the understanding of the stakeholders' positions in the debate. The value trees appear to be the main vehicle for improving understanding and it is not uncommon to find stakeholders express surprise when confronted with other organizations' values. In the oil application, for example, the concern with air pollution expressed by several groups was surprising to some. Similarly, the clear distinctions between fishermen's and environmentalists' values was not completely anticipated. The fishermen, in particular, had been concerned with a lack of empathy on the side of the oil companies and felt that "they just don't listen to our concerns". Value trees thus could provide a useful mechanism for exchanging value-relevant information.

Often the value trees lead to improved understanding and clarity of values and concerns *within* an organization. After interviews carried out to structure the FRG's energy objectives (Keeney *et al.*, in press) almost all respondents expressed appreciation of the process in which their values were brought out and they felt that they had learned something.

The joint value tree, while necessarily an abstraction, can be useful in defining the scope of the discourse about a controversial problem, and for identifying areas of agreement and disagreement. It also provides a legitimate structure for discussions and defines agendas. Missing elements in a value tree can be just as interesting as the listed ones. For example, the joint FRG value tree had no values related to national image and national prestige, and the joint value tree in the oil study includes aesthetic concerns only in a very minor way, in spite of the analysts' probing.

In the past, most conflicts were found in the value structures or, when the joint value tree was considered, in the weights attached to its branches. This was not only true in the offshore oil study, but also in

Edwards' (1980) school desegregation study, Rozelle's (1982) Arizona water project study, and in a follow-up study to the FRG energy analysis. If weights are in conflict, there are no simple solutions to the problem, since weights are expressions of legitimate value differences. The formal sensitivity analysis can merely indicate the extent to which a particular stakeholder may be "swayed" either by option invention or by changes in the weights.

While the usefulness of value tree analysis for conflict diagnosis has been established, its use for face-to-face conflict resolution has still to be proven. The main use of the analysis may be in facilitating communication in mediation processes, in setting agendas, and for investing compromise options or packaging options that are attractive to the opposing stakeholders. There is, however, one major obstacle to this kind of analysis: very few arenas can accommodate this type of rational display of facts, values, and conflicts. The adversarial nature of the courts appears to hinder analytical processes more than it facilitates them. Arbitration is often aided by power brokerage, and rational analysis may not be the most useful activity. Only mediation arenas, which are still fairly rare in the kinds of disputes discussed here, seem directly amenable to implementing value tree analysis.

11.3.2. Pitfalls and recommendation

The lack of an appropriate arena often couples with lack of stakeholder interest to create an atmosphere of benign neglect of value tree analysis. One of the first pitfalls of value tree analysis is this inability to generate involvement and problem focus for the stakeholders. The oil study clearly suffered in this respect, and to a lesser degree the FRG energy study as well. On the other hand, Rozelle's (1982) study of alternative flood control projects in Arizona appeared to have generated much involvement. We believe that this project was more successful in this respect because it created its own arena, because there was a responsible decision-maker who was committed to taking the analysis results seriously, and because it was sponsored by a neutral government agency.

The following precautions are recommended to users of value tree analysis in order to avoid the "irrelevance" pitfall:

(1) Assure that the analysis is linked to a real arena and a real decision.
(2) Obtain sponsorship from agencies that are perceived to be neutral in the issues, or obtain a balanced sponsorship from several agencies.

(3) Work in close consultation with the real decision-making agency whether or not it is the sponsoring agency.

The second most common pitfall in value tree analysis is the refusal of some stakeholders to cooperate, or the loss of some stakeholders after initial cooperation. In Edwards' school desegregation study, the two most extreme groups (one antibussing and one probussing group) refused to participate at all, and some individuals did not provide weights for the final value tree. In the FRG energy study, an association of environmental and peace activists, after initially agreeing to a meeting, did not further participate in the value tree process after this initial meeting. Usually the refusal is a result of strong ideological positions and assumptions that their involvement could be construed as co-option.

To make participation appealing to a broad range of stakeholders, it helps to:

(1) Obtain sponsorship from several sides in the debate and stress the fact that all sides can learn something from the process.
(2) Create a broad range of alternatives that accommodate even radical views.
(3) Be very frank about one's own and one's sponsors' objectives in pursuing the analysis.
(4) Assure confidentiality at each stage of the process and allow stakeholders to reassess the value of their participation at each stage, before their responses are made public.

While the foregoing pitfalls had to do with the process of the analysis, the next two address pitfalls in the techniques themselves. The first is well known to analysts: settling too quickly on a problem definition and structure. It cannot be stressed too much that the initial structuring stages of value construction of the separate value trees are crucial and should be carried out with many iterations and with much flexibility toward restructuring if necessary. In particular, the definition of inappropriate alternatives can become a major stumbling block. Analysts should keep an open mind and define a broad range of initial options. As a rule, the options should include the most preferred alternative of each stakeholder.

The second class of technical pitfalls involves biases in value tree structuring and weight elicitation. One such bias is a result of "selfishness", which would eliminate any values and concerns that favor an alternative that the stakeholder group does not like. The same bias may lead to a lower weighting of those branches of the joint value tree that favor undesirable alternatives. This type of bias has been rare in value tree analysis, but it has been reported in other contexts (see

Dyer and Miles, 1976). The opposite bias seems more common: to express a broader range of values than those that truly represent the stakeholders' position and to state a more balanced set of weights. This "social desirability" bias poses no problem for the purpose of building value trees, but in weighting it may lead to an artificial agreement among stakeholders (see, for example, Stillwell *et al.*, 1981).

To counter these biases the analysts should:

(1) Emphasize that the role of the analysis is to create a realistic portrayal of each stakeholders' value systems.
(2) Stress the diagnostic value of the activity.
(3) Assure confidentiality until the results are accepted by the stakeholder representatives.

Value tree analysis is still in its infancy. Whether or not it becomes an acceptable tool for diagnosing and resolving conflicts about social issues like hazardous materials transport and storage depends largely on the analysts' skills in avoiding the pitfalls described above.

Note

[1] The research reported in this chapter was conducted jointly with Ward Edwards and Richard John of the University of Southern California under the sponsorship of the National Science Foundation (grant No. PRA-8108683). The views and opinions expressed here are solely those of the author and do not necessarily reflect the opinions of the US Government or any of its agencies.

References

Dyer, J.S. and Miles, D. (1976), An actual application of collective choice theory to the selection of trajectories for the Mariner Jupiter/Saturn Project, *Operations Research*, **24**, 220–243.

Edwards, W. (1980), Reflections on and criticism of a highly political multiattribute utility analysis, in L. Cobb and R.M. Thrall (eds), *Mathematical frontiers of behavioral and policy sciences*, pp 157–186 (Westview Press, Boulder, CO).

Keeney, R.L. (1980), *Siting Energy Facilities* (Academic Press, New York).

Keeney, R.L. and Raiffa, H. (1976), *Decisions with Multiple Objectives: Preferences and Value Tradeoffs* (Wiley, New York, NY).

Keeney, R.L., Renn, O., and von Winterfeldt, D. (in press), Structuring Germany's energy objectives, *Energy Science*.

Kunreuther, H., Kleindorfer, P., and Yaksick, R. (1985), *Compensation as a Tool for Improving Risk Communication*, Paper presented at the EPA–NSF workshop "Risk Perception and Risk Communication", Long Beach, CA.

Manheim, M.L. and Hall, F. (1968), Abstract representations of goals: a method for making decisions in complex problems, in *Transportation: A Service* (New York Academy of Sciences, New York).
Marks, G. and von Winterfeldt, D. (1984), "Not in my backyard": the effect of motivational concerns on judgments about a risky technology, *Journal of Applied Psychology*, **69**, 408–415.
Miller, J.R., III. (1970), *Professional Decision Making: A Procedure for Evaluating Complex Alternatives* (Praeger, New York).
Minerals Management Service (1984), *Draft Environmental Impact Statement, Proposed Southern California Lease Offering, February, 1984* (Minerals Management Service, Pacific OCS Region, Los Angeles, CA).
Raiffa, H. (1969), *Preferences for Multiattributed Alternatives*, RM-5868-DOT/RC (The Rand Corporation, Santa Monica, CA).
Raiffa, H. (1982), *The Art and Science of Negotiation* (Harvard University Press, Cambridge, MA).
Rozelle, M.A. (1982), *The Incorporation of Public Values into Public Policy*, Unpublished PhD Thesis (Arizona State University, Tempe, AZ).
Saaty, T.L. (1983), Conflict resolution and the Falkland Island invasion, *Interfaces*, **13**, 68–83.
Stillwell, W., von Winterfeldt, D., and John, R. (1981), *Value Tree Analysis of Energy Alternatives*, Technical Report No. 81-2 (Social Science Research Institute, University of Southern California, Los Angeles, CA).
Ulvila, J.W. and Snider, W.D. (1980), Negotiation on international tanker standards: an application of multiattribute value theory, *Operations Research*, **28**, 81–96.
US Geological Service (1982), *Pacific Summary Report No. 2*, USGS Open File, Report No. 82-1, US Geological Survey.
von Winterfeldt, D. and Edwards, W. (1986), *Decision Analysis and Behavioral Research* (Cambridge University Press, New York).

11.D1. Discussion

H. Otway [1]

Having little experience of writing discussion papers, I did not quite know how to begin this. So I went to the library to see how discussion papers are usually written. I soon realized that they pose a problem for most of us, being a sort of peer review without the benefit of anonymity. The general format seems to be an awkward attempt, such as this, to start on a friendly tone, followed by something vaguely positive about the paper, after which one engages in the serious business of mild criticism.

Fortunately, I came across a discussion paper on the application of risk analysis, buried in the proceedings of a conference held in Germany in 1979, which had solved the problem. It tackled the business of discussion by addressing four astute, and still quite topical, "theses" posed by the discussant. I decided to use the four theses of that paper (von Winterfeldt, 1980, hereinafter referred to as "DvW, 1980") as a basis for this discussion.

However, this proved to be unexpectedly difficult since von Winterfeldt had cleverly anticipated my criticisms and had dealt with them in this chapter (hereinafter referred to as "DvW, 1985"), which I discuss here; therefore, I will quote liberally from DvW (1985) when addressing the four theses put forward by DvW (1980). I will follow this with a general discussion of value tree analysis and conclude with remarks praising both von Winterfeldt and his paper while expressing reservations about the application of value tree analysis in practice.

11.D1.1. Four theses on the application of value trees

Thesis 1

> The first question should be ... for whom and for what purpose? Failure to answer this ... can lead to the most common pitfall of analysis – addressing the wrong problem. (DvW, 1980.)

As far as "for whom?" goes, DvW (1985) recognizes this problem. Cautioning about the "'irrelevance' pitfall" he advises us to:

> (1) Assure that the analysis is linked to a real arena and a real decision.
> (2) Obtain sponsorship from agencies that are perceived to be neutral in the issues, or obtain a balanced sponsorship from several agencies.
> (3) Work in close consultation with the real decision-making agency, whether or not it is the sponsoring agency.

Point 1, at least at a superficial level, is under the control of the analyst if we assume that he can refuse to lend his skills to unreal arenas and nondecision contexts, although this may not always be true in practice.

Points 2 and 3 are more problematical. It can be difficult indeed to find an agency that is perceived as being neutral by all parties to a controversy. Further, the turf-protecting instincts of organizations with overlapping areas of interest and responsibility might outweigh the possible benefits to them of multiple sponsorship of studies. It might be even more difficult to work with the "real decision-making agency" if it is not the sponsoring agency. An organization with decision-making responsibility is unlikely to lend its support and cooperation to a study over which it has no control and where it has not defined the terms of reference. This may be no more than a simple matter of protecting its decision-making prerogatives.

"For what purpose?" is a rather more difficult question. A general problem with multiattribute techniques lies in the elicitation of objectives; it is not that decision-makers do not know what their objectives are, but rather that their true objectives often cannot be stated for organizational or personal reasons. However, the needs and values of the sponsor do help shape the outcome of research. This was foreseen by DvW (1980) who says: "The different purposes of risk assessment for ... clients include satisfying intellectual curiosity, legitimizing already made decisions...". Clearly the legitimization of decisions already taken cannot be stated as the objective of a study. DvW (1985) admits that "very few arenas can accommodate this type of rational display of facts, values, and conflicts.... Only mediation arenas, which are still fairly rare in the kinds of disputes discussed here seem directly amenable to implementing value tree analysis."

On balance, I think it would be difficult for a real decision-making agency to use value tree analysis in practice in a way that would be politically defensible.

Thesis 2

> Risk assessment should aid specific institutions in solving real and complex decision problems. Risk assessments for pure informational or comparative purposes are likely to be irrelevant for decision-making purposes. (DvW, 1980.)

There seems little doubt that value tree analysis, structured to provide information on the issues important to stakeholder groups, is designed to provide information relevant to complex decision problems. However, we must also consider how valid the information obtained actually would be in practice: i.e. could it represent the values of all relevant groups accurately enough to form a basis for policy? Kerry Thomas and I, in a paper which questioned the premises of risk perception research, discussed the need for "depth interviewing with as many different public groupings as is possible", raising "the issue of whether or not the public is willing to cooperate", and speculating that "politically active groups are directly relevant to policy makers (if not policy) ... but that such groups might well be the least inclined to collaborate with researchers who want to investigate their motives" (Otway and Thomas, 1982.)

This concern has also been anticipated by DvW (1985):

> The lack of an appropriate arena often couples with lack of stakeholder interest to create an atmosphere of benign neglect of value tree analysis. One of the first pitfalls ... is this inability to generate involvement and problem focus for the stakeholders.... The second most common pitfall ... is the refusal of some

> stakeholders to cooperate, or the loss of some stakeholders after initial cooperation.... Usually the refusal is a result of strong ideological positions and assumptions that their involvement could be construed as co-option.

DvW (1985) then goes on to identify another problem which can affect the validity of the data:

> One ... bias is a result of "selfishness", which would eliminate any values and concerns that favor an alternative that the stakeholder group does not like. The same bias may lead to a lower weighting of those branches of the joint value tree that favor undesirable alternatives.

Von Winterfeldt recommends ways in which the analyst can try to avoid these pitfalls, but they are not very convincing to me, primarily because they mostly involve advice on how to win the confidence of stakeholder groups and thus are not really "implementable" in the usual sense.

In summary, I think that value tree analysis, in principle, can generate exactly the kind of information that is most relevant to the evolution of policy on controversial topics. Unfortunately, I am afraid that it has inherent limitations in that it is precisely when there is controversy that full and honest stakeholder participation will be most difficult to obtain. If this is indeed the case, then the results of value tree analysis are primarily for "pure informational purposes", failing the test of the second thesis.

Thesis 3

> Solving complex decision problems requires a comprehensive approach which carefully defines the available alternatives and assesses the direct and indirect costs, risks, and benefits of these alternatives in light of the objectives of the decision making institution ... rather than starting with the question "what are the risks?" a comprehensive analysis would start with the questions "what is the decision problem, what are the alternatives, and what are the objectives?" (DvW, 1980.)

Here value tree analysis stacks up pretty well. It avoids the near-sighted risk focus sometimes encountered (e.g. the simple-minded "acceptable risk" notion), it is tied to real decisions, it prefers real arenas, and its primary goal is to get stakeholders to participate in the problem of generating alternatives. This is a significant improvement on most risk perception research, which is not well suited to eliciting responses on salient policy alternatives. In theory, at least, value tree analysis is a good method.

Thesis 4

> Decision analysis is the only comprehensive and practical methodology for aiding complex decisions. However, decision analysis needs to be adapted to the political and institutional realities of decision making on problems of technological risks. (DvW, 1980.)

The fourth thesis shows the author's sense of dissatisfaction, in 1980, with the constraints of traditional decision analysis and his desire to improve it:

> The application of decision analysis to problems involving large technological risks is still in its infancy ... requir[ing] an adaptation of the traditional decision analysis process. For example, risk problems typically involve negotiations between conflicting interest groups. Decision analysis ... has been mainly developed within the single decision maker paradigm. But in spite of its present shortcomings decision analysis ... has the potential of meeting the above requirements for a sound application of risk assessment. (DvW, 1980.)

In retrospect, we see that this was an important historical moment, the first expression of the idea that a value tree could be created by grafting stakeholder group values onto the still immature decision tree.

11.D1.2. Discussion

Von Winterfeldt gives an excellent overview of value tree analysis, identifying its roots in multiattribute decision techniques, explaining clearly how one actually goes about using it, and outlining the common pitfalls which must be avoided. One of the more appealing aspects of value tree analysis is that it is based on the values held by real people, as opposed to those dreary and sterile analyses which go on about what "rational" people would do if they were only wise enough to accept the values of the analyst instead of stubbornly striving to achieve their own objectives based on their own values.

Another appealing aspect of value tree analysis is that it deals with the values of policy-relevant groups. Most risk perception studies, partly owing to financial constraints, have used opportunistic samples, such as club members or psychology students. Samples of the general public, when they can be had, usually require that the data be analyzed on the basis of what could be called involuntary group memberships, i.e. sorted on the basis of demographic variables. These differences may be significant, but are difficult to interpret in a politically meaningful way. Value tree analysis, in contrast, deals with groups that have voluntary, real-world memberships, allowing it to more effectively model real political processes.

Yet another appealing aspect of value tree analysis is the close relationship between the analyst and stakeholder groups. This two-way relationship undoubtedly facilitates a clearer understanding on the part of the analyst of how those affected perceive the problem under investigation. This is a drawback of psychometric pencil and paper surveys: there is little contact with the respondents, generally no feedback to them, and often no way to tell whether all salient items have been addressed or whether the respondents have understood the items as was intended. However, I wonder if we might have a bit of "new toolism" here, an attempt by the decision analysis community to get a piece of the risk action. This raises a few rhetorical questions.

Is value tree analysis really the best framework for establishing what sounds like an analyst–respondent relationship based on trust and mutual respect? Could a psychiatrist or a sociologist, not equipped with the value tree methodology and equations, obtain an equally valid understanding of people's concerns just by listening to them? Could it be that the real function of value trees is to give decision analysts something to do with their hands while they talk to people? Are the good results obtained from value tree analysis a confirmation of the method or a tribute to the interpersonal skills of the analysts? If this be the case, the results may not be replicable by other investigators.

In summary, in a perfect world, where unbiased organizations could be found, where the agencies managing the process and stakeholders holding extreme positions were truly willing to compromise, and where all parties would participate wholeheartedly in the process of fully and accurately defining their value trees, the value tree methodology might be more relevant to policy formation than conventional risk analysis or risk perception studies. But its utility will also depend on the degree to which interest groups are willing to trust unelected analysts to solve political problems in preference to the "smoke-filled room" dealings of elected politicians. This is an important point because value tree analysis is basically the use of the decision analysis framework to model political processes which, although imperfect, do function. In practice, the contribution of value trees to the solution of controversies will ultimately depend upon the extent to which ideal conditions for its application can be found.

11.D1.3. Concluding remarks

I found von Winterfeldt's paper to be very good. It is well written, clear, logical, complete and illustrated with a real case study. Moreover, as we have seen, it quite carefully outlines the shortcomings of value tree analysis as well as its strengths; in all respects it is a nice piece of craftsmanship and the author is to be congratulated.

Value tree analysis itself also deserves favorable mention. It is descriptive, recognizes the political importance of stakeholder groups, is structured around the evaluation of policy alternatives and, as such, is well suited to analyze the political processes that are of most importance to policy decisions on controversial topics.

I am somewhat less optimistic about the prospects for using value tree analysis in practice. Its success in real policy arenas seems too dependent on a degree of altruistic behavior that is rare in interest groups or any other organization. DvW (1985) concludes by saying that "Value tree analysis is still in its infancy. Whether or not it becomes an acceptable tool for diagnosing and resolving conflicts about social issues ... depends largely on the analysts' skills in avoiding the pitfalls...". It will also depend upon finding special arenas that can best accommodate what appear to be inherent limitations of the method. Von Winterfeldt suggests mediation as one possible application; another might be to aid negotiation of terms of implementation once stakeholders have agreed to accept a particular course of action. Despite these reservations, I think the "infant" value tree is worthy of fertilization.

Note

[1] I am grateful to Detlof von Winterfeldt himself for his comments on an earlier draft of this discussion.

References

Otway, H. and Thomas, K. (1982), Reflections on risk perception and policy, *Risk Analysis*, **2**, 69–82.

von Winterfeldt, D. (1980), Four theses on the application of risk assessment methods: Discussion paper on Rowe, W.D. (1980), Risk assessment approaches and methods, in J. Conrad (ed), *Society, Technology and Risk Assessment*, pp 35–38 (Academic Press, London).

11.D2. Discussion

I. Rosenthal

Von Winterfeldt has presented his study on the application of value tree analysis to offshore oil drilling clearly and objectively. In commenting on the problems that arose in this particular case study, von Winterfeldt implied that, while these problems might be of generic nature, resolution could be obtained by changes in methodology. From my viewpoint, that of the manager of a chemical company's Safety, Health, and Environmental Affairs organization, the difficulties he has experienced were not unexpected. However, I am not optimistic that refinement of the techniques can be expected to resolve the problems described. The fundamental problem arises when one moves MAUT or value tree analysis from a decision tool used by a single individual who must reconcile his different interests and values to the other extreme of a group of stakeholders who may feel no need to take each others interests into account.

It is difficult to see how value tree analysis can be successful unless the stakeholders involved accept, at least implicitly, a common utilitarian index of "goodness" and also agree, that once this "goodness" function has been maximized and elucidated (by value tree analysis), they will endorse it. Von Winterfeldt himself points out that value differences and "selfishness" both stand in the way of an agreement on the "optimum goodness". Put another way, there appears to be an unwillingness to voluntarily accept an optimum "goodness" solution if it affects either a stakeholder's fundamental values or his basic material interests.

Von Winterfeldt implicitly recognizes this problem of voluntary action or selfishness when he recommends that the work be done in "close consultation with the real decision making agency". Under such an arrangement, one is essentially dealing with a situation that is similar to the use of MAUT in arriving at an individual decision. The value tree process will make the decision-making agency aware of the degree of support/opposition it faces from the different stakeholders depending on which variant of the proposition it executes. The agency and/or decision-maker can then choose the proposition that will optimize its own utilitarian goodness equation. It can decide which stakeholders it can afford to antagonize, which it must carry along, and where it wants to position itself with the remainder. In essence, value tree analysis then becomes a tool that assists a ratioinal political decision by the group or individual having the power to make the decision. It can promote an understanding of stakeholder concerns and stakeholders' feelings about the level of benefits that the stakeholders believe they are entitled to, and such understanding promotes the ease with which the agency can promote "deals". It also clearly points out which are the single issue stakeholders with whom there is no dealing. For example, why should one expect a group that feels that expanded energy usage is inherently bad to accept any risk associated with energy expansion? There are no benefits to compensate for any level of risk.

In conclusion, I believe that the techniques and approaches put forward by von Winterfeldt are a valuable addition to our search for better techniques of resolving conflict. However, I believe they are most valuable when used by the decision-making unit (person, agency, local council) in assessing stakeholder positions as part of the process by which the decision-making unit optimizes its course of action (a political process) rather than in promoting a decision commonly arrived at by all stakeholders.

PART FOUR

Risk Management and Insurance

CHAPTER 12

Regulating Environmental Risks: A Comparative Perspective

T. O'Riordan and B. Wynne

12.1. Introduction

In this chapter we offer a comparative look at regulation of environmental risk in industrialized economies. We argue that risk regulation is part of a national style of government. But, while there may be a degree of national idiosyncrasy, there are also important and growing elements of common ground underlying contemporary developments in risk regulation. The common features include some dependence upon self-policing by the risk creator, an element of more formal intervention through structured and often laborious consultation, a reliance on expertise, and an uneasy relationship between specific technical advice and judgment on the one hand, and the need to command public confidence and credibility on the other.

These developments are leading to a degree of convergence of regulatory styles: towards more open, externally accountable procedures, towards a greater element of politically as well as scientifically credible safety standards, and towards a more formal approach to enforcement even when there is a greater dependency on self-policing or at least a more collegiate approach to regulator–client relations. These trends will not eclipse the idiosyncrasies of national regulatory styles; but we suggest that risk regulation generally is passing through an important transition built around the practical reconciliation of general needs for "third party" reassurance (public credibility) and the need to define and control specific risks.

This transition is exposing many tensions. More difficulties are in store for the regulation and management of radioactive and other hazardous materials. These difficulties relate to mechanisms to widen even further public participation in both the standard setting and the disposal strategy process, to strengthen the external accountability of monitoring and enforcement, and to ensure that "hazard face" regulatory action is consistent with both organizational policy and the many variables associated with particular circumstances. This latter problem is likely to become most observable in the management of hazardous wastes. Central control and homogeneity of regulation may not fit easily with the realities of hazardous waste management, where flexibility and discretion are important practical factors.

The usual approach to analyzing the impact of political and cultural factors upon regulation has tended to imply that facets of national style make a uniform local imprint upon regulation whatever the issue being regulated. While not wishing to deny the importance of comparative cultural analysis of decision-making and public credibility, we wish to suggest that these cultural factors create different effects on issues according to intrinsic features of those issues. We will illustrate this point in outline by reference to the case of hazardous wastes.

The policy implications of this comparative review are that in all political cultures the general problems of institutional credibility will continue to gain in importance, and implementation failures will only exacerbate them. Therefore local enforcement will be subject to increasingly stringent public attention, and new means of developing shared authority between regulators, regulated, and the public in the face of scientific uncertainty will have to be produced.

12.2. Environmental Risk and Regulation

To regulate means to control by authoritative rules, and to impose consistent and, ideally, predictable restrictions on an activity in accordance with these rules. "Authority" does not mean coercion, but varying degrees of persuasion. Environmental regulation applies to controls over activities which could affect the well-being of third parties who may not be in a position either to know about or to avoid the dangers to which they are involuntarily exposed.

Risk regulation can be divided into four components (as illustrated in *Figure 12.1*):

(1) *Standard setting*: the determination of suitable levels of emission, general environmental safety, and manufacture and maintenance of equipment, judged to be acceptable to the most exposed populations.

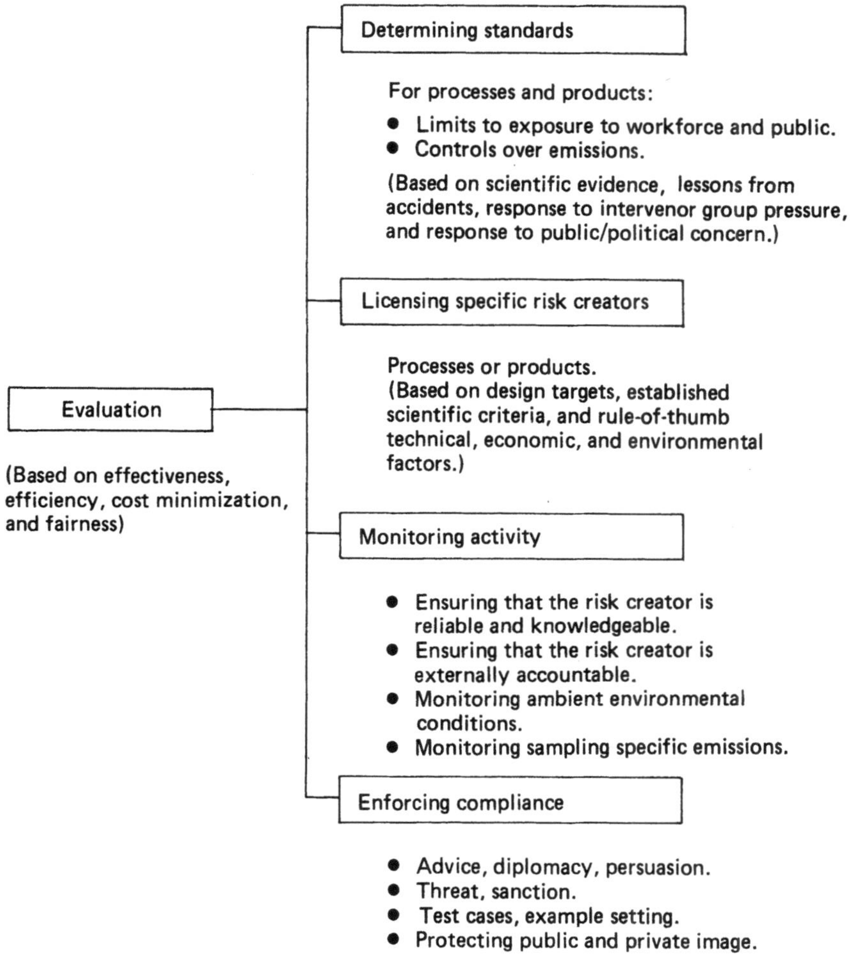

Figure 12.1. The practice of risk regulation.

(2) *Licensing*: the application of specific requirements of health and safety to particular plants, processes, or activities. Licensing is the process of applying the general to the particular on the basis of agreed rules. The application of these rules is, however, rarely predictable or clear cut.

(3) *Monitoring and compliance*: monitoring of regulated activities to ensure that they conform to agreed standards, targets, and licensing conditions. Enforcement of licensing conditions occurs through advice, negotiation, threats, legal action, and penalties. There are important rules and agreed forms of behavior in the enforcement process which are subtly changing as a result of increased public scrutiny.

(4) *Evaluation* of the efficiency, effectiveness, and fairness of regulation to refine and improve the whole regulatory process. This is usually relatively unstructured, although official inquiries, especially following failures, are a formal means of evaluation.

Of these four functions, the evaluative function is the least well understood and followed. There are no agreed criteria for regulation, for much of contemporary regulation has grown organically in response to technical developments and political pressures neither of which are always predictable. Indeed, it is important to see formal regulatory innovations as an *addition* to the informal regulation which is part and parcel of ordinary social and economic interaction. In this way one can more clearly see formal regulation (e.g. via specific legislation and statutes) as rooted in prevailing national processes of political economy and culture, though affected also by the specific history of given issues and regulatory institutions.

Four overall yardsticks for regulation are normally used: effectiveness; efficiency; cost-effectiveness; and fairness. Even without the problem of public justification and credibility, it is all but impossible to reconcile these criteria in environmental risk regulation. Indeed, as Kasperson points out in Chapter 7 of this volume, there are as yet no agreed rules as to fairness in environmental risk regulation and no useful studies of how far equity considerations influence regulation or are set against tests of administrative efficiency and cost-effectiveness.

In 1976 the UK Government agreed that regulation should be so designed that the benefits of control should be more clearly related to costs of regulation (HM Government, 1980, pp 40–41). The aim was to make environmental regulation more economically justifiable so as to reduce any unnecessary financial burden either on industry or on publicly funded activity. The "deregulation" thrust of the Reagan administration in the USA after 1979 was in the same direction. Subsequently, in 1980–1982 an attempt was made to streamline the administrative efficiency of the UK regulatory inspectorates, but initiatives proved unsuccessful. Baram (see Chapter 13) also shows that in the USA, the courts have ruled that cost–benefit criteria do not form a permissible basis for regulatory action. Risk benefit justification is therefore still in confused infancy. Moreover, administrative efficiency is not related to the effectiveness of regulation. Equity considerations do not yet appear on the evaluative agenda at all.

12.2.1. Difficulties in establishing a sound risk regulatory process

Other contributions to this book suggest why environmental risk is such a difficult phenomenon to regulate (see especially Chapter 10 by

Slovic and the ensuing Discussions by Lee and Renn). We summarize the problem as follows:

(1) There is no agreement as to what constitutes "acceptable" risk. The *process* of determining that level of risk is just as important as the actual safety standard finally reached. Acceptability of regulation is about style, and trust, and confidence building (i.e. about institutional relationships of regulation). This is why broad-based public support for the authority of regulatory institutions is such a critical factor.

(2) The whole regulatory process is characterized by uncertainty – uncertainties about the nature and distribution of risk, as well as about the distribution of responsibility and its possible consequences. The models used for identifying and predicting hazard are subject not only to "marginal" imprecisions (the conventional framework of risk management) but also to fundamental ignorance. Regulatory processes artificially process the latter into the former.

(3) The increasing use of formal models – both statistical (in the form of mathematical relationships based on rigorous laws) and conceptual (in the form of imagined combinations of circumstances and events building up a picture of possible faults) – can be at odds with more conventional approaches to regulation based upon experience, tested judgment, and intuition. These conventional approaches depend upon a regulator's empirical knowledge of the process – social as well as technical – his/her understanding of what technical developments could be applied to improve that process, a sense of what is managerially and financially practicable for the risk creator to cope with, and a diplomatic recognition of when and how to put on the pressure. Note that none of these factors involve particular models of risk: they relate to what the regulator "thinks is right" and are often highly flexible to particular conditions.

(4) Public interpretations or riskiness may be at variance with the judgments of professional regulators. This is probably the most difficult problem facing the modern regulatory official and the politician, nowadays responsible for determining final standards. People do not judge risks solely in terms of physical dangers, but regard them as part of a technology or a decision-making process with which they feel comfortable or uncomfortable. People are also confused by the political and technical mystification associated with expertise; in their confusion, they cling to the views of credible authorities. Given the absence of genuinely independent but informed commentators, these "credible authorities" may be "public interest" pressure groups, the media, their trade union,

or other trusted social reference groups. As science has been increasingly (over)used as an attempted means of regulatory authority, so it has been correspondingly "deconstructed" (exposed and found wanting) in public by critical scientists (and well-briefed lawyers) demonstrating the lack of any watertight logic-bound or evidence-bound basis of official regulatory claims for scientific justification of standards and policies. Authority and expertise have therefore been dispersed into a confusing tangle of competing claims and allegiances. This has occurred to different degrees in different political cultures, depending upon the style or social use of science, and the institutional means of reviewing its claims. We illustrate this point later by contrasting the US and UK cases (pp 323–325, 329–330).

(5) "The public" is not a monolith of similarly behaving individuals. There are a great variety of "publics" responding to environmental situations. What appears evident from research (see, for example, Otway and Thomas, 1982; van der Pligt, 1985) is that two polarized groups stand apart: one generally favors technology and economic progress through current mechanisms of innovation, and is quite supportive of established expertise and regulation; the other group is suspicious of complex technology or, perhaps more accurately, of the corporations that devise and promote it, frustrated that they cannot halt or at least alter the development of that technology, and determined to be part of the process of regulation which they regard as much a political as a technical exercise. It is wise to be cautious about any firm figures, though it is estimated that about 30% of the population of Western nations fall into the former category and about 20% into the latter, with about half the population in the middle seeking reassurance. The latter may be sceptical, but are capable of being moved in either direction (see Milbrath, 1984). One of the most important issues in contemporary environmental risk management is the battle for public credibility, a major reason why risk regulation is so much more politicized than it used to be.

(6) What emerges from all this is escalating uncertainty as to how to proceed with the regulation of environmental risks. The feasibility of reconciling all the conflicts appears to be diminishing fast. The technical people prefer *quantifiable* and *predictable* criteria upon which to base judgements – that is the stuff of science and the scientific method. The risk-creating industries likewise seek clear targets and standards which lead to consistent licensing procedures so that they can embark on major investments with very long time horizons with a high degree of managerial confidence. Thus industry laments the constant alteration of regulatory requirements in response to unpredictable political

reactions, adverse publicity over accidents and near accidents (some of which are unsuccessfully covered up), and developments in scientific understanding and technological innovation.

Examples abound. In the aftermath of every environmental disaster there are significant shifts in regulatory policing and licensing conditions. Following the Love Canal case in 1978, the emphasis of the Resource Conservation and Recovery Act (RCRA) 1976 switched from energy and materials conservation to toxic waste disposal and (in theory) considerably toughened regulation over disused and operational chemical waste sites (see Chapter 5 by Kleindorfer and Kunreuther and Chapter 13 by Baram in this volume). Even so, RCRA had to be radically tightened by congressional reauthorization in 1984. In the wake of Three Mile Island, the worldwide nuclear industry passed through a quantum leap in regulatory safeguards (see Chapter 7) with an estimated $550 million invested in retrofitting new safety equipment (Evans and Hope, 1984). Both Lagadec (Chapter 1) and Naschi (Chapter 2) address some of the responses likely to confront the multinational chemical industry in the tightened regulatory climate following the Mexico City gas explosion and the Bhopal fertilizer plant disaster. The Seveso tragedy of 1976 has left its legacy in a European Community Seveso Directive (European Communities Council, 1982) which requires much more formal justification of safety and evacuation measures from various classes of chemical works, a more open risk assessment undertaken by industry, and more formal regulatory review of both evacuation plans and information availability both to workers and to the public (see also Chapter 13 by Baram). An important point about many of these reactions and innovations is that they tend increasingly to be of international scope or interest, setting precedents for attempts at standardization across national regulatory arenas.

12.3. Risk Regulation and Governing Styles

Risk regulation is set in institutions and instruments of government which reflect established national traditions and customary practices. Nevertheless, there are certain common characteristics to all forms of regulation, characteristics which become molded by national convention:

(1) Consultation is a vital part of government. It is not only wise but politically necessary to discuss likely policy proposals with key affected parties. This not only helps to smooth out disagreements, but signals to decision-makers what is achievable, if not

always acceptable. There will be notable differences as to *what* is discussed and *when*, but some form of dialogue is a feature of all regulatory approaches. What is likely is that this dialogue will become more comprehensive, extending to greater use of advisory committees and to well-staged forms of public participation (see notably Chapter 11 by von Winterfeldt and the Discussion of Chapter 10 by Renn). Nowadays many regulatory procedures require affirmative evidence of consultation (e.g. the Seveso Directive), while, as Cohen points out in his Discussion of Chapter 13, UK regulatory practice depends on a round of discussion and commentary from interested parties.

(2) A dependence upon expertise is also necessary, though with regard to environmental risk "expertise" has a changing connotation. As already noted, credibility and reassurance become equally vital as components of advice as technical competence. Two interesting developments stem from this. One is a more explicit recognition that experts have to become familiar with expertise in other areas. Engineers, for example, need to know a lot more than their predecessors about toxicology or medical diagnosis, while economists require a sound understanding of fault tree analysis and cognitive psychology. This point is developed by Covello and Merkhofer in Chapter 8, though the connection between engineering specialisms and creative economic accounting is far from well established. Risk regulation will have to depend more and more upon "synthetic expertise", with a truly international flavor.

The second development is that the expert has to devote time and effort to the skill of being able to translate specialisms into a language that is readily comprehensible to affected parties, in a form that allows them to react in a way in which they feel most effective and comfortable. This means conveying *faithfully* to intelligent but lay people both technical substance and concepts of uncertainty. This extraordinarily difficult task is not one for which either the pure or the synthetic expert is appropriately trained. Nor are regulatory bodies yet very adept at representing uncertainty. These points are addressed by Slovic in Chapter 10 and Lee in his Discussion of Chapter 10, though neither offers specific prescriptions as to how these problems can be overcome in practice.

(3) Regulatory compliance depends upon some degree of self-policing. This is partly because resources are simply not available for comprehensive scrutiny. But it is also necessary because self-policing forms a vital element of compliance through streamlining consultation, developing a collegiate sense of cooperation, and sharing expertise. Without a fairly high degree of self-policing,

enforcement could not be cost-effective: the contemporary problem is to seek a new balance between scrutiny and client dependency to complement the requirements of political acceptance and public reassurance. What is evident, however, is that the relative degree of self-policing is a highly variable commodity. It is much more likely to be found among large corporations with a genuine interest in "good neighbor" relations, or where the consumer may react to adverse publicity over certain products, and in industries where there is a publicly recognized danger (e.g. certain classes of the chemical industry). It is least likely to take place among small industries which are not perceived to create a hazard or where the regulatory arm is too short to ensure adequate surveillance.

(4) As has already been mentioned, the technical aspects of risk regulation are increasingly being infiltrated by political considerations. The determination of "acceptable risk" means the adoption of standards that technical people regard as unnecessarily strict. The so-called "gross disproportion" yardstick of determining at what point an additional safety measure is no longer justified by the social benefits gained is widening. A pervasive feature of hazardous material regulation is the growing strictness of safety standards at ever greater cost per statistical life saved. For example, at the margin, the cost of saving a statistical life from, say, cancer or an especially severe explosion can be as high as $20 million. Typical values of life in the insurance industry rarely exceed $250 000 and, even in notoriously high court cases of compensation over liability suits, awards rarely exceed $3 million. Mummery (1985) suggests that the cost of saving a life in the recent decision to install 150 million pounds sterling of pollution control equipment at the Sellafield nuclear fuel reprocessing plant in Cumbria, England, could be as high as $100 million per possible life lost. Even then, however, elements of the public remain dissatisfied. This is why the "residual risk" phenomenon is becoming such a thorny issue.

These four common elements are, however, filtered and structured by governing styles in different ways. For the purpose of simplicity and illustration four differing regulatory approaches can be identified. One is based on ***adversarial principles***, another on ***consensual lines***, a third upon ***authoritative procedures***, and a fourth upon ***corporate characteristics***. In no country is the "pure" form of any of these approaches found, but it is interesting to note that each of these approaches is associated with distinctive characteristics in regulatory institutions and regulatory decision-making. Some institutions are designed to be authoritative: they have executive powers, they are controlled by nonaccountable administrators and technical advisers, and they minimize the practices of consultation and negotiation. Some regulatory bodies are so structured as to build in representative opinion, through selective and informal procedures (this is common in the UK, for example – see Cohen's comments on Chapter 13). Others rely extensively on varying forms of advisory committees with more or less representation from wider interests. Others still combine technical expertise and interest group representation so that a more corporatist approach is pursued. Governing style, regulatory approach, and regulatory structure interconnect to produce patterns of commonality within other patterns of idiosyncrasy.

We caution against too ambitious an interpretation of these "styles". In almost all regulation, various combinations of approaches are tried (see Brickman *et al.*, 1982). Indeed, examination of regulatory approaches within each "style", such as that conducted by Wilson (1981) in the USA, demonstrates a marked diversity of approach between agencies depending upon such factors as their "age" in the administrative order, their regulatory experiences, related interest groups, and, critically, supporting regulation and statute. An ex-deputy director of the US Food and Drug Administration (Hutt, 1983) confirms that even within formalistic rule-making procedures a federal regulatory agency can exert informal judgment and discretion. So, while styles are important, in that they set contexts and establish custom and tradition, there are important variations.

Another critical reason why regulatory style will vary is because of the structure, political relations, and modes of accountability of the institutions being regulated. Take, for instance, the question of nuclear power. In the USA, the utilities are disaggregated, customer sovereignty is powerful, state bodies control the prices, and partial competition is rife. In France, the opposite is true: the industry is highly centralized, and it commands a high degree of state support, and, generally speaking, public loyalty. In the UK, the industry is also monolithic, but it is subject to far greater external accountability both in the setting of safety and in pricing. We observe that regulatory issues may have intrinsic properties that affect the optimal style of regulation, but that there is no universal distinction between regulator and regulated institutions, because both are being evaluated by critical expertise and public prying.

12.3.1. The adversarial approach

The adversarial approach to regulation is particularly associated with the USA, but elements are becoming evident elsewhere. Its principal characteristics are the establishment by large executive agencies of precise standards and detailed rules, written down so as to be capable of precise legal interpretation. Regulation, especially in standard setting and licensing, is very formal and elaborate. This is because the agencies are constitutionally independent of the legislature (Congress), while the judiciary is able to review the legitimacy of agency decisions. The scope of judicial review has waxed and waned under various doctrines, but regardless of these fluctuations litigation against administrative decisions is constitutionally far more possible than in arrangements (e.g. those in the UK) where executive and legislature are combined via the doctrine of parliamentary accountability of administrative decisions, and through collective ministerial responsibility. The adversarial approach is also characterized by high levels of openness of documentation and hence exposure to criticism and lobbying from a wide array of interests, which usually have many points of entry. The regulatory process is therefore subject to incessant conflict, bargaining through scientific claim and counterclaim, and formal adjudication. This occurs most notably in the standard setting and licensing procedures.

An important distinguishing feature of the adversarial approach to risk regulation lies in the control of safety and design by the regulatory authority. The client, or risk creator, is expected to meet predetermined conditions and to obey preset rules. The responsibility for safety lies mostly in compliance. This is judged to be necessary because no single risk creator is deemed big enough to have the resources to undertake its own safety assessment. This may be true of many of the nuclear power utilities, but it is not true of chemical corporations. There is an uneasy lack of consistency in precise location of regulatory control from one risk creator to another. This is particularly evident for hazardous waste management, though less so for radioactive waste.

Critics of this approach point out that the regulated actor is not encouraged to be responsible for thinking about safety at all stages of project management – concept, standard setting, design, fabrication, and execution. As an externally imposed concept, safety does not become part of the "mindset" of the whole process of project conception, birth, and life.

The adversarial approach tends to be found in governing systems where a strong central institution is expected to regulate very diverse and geographically dispersed client groups, where patterns of ownership, management skills, and environmental circumstances vary enormously. It also reflects a political culture of institutionalized distrust (the separation of powers) which is suspicious of self-regulation and expects accountability. Cultural anthropologists (see for example, Thompson, 1982; Douglas, 1986) have also argued persuasively that the adversarial style is most consistent with political cultures where pluralistic groups have a strong influence on public life. This is true of populist political culture in the USA. The adversarial approach can be very time-consuming and procedurally costly, and does not necessarily result in more equitable regulations. The corresponding tendency of regulatory bodies to try to regulate everything in standard fashion ab initio, rather than proceed incrementally and pragmatically, tends to create "all or nothing" regulatory syndromes, with dramatic ups and downs, or successes and failures. Wynne (1986) illustrates this in his study of the regulation of hazardous wastes in the USA.

The adversarial process is also arguably more prone to instability, so that the risk creator is often unsure whether any particular technology or management decision, which may depend upon long lead time and much financial investment, will be acceptable in a relatively short time scale.

12.3.2. The consensual approach

The consensual approach to risk regulation bears the most marked contrast to the adversarial. It is based on a high level of trust between institutional actors, and on collaborative relationships between regulated and regulator. As a result confidential arrangements develop, ostensibly to protect commercial information, but also to exclude the inflexibilities of third party access. In addition, the consensual approach places a lot of emphasis upon the freedom of the regulator to exercise variable judgment. Rules are expressed so as to be deliberately imprecise and flexible from case to case. Technical uncertainty and imprecision are designed into regulations but are compensated for by a sense of mutual trust and an expressed belief in the basic competence and high ideals of all those involved. The regulator is expected to act in part as adviser, in part as consultant, and in part as policeman. As a consequence, the consensual approach relies more heavily on self-regulation: indeed, the ultimate responsibility for safety usually rests with the regulated, not the regulator. Technical norms are usually "advisory" and carry no statutory power when stated in precise form.

One often-stated advantage of the consensual approach is that the risk creator is in a better position to be assured of some stability in the pattern of regulation. He can expect gradual, negotiated evolution of regulatory standards and licensing procedures. Long-term technological and managerial planning and commitment are therefore more possible. However, this can create problems for the regulator when attempting to enforce compliance if there is a political need to change regulatory standards.

Critics argue that the Achilles' heel of the consensual approach is an overdependence on self-policing. This often means that practices become established which are not easily improved without a lot of external conflict and pressure, organizational change, and financial investment. A classic example can be found in the recently published report of the UK Hazardous Waste Inspectorate (1985). This points out that many waste sites are not properly managed, that illegal dumping is commonplace, that regulations about the practice of disposal are often flouted, and that some waste disposal authorities are clearly incompetent. It will require an enormous investment in managerial capacity and public money to improve this lamentable state of affairs, which has been allowed to develop as a result of complacency fostered by a hitherto undisturbed consensual approach. It is possible that the normal consensual UK approach will be ill equipped to cope with this and that more formal procedures will be required, implying a more adversarial relationship with industry (see Wynne, 1986, Chapter 6).

In the consensual approach, regulatory officials tend to regard themselves as professionally competent and thoroughly expert. They like to be thought of as part of an elite who share expertise, specialized information, and a privileged position with their industrial (regulated) colleagues and who take pride in the quality of their work. This point is well illustrated by Richardson *et al.* (1983) in their study of water pollution control in the UK. Enforcement by recourse to the courts is a rare occurrence because an inspector believes it is a sign of failure to have to resort to litigation rather than persuasion. In his Discussion of Chapter 13, Cohen endorses this line. Regulators also like to believe that they are armed with the best information available, that they can seek advice from colleagues or independent consultants so as to be, in their opinion, in the best position to advise and to nag. They particularly dislike external criticism, especially from those whom they do not regard as members of their "club", and resent political interference. There is therefore a constant danger of complacency and an insular sense of self-satisfaction. In the contemporary politics of environmental risk regulation, these attitudes are likely to be increasingly counterproductive and damaging, a point which is implied by the Tenth Report of the Royal Commission on Environmental Pollution (1984).

The consensual approach is also characterized by a flexible approach to standard setting and licensing. In some important instances maximum permissible exposures or discharges are precisely laid down and universally applied – notably with respect to radioactive exposure for workers and the general public. But these are regarded as backstop outer limits, not to be even approached in normal circumstances. Discretionary practical standards of "best practical means" or "as low as reasonably achievable" doses are supposed to be negotiated and fulfilled well within these precise maxima.

In most aspects of environmental risk, levels of exposure, standards of monitoring, and conditions for maintaining equipment are not precisely laid down. The guiding principle is what is practicable – namely, what seems appropriate to the circumstances of the risk creator within general guidelines, having regard to the quality of the environment in the locality, the state of technology, the cost of meeting regulations, and public opinion. There is no legal definition of public opinion: the courts usually interpret it as what reasonable people ought to expect. "Reasonable people" are those who apply "common sense and balanced judgement". It follows from all this that litigation in the regulatory process by intervenor groups is unusual. Indeed, it is made almost impossible on formal legal grounds, since regulations are administrative decisions which in the UK are theoretically accountable to parliament, not the courts, unless a minister is thought to have overstepped *legal* (but not technical or social) propriety.

This is a key issue for trade union representatives and for public spirited environmental groups anxious about exposure to low levels of possibly carcinogenic materials. This point is well documented by Frankel (1984) who is the lead campaigner for a statutory right to know about all forms of pollution risk in the UK and who is one of the drafters of parliamentary legislation aimed at guaranteeing a right to information. Note, however, that the purpose is not to increase the scope for litigation, which is not particularly relished in the consensual culture, but to ensure a better surveillance of administration action by scrutiny and political pressure (see Macrory, 1983).

12.3.3. The authoritative approach

In the authoritative approach, regulators are granted considerable freedom to set standards and enforce compliance, with minimal (though often important) consultation either with the risk creators or with the public. The formal scope for legal redress is limited, except where regulators can be shown to have acted arbitrarily or contrary to statutory procedures.

This approach is most likely to be found in countries with strong central government but weak legislatures, where local or regional government is constitutionally limited to executing commands from the center, and where the public have little tradition of militancy or distrust. The closest example to this approach is to be found in France, where the regulatory official is granted considerable freedom to act, where rights of appeal are restricted, and where public consultation is formal, preemptory, and one-sided (see Macrory, 1982). The most elaborate aspect of this style is usually negotiation in private between government agencies, and with industry, though even the latter is performed in a climate of strong governmental elitism and paternalism.

The authoritative style is regarded as efficient and cost-effective by its supporters, but unjust and inequitable by its critics. It is not an approach which commands intrinsic public confidence, except by default of issues being brought to public attention. Hitherto, it is this ability to keep issues from the public agenda which has been its strength, but in the area of environmental risk regulation this is already becoming eroded. Its mode of operation is so at odds with the public requirements increasingly demanded of risk regulation already summarized above that it is likely to undergo further change. Nevertheless, the point about the interconnectedness of style, and approach and structure must not be forgotten. In France, for example, regulatory officials in the nuclear industry believe very much in their exclusive powers and competence: they will not give way easily to any attempts to open up their standard setting or licensing procedures. In the hazardous waste field too, government agencies have created what is regarded as a successful infrastructure of waste treatment and disposal facilities, and regulatory instruments including economic incentives to enhance their proper use.

For the reasons already stated, there are few examples of pure authoritarianism as associated with professional elitism, and of relatively uncritical support for particular kinds of economic development.

12.3.4. The corporatist approach

In many respects the corporatist approach is an amalgam of the other three operating in a particular structure of relationships. Corporatism refers to collegiate forms of organization in which different interests maneuver to promote their common interests. Corporatism is found where powerful groups have mutual advantage in acting collectively. In the FRG, for example, trade unions and the chemical industries work hand in hand (although somewhat formally) in the standard setting process over chemical product regulation. Corporatism, therefore, is appropriate for consensual and authoritative approaches to regulation where major interests perceive a collective self-interest in establishing alignments both with and against regulators. Corporatist structures are difficult for "footloose" public interest organizations to penetrate, and can also provide a powerful block to international agencies seeking to impose internationally agreed standards on a national interest.

Corporatism as manifested in the FRG system could be regarded as an intermingling of consensual or authoritative elements of institutional structure with embryonic adversary elements. Thus the FRG constitutional structure is formally set up on federal lines, with considerable state autonomy and a strong legal oversight of administrative actions at both state and federal government levels. It was legal recognition of protesters' claims of arbitrary government decision-making, for example, which created a holdup in the FRG nuclear program. Yet historical elements of elitism and informal collaboration between government and industry mean that a strong sense of mutual, unitary responsibility exists, even extending to labor interests. In addition, science is established in a strongly hierarchical way which consolidates almost a monopoly role for experts in science and technology decision making. Thus, even with the embryonic adversary elements of government, scientific advice tends to be consensually authoritative, being challenged far less than in the USA or the UK.

The result is a form of decision-making in which composite interest groups nominally express their positions and negotiate in relatively formal ways compared to the UK, yet not in any way like the completely public, legally ritualized, and extravagantly adversarial ways of the USA. Corporatism means a formal structuring of composite interests in a way alien to the UK, but with a very strong political pull to achieve consensus in a manner alien to an adversary system. A typical example from hazardous waste management in the FRG is that provided by Wynne (1986, Ch 4) where a joint committee of technical experts from all the states, together with a federal representative, developed central regulations – the hazardous waste classification scheme – through which a consensual agreement is expected. Only then is the proposed scheme revealed to industry, who negotiate their own revisions relatively formally, though not in public. In the UK, industry would have been represented at the outset had such a committee even been formed; in the USA, "negotiation" would have been via litigation, or via formal hearings procedures for "de-listing", akin to litigation.

12.4. Convergence and Coalescence of Regulatory Styles

This discussion of the corporatist regulatory styles emphasizes the qualification made at the outset, that these four comparative typologies should be regarded as *ideal types*, none of which exists in pure form in any system. Nevertheless they are useful in two related respects.

First, they can be treated as patterned indicators, providing descriptive categories which point to unidentified or unclarified underlying structural features of political culture that affect styles of government. The work of Douglas and Wildavsky (1982) and Thompson (1982) provides a promising analytical framework, using the grid group anthropological terms of Douglas (1986) to characterize different types of political regime or political cultural style. Even so, an ambiguity exists as to whether even such a fruitful framework as this can ever hope to clarify "intrinsic" structural features as explanatory factors of regulation; our view would be that "structural" in this sense is inherently ill-definable because the *intrinsic* nature of "structure" in social terms is always a composition of heterogeneous, partly conflicting, dynamic tensions and tendencies. Therefore such typologies of comparison, even theoretically embedded ones, will only ever be approximations.

Second, these ideal types provide a framework within which different substantive elements of local "style" can be identified and separately treated. For example, once these categories are treated as useful analytical constraints, we can see that some features of regulatory style may be more identifiable than others. To illustrate this, the Netherlands exhibits elements of "pillared" interest group pluralism and open adversary interest conflict right into the heart of cabinet decision-making; as such it arguably shows more approximation to the US model of the adversary style of regulation than any other European comparison. As discussed in Wynne (1986), the facts that the Dutch experience probably shows a more open information circulation, that the Dutch use highly elaborated precise numerical standards, and that the Netherlands was the only European country to react to past hazardous waste dumps crises by specific emergency legislation like in the USA, are all circumstantial indicators of an adversarial – or what Douglas, Wildavsky, and Thompson call "sect-dominated, anti-hierarchical" – political culture. Yet, when one looks at other elements of Dutch regulation, this identity fails. Central government–industry relations look more consensual, or at least corporatist, than those in the USA. There is more effective pressure to reach inter-interest compromise among like-minded interests. The open pluralism of Dutch decision-making is arguably more to do with a long history of coalition government and lack of collective cabinet responsibility than it is a function of anything akin to the US history of constitutionally established mistrust and mutual limitation of powers.

Likewise, one can take points of similarity between the Netherlands and the UK (e.g. close informal links between government and industry, and informally close-knit industrial lobbying) and still find strong differences, as on the openness of information and third party access, and on the role of science. In the UK, for example, scientific judgment and advice are normally in a privileged discretionary position within or very close to government, and are incorporated with all kinds of unquestioned value judgments into the heart of policymaking. In at least some Dutch regulatory activity, science plays a more open and directing role, being part of legally inflexible regulatory implementation, via "mechanical" decision rules (e.g. precise concentration limits in hazardous waste definitions). The institutional roles of science reflect aspects of the history of scientific professionalization generally, yet are also consistent with other features of local political culture.

The above discussion was intended to point to the pitfalls of simplistic readings of comparative frameworks for analyzing regulatory practices in different settings. At the very least, outcomes of regulation are not predictable by comparing formal regulations; and indeed such comparison is often impossible when some systems do not in their nature even express any formal standards. This might imply that one should confine analysis to comparison of actual enforcement practices. Such practices are still subject to gross variations and uncertainties despite the recent growth of attention devoted to them. However, the wider and more complicated comparison should be attempted.

Narrowing the parameters of comparison in regulatory practice may give a misleading picture. A quick look at the hazardous waste classification systems between, say, the US and the UK suggests an enormous discrepancy: whereas the US system has over 400 items, the UK system has about 30. Closer inspection, however, reveals that the classification patterns systems are not comparable. Each UK category is a composite (e.g. "mercury-containing compounds") which may equal tens of the specific chemicals listed in the US counterpart. Even when this discrepancy of method is clarified, it appears that the UK list serves a different regulatory function from that served by its US counterpart. Wastes on the UK list can be exempted by producers if their waste does not meet several other criteria; it is up to the local regulatory agency legally to prove that all relevant criteria have been met. Wastes on the US list are controlled *in addition* to any material not so listed, which are *also* controlled if they meet certain specified (but different from the UK) tests (which it is up to the *producer* to perform and prove). Not only this, but the UK hazardous waste list is only meant to be for *registration* purposes, the locus of official *control* being site licensing of treatment and disposal facilities. The US and FRG hazardous waste management programs are meant to restrict the options available for treatment and disposal: i.e. the restriction applies to a given waste, not a given facility as in the UK.

From all of this one can see that in many cases it is very difficult to compare regulations in a meaningful way. Of course, in cases where decisions to ban, for example, aldrin/dieldrin have occurred in the US, but not the UK, a comparative analysis is easier to undertake (Gillespie *et al.*, 1979). However, the point remains that in many cases what is more comparable is how the regulatory process *justifies* or *attempts to justify* its decisions, i.e. how it tries to cultivate social authority, rather than the substantial content or effects of those decisions.

In addition to these complications in attempting comparative analysis of regulatory styles, it is important to remember that different categories of risk are not necessarily capable of being adapted to similar regulatory mechanisms. Hazardous waste regulation is a very different issue from, say, the regulation of food additives or contraceptive devices. In the latter cases, consumer interests are much more apparent and will affect regulation; for hazardous wastes consumer interest is at best indirect. Public interest enters at an entirely different point, namely the proposed transportation methods and disposal facilities. Furthermore, the role of national comparative analysis is different. In the food additive and contraceptive pill issues, international comparison may be stimulated by industry if an international production or marketing structure exists, in order to generate a uniform regulatory environment. In the hazardous waste case, national or international industries may prefer a state of inconsistency in national approaches, because this may allow them to trade in wastes, in order that cheaper disposal options might be found.

Yet international standardization may be encouraged by public pressure reacting to loopholes. Wynne (1986) has described several specific properties of hazardous waste which render it unsuitable for conventional "dispersive" environmental regulation. These properties – notably the complex physical behavioral life cycle of hazardous wastes, the ambiguity in defining the key terms "hazard" and "waste", the extreme heterogeneity of risk-generating materials and circumstances, and the fact that regulation per se also means cultivating a new, hazardous industry – all mean that standardized, precise, and inflexible regulatory approaches and techniques are unrealistic. Consequently, a variety of institutional mechanisms have to be established. These considerations cut across local elements of regulatory style and, from country to country, will harmonize more in some systems than in others. The same general point, though with different specific factors and conclusions, can be made for other categories of risk. In sum, we argue that circumstances surrounding the management of hazardous wastes are creating pressures that tend toward convergence of regulatory approaches while the realities of regulation are demanding a much more varied regulatory response in actual practice. This important divergence has not been properly analyzed.

12.5. Conclusion: Convergence versus Differentiation

We have identified common currents underlying environmental risk regulation regardless of the context in which it is practiced, and have discussed the different elements and interactions of local regulatory styles, embedded in political cultures, which sift and shape these common properties. This mix of factors is further complicated by the increasing significance of certain "languages" used for public justification and to enhance credibility. The outcome is a form of double discourse. There is the partly judgmental, flexible task of describing and controlling countless actual risk phenomena. This clashes with the need to project reassuring images of universal and formally accountable scientific methods coupled with precise control instruments and standards – in short, a public image of management competence, fairness, and trustworthiness.

This potential dualism exists in all social encounters, but it becomes more significant in some political cultures, notably the consensual and corporatist styles. Other universal factors underlying regulatory change and diversity have already been described: the pressure for accountability and explication of "internal" regulatory decision rules; the trend in modern economic conditions towards cost–benefit scrutiny of extending regulatory scope or rigor; the decreasing ability of scientific evidence to command authority (i.e. the dismantling of the mysteries surrounding scientific knowledge; Rip, 1985; Wynne, 1986) and the corresponding need to act in advance of conclusive scientific opinion. All this tends to increase dependence upon, or to encourage the development of a more justifiable *overt defence* of, informal self-regulation.

These common factors also have to be interpreted in local settings. Why is it, for example, that the interest in exposing the conventions underlying regulatory science in the US Environmental Protection Agency (EPA) arises from interests sympathetic to deregulation, whereas in the UK it is associated with a demand for stricter regulation (Millstone, 1985)? Would one find the same relationship if one examined the regulatory science of an older, more established, and less "environmental"-oriented US agency than the EPA – say, the Food and Drug Administration? This double discourse around science takes a different form in the UK from that in the USA, because, in the UK, a different image of what science really *is* dominates the process of legitimation and social authority. In contrast to the formalistic rule-bound image of science in the USA, in the UK the dominant symbol remains that of a "priesthood" – of craft skill, of tacit judgment, of intuition finely honed over years of experience, where ability to follow formal rules of scientific method is merely competence, not *expertise*. This corresponds with the central UK role of discretion, informal negotiation in private, and nonspecified standards. Yet, even in the UK, "third party" pressure, from such as the European Commission, parliamentary committees, more analytically competent environmental and labor groups, and even local authorities, is increasing the demand for more formal justification. So formal public accounts of decision rules are required even when these rules have not been subject to thorough external review.

Overall, the regulatory response to rising public mistrust has been to intensify traditional forms of symbolic action, in order to reiterate and rephrase scientific statements of justification and reassurance. This tends to heighten inflexibility and may inhibit the ability of regulatory bodies to respond flexibly to variations in the realities of risk creation and control. In many cases, the most potentially significant dislocations exist between central and peripheral parts of the same regulatory network. The problem here is that the center responds to the needs of providing symbolic reassurance, while the periphery (i.e. the inspectors) actually attempts to regulate, mindful of the need to be flexible and discreet. Add to this the growing sophistication and intensity of specialist "public interest" bodies and it becomes evident that stability and common ground in environmental risk management are unlikely to be allowed to flourish.

We conclude, therefore, that as "official" institutional means of regulation fail to meet political demands, so unofficial methods will increasingly take their place alongside formal regulation. Idiosyncrasy, uncertainty, and heterogeneity are likely to increase in practical regulatory issues and settings, even if, as is also likely, the official international language of regulation becomes increasingly uniform. The dangers are that *environmental* risk regulation will try to conceal its own problems by accepting for supposed management and control almost whatever ends up in the "back-end" of the chain, production → waste → disposal → residual risk. Tensions, uncertainties, and disorientations are likely to grow while "environmental risk management" accepts a wildly overflowing agenda of responsibilities and attendant uncertainties without looking for the means to reflect these into their proper arenas. We refer here to the measured and consistent social control of all aspects of industrial production and the more rounded justification of technological change. It is only in that "whole" context that risk regulation can be fairly evaluated, for hazardous waste management is really a "last frontier" in pollution control. Attempts to internalize the full costs of production have never entirely succeeded because, in the past, a relatively uncontrolled outlet existed. The current challenge is to close this last externality. Whether this can be done sporadically and inconsistently, or in more measured ways, will test the ability of hazardous waste management to lead the regulatory field.

References

Brickman, R., Jasanoff, S., and Ilgen, T. (1982), *Chemical Regulation and Cancer: A Cross-national Study of Policy and Politics*, Report for the National Science Foundation, National Technical Information

Service (NTIS), Accession No. PB83-206771, November (NTIS, Washington, DC).
Douglas, M. (1986), *Risk Acceptability According to the Social Sciences* (Russell Sage Foundation, New York).
Douglas, M. and Wildavsky, A. (1982), *Risk and Culture* (University of California Press, Berkeley, CA).
European Communities Council (1982), Council Directive of 24 June 1982 on the major-accident hazards of certain industrial activities, *Official Journal of the European Communities*, 5 August.
Evans, N. and Hope, C. (1984), *Nuclear Power: Futures, Costs and Benefits* (Cambridge University Press, Cambridge, UK).
Frankel, M. (1984), *Protecting the Polluter: A Report on Environmental Secrecy* (Campaign for Freedom of Information, London).
Gillespie, B., Eva, D., and Johnston, R. (1979), Carcinogenic risk assessment in US and UK. A case of aldrin–dieldrin, *Social Studies of Science*, **9**, 265–301.
HM Government (1980), *Report on Non Departmental Public Bodies*, Cmnd. 7797 (HMSO, London).
Hutt, T. (1983), *Consensus without Litigation*, Paper presented to Bellagio Conference on Regulatory Policies for Carcinogens, August 1983, Bellagio, Italy.
Macrory, P. (1982), *Public Inquiry and Enquete Publique* (Institute for European Environmental Policy, London).
Macrory, R. (1983), Environmentalism in the courts, in T.O'Riordan and R.K. Turner (eds), *Progress in Resource Management and Environmental Planning*, pp 153–168 (Wiley, Chichester, UK).
Milbrath, L. (1984), *Environmentalists: Vanguard for a New Society* (State University of New York Press, Buffalo, NY).
Millstone, E. (1985), Personal communication to B. Wynne, September.
Mummery, P.W. (1985), *Reduction of BNFL Discharges: the ALARA Principle* (British Nuclear Fuels, plc, Warrington, Cheshire, UK).
Otway, H. and Thomas, K. (1982), Reflections on risk perception and policy, *Risk Analyisis*, **2** (2), 69–82.
Richardson, G., Ogus, A., and Burrows, P. (1983), *Policing Pollution: A Study of Regulation and Enforcement* (Oxford University Press, Oxford, UK).
Rip, A. (1985), Experts in public arenas, in H. Otway and M. Peltu (eds), *Regulation of Industrial Risks* (Butterworths, London).
Royal Commission on Environmental Pollution (1984), *Tackling Pollution: Experience and Prospects*, Cmnd. 9149 (HMSO, London).
Thompson, M. (1982), Postscript: a cultural basis for comparison, in H. Kunreuther and J. Linnerooth (eds), *Risk and Decision Processes: The Siting of LEG Facilities in Five Countries* (Springer Verlag, New York and Berlin).
UK Hazardous Waste Inspectorate (1985), *First Report* (Department of the Environment, London).
Van der Plight (1985), Public attitudes to nuclear energy: science and anxiety, *Journal of Environmental Psychology*, **5** (1), 87–98.
Wynne, B. (1986), *The Hazardous Management of Risk – Comparative Institutional Perspectives* (Springer Verlag, New York and Berlin).

12.D. Discussion

T.J. Jones and S.M. Swanson [1]

In their comparative analysis of regulation of hazardous materials across several countries, O'Riordan and Wynne posit four distinct regulatory styles. These are:

(1) Adversarial.
(2) Consensual.
(3) Authoritative.
(4) Corporatist.

Without considering whether these differences in style manifest themselves in differences in regulatory outcome, they argue that the systems are converging. The European systems are beginning to exhibit some characteristics of the adversarial systems: "more open, externally accountable procedures, ... a greater element of politically as well as scientifically credible safety standards...". At the same time, there are some signs of less adversarial regulatory decision-making in the USA, which they characterize as "mediation" efforts.

In our comments, we would like to discuss our experience with the adversarial systems in the USA as well as with less adversarial mediation efforts.

In preparing to comment on this chapter, we discovered a recent comparative analysis by Brickman *et al.* (1982) which examined national policies for regulating pesticides, food additives, chemicals in the workplace, and industrial chemicals subject to premarket or premanufacture notification. This study identifies regulatory styles similar to O'Riordan and Wynne's. Brickman *et al.* go on to observe that the differences in regulatory styles do not seem to result in significantly different regulatory outcomes for the class of problems the study addressed. That means it does not appear to be useful to distinguish between regulatory styles for forecasting differences in regulatory outcomes or for predicting which system is likely to lead to the "best" regulatory decisions.

Since we cannot promote one system over the other based on the quality of the decisions made by that system, we must search for other factors to help us compare the systems. Here O'Riordan and Wynne as well as Brickman *et al.* suggest that the US adversarial system is much more time-consuming and costly than its European counterparts. Obviously we Americans see some value in the adversarial system; or else why would we use it at considerable costs both in money and time to make decisions which are remarkably similar to those made with less resource-intensive systems in Europe? This question is all the more important for Europeans to consider if one accepts O'Riordan and Wynne's hypothesis that Europe is moving toward more adversarial decision-making processes.

We would like to make two principal points regarding the convergence of regulatory styles toward the adversarial – and less efficient – US system. The first point answers the question just posed: Why does the US use the inefficient adversarial system? In addition to being an important part of the US political tradition, the adversarial system is well suited for making certain kinds of complex regulatory policy choices that involve trade-offs between risks associated with hazardous materials in commerce and the economic benefits of that commerce. Our second point is that, after the broad and sweeping questions of policy have been resolved and precedents have been set, it is possible under the adversarial system to make many decisions in a nonadversarial manner.

O'Riordan and Wynne identify situations when it is difficult to judge whether a particular regulatory process is "just, efficient, or cost-effective". It is in just these situations where the adversarial system is at its best. The situations they describe are:

(1) There is often little agreement on what constitutes an acceptable risk.
(2) The whole regulatory process is characterized by uncertainty.
(3) The increasing use of models can be at odds with more conventional approaches to regulation based on experience, tested judgment, and intuition.
(4) Public interpretations of riskiness may be at odds with expert judgment.

Obviously a regulatory system is in a state of transition when the above conditions exist. The old rules do not seem to apply anymore, uncertainties abound, and "credible experts" are questioned at every turn. While the adversary system may not be efficient in terms of transaction costs, it is effective in forcing decisions about the acceptability of risk, resolving uncertainties, and restoring credibility to the regulatory process – if not to regulatory authorities.

Periods of transition are signaled by disagreements among experts. When experts disagree, one is usually confronting issues on the edge of science which cannot be resolved by science or expertise alone. Nevertheless, science and expertise can often inform resolution of these issues.

As an example, the authors have spent considerable portions of the last seven years involved in an adversarial process aimed at developing a new occupational exposure standard for benzene. A major issue is whether workers exposed at the current permissible exposure limit of 10 ppm are at significant risk of experiencing adverse health effects in the future. Scientists and experts from many different backgrounds and representing many different interests cannot agree what risk benzene-exposed workers are incurring, much less whether a particular level of risk is acceptable or not. The critical issues which are capable of resolution are not scientific. They are rather socio-economic and political ones that revolve around what constitutes prudent public policy in the face of considerable uncertainty. In the USA, choices such as these are reserved for the adversarial process where both the scientist–expert and the well-informed citizen have a part in the decision. After all, this type of decision involves value judgments where the ordinary citizen has as much to contribute as the scientist–expert.

Admitting that seven years is a long time for a society to be considering whether to change an occupational benzene standard, we would argue that once several landmark cases such as benzene are resolved, the system will revert to a less adversarial mode to decide on appropriate exposure standards for other substances. This will be possible since many of the important policy judgments regarding estimation of risk and its significance will apply to subsequent compounds.

The benzene example also supports our argument that, over time, less adversarial means for resolving regulatory issues will be adopted. Recently the authors participated in an attempt, albeit unsuccessful, to develop a joint labor–management recommendation to the Occupational Safety and Health Administration for a new benzene standard. While the attempt was unsuccessful all parties to the discussions felt that they were useful and might eventually contribute positively to resolution of the issue through the traditional rule-making process. In this instance, some of the broad policy issues had been resolved but significant issues remained.

Ironically, resolution of one major issue in the 1980 Supreme Court benzene decision contributed to the breakdown of the mediation efforts. The court decided that the Occupational Safety and Health Administration must make a finding that a significant risk exists under the current standard before it can change the standard. The court did not address the issue of how large the risk was or what would constitute a significant risk. The recent mediation foundered partly because of remaining divergence of opinion about the size and significance of risk at current exposure levels.

In general it is very difficult for industry to agree that risks exist at current exposure levels based on theoretical models. Accepting that some theoretical risk exists could subject industry to a flood of product liability suits. While industry is willing to take prudent steps to reduce risks by reducing exposure, it may find it difficult to participate in a negotiation process which assumes that a significant risk exists – the condition precedent to the need for a new exposure standard. Mediation efforts will be undertaken more frequently if a means can be found to divorce the significant risk finding from the development of the standard itself. This problem highlights a clear advantage of the European systems. Since a rationale for a new standard may often be unstated, industry can take prudent measures to reduce risk without the specter of a finding of significant risk.

In conclusion, the adversarial system has its advantages for resolving some of the complex societal trade-off issues that will confront us in the regulation of hazardous material in the next decade. While more adversarial systems will undoubtedly disturb the status quo in Europe, we expect that in some years hence Europeans will have a sanguine attitude about their utility. In the USA, the pendulum will swing back toward less adversarial systems as the broad policy issues become resolved and attention turns toward applying these policies to individual hazardous materials.

Note

[1] The authors wish to thank Camilla Hegeler for research assistance in the preparation of this Discussion. The views expressed in this comment are the authors' and do not necessarily represent those of their respective institutions.

Reference

Brickman, R., Jasanoff, S., and Ilgen, T. (1982), *Chemical Regulation and Cancer: A Cross-national Study of Policy and Politics*, Report for the National Science Foundation, National Technical Information Service (NTIS) Accession No. PB83-206771, November (NTIS, Washington, DC).

CHAPTER 13

Chemical Industry Hazards: Liability, Insurance, and the Role of Risk Analysis

M.S. Baram [1]

13.1. Introduction

Over the last decade, *toxic chemicals and other hazardous materials* have become essential features of economic growth in the USA, Western Europe, and the "Third World". Production and trade have increased to meet the demands of new users.

For example, the rapidly growing computer and electronics industry is a major new user: a semiconductor firm may use over 2500 chemicals, many of them highly toxic, in the manufacture of chips and other computer parts. Hospitals have become major users of various radiopharmaceuticals and radiochemicals in providing health care. Agricultural productivity is increasingly reliant on the use of chemical pesticides. Thus, toxic chemicals and other hazardous materials now pervade industrial and agrarian societies.

These hazardous materials pose *risks to health, safety, and the environment* throughout their "life cycle", including stages of production, shipment, storage, interim use in industrial processes, end product use by industry and consumers, and waste discharge and disposal.

The risks include *accidental releases and spills* which contaminate the environment (e.g. the dioxin accident at the Hoffman-La Roche plant at Seveso, Italy; the kepone spill by an Allied Chemical subcontractor at James Bay, Virginia; and other major accidents at Beek, the Netherlands, and Velbert, the FRG. In some cases, accidents inflict immediate harms to the health of workers and community residents (e.g. the methyl isocyanate accident at the Union Carbide subsidiary in Bhopal, India, which caused several thousand deaths and permanently impaired the health of many other people; the explosion at a plant in Flixborough, England, which killed 28 people).

The risks also include numerous latent disease or chronic health hazards for consumers of various industrial products, and for workers and community residents exposed to ***gradual unintended releases*** (e.g. from leaking storage tanks or waste disposal sites) or to ***routine industrial emissions or discharges*** which have been permitted by government authorities. Finally, latent disease risks have so impressed the public consciousness that emotional distress is now increasingly claimed as a health hazard, even in the absence of any medical findings of physical illness or scientific proof of increased risk (e.g. no evidence of biological changes as precursors of disease).

Many persons in the USA are now seeking compensation and other remedies from industry and insurers, for actual harms and perceived risks. Claims that various illnesses such as cancer, birth defects, and emotional distress were caused by, or are likely to result from, prior exposure to asbestos, herbicides, toxic wastes, ionizing radiation, and other hazardous agents are now commonplace in federal and state courts.

Thus, the increase in commercial production, distribution and use of hazardous materials has led to an increase in actual and perceived risks, and to a rising spiral of claims against industry and insurers in the USA. In addition, it has stimulated increasing government regulation and litigation to secure "clean-up" costs from industry in the USA and the European Economic Community (EEC), and new demands by labor unions and local communities for more effective risk reduction efforts by industrial managers.

The result is that all sectors of industry producing and using hazardous substances, and their insurers, are ***increasingly vulnerable to economic losses*** from liability determinations in the USA and from more stringent government actions in all industrial nations.

13.2. The Challenge to Industry and its Insurers

How to deal with increasing economic vulnerability has therefore become an issue of critical importance to all manufacturers, shippers, users, and disposers of hazardous materials, and their insurers. In addition, persons exposed to the hazards and government authorities have obvious concerns about this problem, concerns which include the quality of life and the protection of their physical and emotional health, their jobs, economic growth, and the availability of funds to compensate injured parties.

Firms doing business in the USA face the greatest degree of economic vulnerability. The US legal system enables injured persons and persons at risk to readily bring "toxic tort" actions against industry in state courts, with increasing success in securing high monetary compensation awards and other costly remedies from industry for harms to their health and environmental interests (see Nothstein, 1984).

The implications of this high degree of economic vulnerability in the USA *are not confined to US firms and the US economy*, however. Many of the firms doing business in the USA are multinational corporations, and losses in their US operations affect their branches in other countries. Further, insurance coverage for firms doing business in the USA is provided by an international network of insurers and reinsurers, many of whom are based in England, Switzerland, the Federal Republic of Germany, Scandinavia, and Japan. Losses which occur in the USA are therefore borne in most instances by this international network of insurers, and affect their ability to provide insurance coverage for industrial activities in all nations [2].

For example, the enormous liability awards against US firms for injuries due to their sales of asbestos without disclosures of the health risks, their careless disposal of hazardous wastes across the USA, the Three Mile Island nuclear plant accident, and other accidents and spills, have had severe economic consequences for many insurers and reinsurers in Western Europe, and consequently on industrial firms in many nations which rely on these same insurers and reinsurers for adequate and affordable coverage. If persons injured by the 1984 Bhopal accident are permitted to secure compensation in US courts from Union Carbide, a US firm, the economic implications will be severe and felt in all nations. Thus, the vulnerability of a single firm subject to the US legal system may affect industry, insurers, and economic conditions across most political boundaries [3].

The challenge for industry and its insurers is to develop the ability to overtake increasing risk and economic vulnerability, particularly for firms doing business in the USA, where vulnerability is greatest, and bring about the control and abatement of these dynamic forces. This view, cogently expressed by Dr Orio Giarini of the Geneva Association [4], calls for a proactive response by industry and insurers. It is not intended to support the "negative" approach of those industrial firms and insurers who have traditionally sought to reduce their economic vulnerability by lobbying the US Congress and state governments to change the legal system to lessen the ability of injured persons to secure remedies, or by increased reliance on defensive tactics in litigation. These negative approaches have severely damaged corporate credibility, infected corporate personnel with irresponsible attitudes, and failed to stop the spiral of liability.

The challenge is considerable, but it can be met if the manufacturers, shippers, users, and disposers of hazardous materials ***develop and use improved methods of risk analysis, health and environmental monitoring, and safety engineering*** to reduce hazards and prevent risks and consequent liability. Insurers must participate in meeting this challenge by using their considerable expertise and influence over industrial firms, which need affordable insurance, to force positive responses from industry [5].

Meeting this challenge is more than an economic necessity for industry and insurers in the private sector: it is also supported by many who seek *public policy reforms* to improve protection of health, safety, and environment. There is growing support in the USA for the *concept of using private insurance as an alternative to government regulation* – for insurers to play the leadership role in forcing safer industrial practices and thereby preclude the need for more government regulation (Baram *et al.*, 1981).

According to the Cato Institute (1985, p 1):

> There are many business activities that place third parties, employees, or consumers at risk. The traditional response ... has been government regulation.... There remains a need to find a better way to control business behavior ... we would raise the hypothesis that private insurers should provide a more efficient, effective alternative.
>
> ... insurers who protect third parties have strong incentives to police the behavior of firms whose actions directly affect the financial exposure of the insurer. Consequently, we would expect insurance companies to develop risk assessment procedures and to use these to mandate risk management strategies for their customers.

This chapter is an attempt to develop a framework for the insurance industry to meet the challenges before it – to force the chemical industry to overtake and reduce its economic vulnerability, and, consistent with this, to demonstrate that insurance is a viable private sector alternative to government regulation of hazardous materials and their risks.

Some elements of this framework for insurers are set forth in the following sections. They include evaluations of current legal developments in the USA which impose liability on industry and insurers, emerging industrial responses for reducing risks and economic vulnerability, and several concepts and methods of risk analysis which, if reduced to practice and applied, could benefit industry, insurers, and the general public.

13.3. Legal Developments Promoting the Economic Vulnerability of Industry

In the USA, persons who claim injuries to their health, property, and environment from industrial activities (e.g. hazardous waste disposal) can use the tort theories of the common law to secure monetary remedies from the industrial firms which were responsible for the injuries. By bringing a tort action in a state court, the injured person (plaintiff) has the opportunity to prove the responsibility of one or more industrial firms (defendants) and to secure a decision holding such firms liable for the monetary compensation (compensatory damages; see generally Prosser, 1984; Nothstein, 1984).

The plaintiff in a tort action may also seek additional compensation to "punish" a defendant whose harmful activities involved intent, malice, fraud, or gross negligence (punitive damages). This punitive damage remedy, if secured, can be extremely costly to the defendants, since it is often considerably larger than the monetary level of the preceding compensatory damage award, and is not insurable in many states [6].

The plaintiff (or group of plaintiffs joined in a single court action) may also seek additional remedies with costs to be borne by the defendant (which may be one or more industrial firms joined in the single court action). These additional remedies may include court orders for ongoing medical surveillance of the plaintiffs and their families, for ongoing environmental monitoring, for restoration of contaminated environmental conditions or property (e.g. groundwater), and for the abatement of the industrial activities creating the hazards.

A case now before the state courts in New Jersey demonstrates the array of remedies. In November 1983, the Superior Court in Tom's River, New Jersey, awarded 350 residents of Jackson Township $16 million in damages to compensate them for the contamination of their water supply system (well water) caused by the operation of a municipal waste disposal facility [7]. The contaminants included benzene, acetone, and residual aircraft fuels.

At the trial, no illnesses or injuries to health were established by the plaintiffs who had unknowingly used the contaminated water supply for several years, but it was established that they were at risk of future illnesses. The jury therefore awarded the plaintiffs $8.2 million in medical surveillance costs to detect and treat any future medical problems that might arise as a result of their exposure to the pollutants, $5.4 million for impairment of their quality of life, $2 million for emotional distress, $104 000 for their expenses in securing new water supplies, and $92 000 for nuisance damages. In addition, the court ruled that the plaintiffs would be allowed to return to court in the future to seek additional damages, should evidence become available linking any specific illnesses experienced by the residents to their exposure to the contaminated water.

This decision was appealed by the defendants, and a state appeals court has recently reduced the damage award by nullifying the awards for medical surveillance and emotional distress. It left the $5.6 million remainder of the award by the lower court intact, however. This decision is now being taken on further appeal to the state's Supreme Court, and possibly into a federal court, by the plaintiffs who seek restoration of the original award [8].

Irrespective of the final outcome of this case, many tort law experts now feel that US courts will soon allow plaintiffs to secure remedies, such as the recovery of damages and the provision of medical surveillance funds, for possible future injury on the theory that increased risk of an illness (e.g. cancer) constitutes a present injury to the plaintiff. Thus, many expect that plaintiffs who provide sufficient evidence as to their being at some appreciable increased risk (e.g. statistical evidence, biological monitoring evidence) will soon prevail in such cases. A similar trend is expected with regard to recovery for emotional distress [9].

In related litigation, a New Jersey court ruled on the coverage issue for one of the insurers. It found against the insurer by holding that the gradual nature of the contamination did not exclude its coverage, that coverage for "sudden and accidental" releases provided by the policies was not to be limited only to instantaneous releases, and that unintended pollution damage from gradual contamination constituted an accident and qualified for coverage [10].

In addition, federal, state, and local governments are now suing industrial polluters for millions of dollars to clean up hazardous waste sites which threaten public interests in health and environment. This development is also occurring in other nations [11].

Such toxic tort and other legal actions have been increasingly common in industrial states such as New Jersey. The outcomes indicate clearly the willingness of US courts to adapt the common law and statutory principles for humanitarian purposes – to provide remedies to persons who have been exposed to hazardous materials and who are either ill or at increased risk of illness as a result of industrial wrongdoing.

The courts have also been motivated to adapt the common law because they are now faced with clusters of cases with similar and distinctive factual circumstances (e.g. workers with asbestosis or mesothelioma, both rare diseases; residents of communities with unusually high rates of cancer or birth defects, who have been exposed to hazardous wastes improperly disposed). The cluster phenomenon induces a "commonsense" response by the courts that something is wrong and remedies are in order.

Another explanation as to why the courts now readily provide such remedies is that the judges and juries involved in these cases are also motivated by the desire to set forth qualitative criteria for responsible corporate behavior in the new industrial society with its ubiquitous industrial hazards. The desire for qualitative concepts of responsible behavior may have been stimulated by the inability of government regulatory actions, usually in the form of quantitative (numerical) standards, to prevent the harms, and by the use of cost–benefit analysis by US agencies to set standards, an approach which brings only economic values to bear on health and environmental problems [12].

Thus, the common law of "toxic torts" has essentially set forth a "moral code" for industry by defining three moral principles as legally enforceable duties for corporate managers: the duty to make a reasonable effort to identify hazards to health and the environment; the duty to warn persons at risk (e.g. workers, consumers, community residents) of hazards they would otherwise not reasonably be expected to know of (e.g. latent disease risks from asbestos); and the duty to act to abate the hazards in a diligent and responsible manner (see Prosser, 1984; Nothstein, 1984; Baram, 1984).

Finally, there is the influence of new scientific findings on the courts. Studies by toxicologists, epidemiologists, statisticians, medical researchers, and others have produced a wide variety of findings as to health and environmental risks and disease causation. In a given case, usually no single study provides conclusive evidence of risk, causation and industrial responsibility. But, in the aggregate, these studies comprise a mosaic of information from different sources which can more readily be interpreted by fact finders (e.g. the jury) as sufficient evidence to establish the liability of the industrial defendants [13].

For these and other reasons, state courts have adapted the common law to enable plaintiffs to more readily secure remedies and impose liability on industry. The adaptations take place on a state by state basis, but are usually stimulated by initial actions in the courts of "leading states" (e.g. New Jersey, California, Michigan, Massachusetts). The adaptations have had one major policy result: they have reduced many of the "obstacles" in the tort system which had previously kept most plaintiffs from successfully securing compensatory and punitive damages and other remedies for injuries and risks arising from industrial hazards.

For example, consider the following adaptations which have increased the economic vulnerability of industrial firms using hazardous materials:

(1) *Statutes of limitations.* These state laws define the period during which an injured person may bring a tort action (e.g. in New Jersey, up to six years following injury to property, up to two years following injury to health or other personal injury: NJSA 2A, 14-1, 2). Thus, victims of long-term latent diseases, which usually become apparent to the victim some 10–30 years following the injurious exposure, would be unable to bring tort actions after the illness symptoms have appeared so late.

To reduce this obstacle, courts in New Jersey and other states have adopted the "discovery" rule for purposes of interpreting the statutes of limitations. The "discovery rule" provides that the period during which tort action can be brought does not begin to run until the injured party discovers, or by reasonable diligence and intelligence should have discovered, the illness and other related facts which form the basis for the tort action. Thus, the short periods during which a tort action can be brought have now been considerably extended, and permit most tort actions for long-term latent disease and property damage to proceed [14].

(2) *Theories of liability.* The tort system in each state provides several theories of liability which are of potential usefulness for the victim of exposure to hazardous substances. These include negligence, nuisance, trespass, and strict liability theories, with strict liability being the newest and most favorable for plaintiffs because of its easier evidentiary requirements (Prosser, 1984; Nothstein, 1984).

Whereas negligence generally requires the plaintiff to produce sufficient evidence of disease causation and the defendant's failure to meet a standard of due care, and nuisance generally requires a finding that the benefits of the defendant's harmful activities are outweighed by the harms to the plaintiff, all that is required of the plaintiff who relies on strict liability theory is proof that the industrial activity was abnormally dangerous and was the cause in fact of the harm to the plaintiff. Since the issue of whether an activity is abnormally dangerous is now usually deemed to be a question of law for the court, the only factual issue for which the plaintiff must provide sufficient evidence is the causation issue (Rogers, 1977, p 158).

Courts in several leading states such as New Jersey have now adopted the doctrine of strict liability for cases involving toxic chemicals in the environment; and courts in virtually all states have adopted the doctrine for cases involving harmful products which injure consumers or workers. Thus, in *New Jersey v. Ventron*, which involved the seepage of mercury from a processing plant into a river, the Supreme Court of New Jersey stated:

> A landowner is strictly liable to others for harm caused by toxic wastes that are stored on his property and flow onto the property of others ... [T]hose who use, or permit others to use, land for the conduct of abnormally dangerous activities are strictly liable for the resulting damages [15].

In so holding, the court (as have other courts) relied on the formulation of "abnormally dangerous activities" set forth by the American *Restatement of Torts* (2nd edn), a compendium of tort law which reflects the views of leading authorities on tort. The *Restatement* provides that an activity may be considered abnormally dangerous if the following elements are present:

> (a) existence of a high degree of risk of some harm to the person, land or chattels of others;
> (b) likelihood that the harm ... will be great;
> (c) inability to eliminate the risk by the exercise of reasonable care;
> (d) extent to which the activity is not a matter of common usage;
> (e) inappropriateness of the activity to the place...; and
> (f) extent to which its value to the community is outweighed by its dangerous attributes. (*Restatement*, Section 520.)

After considering these criteria, the *Ventron* court (as have others) found the disposal of hazardous wastes to be abnormally dangerous, thereby affording the plaintiff the benefit of using strict liability theory with its reduced evidentiary requirements. Similar results have also been reached in cases involving harms arising from the industrial storage of chemicals that were not wastes [16].

Thus, in New Jersey and certain other states, evidence that an activity involving toxic chemicals cannot be made completely safe, even with the application of state-of-the-art safeguards, does not preclude, and may even expedite, liability and compensation. This is decidedly in the plaintiff's favor, since negligence theory would require a determination as to whether or not safeguards are available and should have been adopted.

Strict liability theory has been extended even further to the benefit of plaintiffs in other cases involving harmful products and the defendant's failure to warn. In the famous *Beshada v. Johns Mansville Products Corp.* case, the New Jersey court applied strict liability to harms resulting from risks that were unknown and even unknowable at the time of the activity, and rejected the state of the art defense argued by the defendants [17]. Thus, in

New Jersey and other states following its lead, whenever a toxic substance is spilled, leaked, discharged, emitted, or otherwise escapes the premises of a commercial facility where it has been stored, plaintiffs can rely on strict liability to secure compensation as well as other remedies for all harms which result.

(3) *Allocation of liability among responsible parties.* Toxic tort cases typically involve multiple pollutants and polluters, but identification of which pollutants and polluters caused the plaintiff's injury may be impossible. Traditionally, this would have thwarted plaintiffs in such cases. But many state courts now allow a toxic tort plaintiff to proceed against several parties whose separate activities led to the convergence of pollutants which caused a resulting "indivisible" harm. Again, New Jersey law illustrates how state courts have reduced this toxic tort "obstacle" for plaintiffs.

The rule in New Jersey in toxic tort cases is that the parties who are in any way responsible for the hazard are "jointly and severally" liable for the consequences of environmental impairment that results. As explained in the *Ventron* decision:

> Damages for a total injury ... are assessable against each of two or more tortfeasors whose wrong was a substantial factor in proximately causing injury ... whenever the total injury ... cannot be subdivided and liability for its several parts attributed and allocated to individual tortfeasors [18].

This rule of "joint and several" liability has two important consequences. First, the plaintiff need not demonstrate what portion of the harm suffered is attributable to any particular responsible party. Second, the plaintiff can proceed against fewer than all responsible parties, and obtain full recovery from one or more of the parties it names as defendants. This leaves to the defendants the task of obtaining contributions from or reallocating liability among others who may also be responsible for the harm. (New Jersey has its Joint Tortfeasors Contribution Law, NJSA 2A, 53A-1, et seq., to provide a framework for apportionment of liability in such cases.)

Even in the rare case where a plaintiff has no responsible party to sue because they are all out of business, or financially unable to provide damages compensation, or no longer subject to the jurisdiction of the state court, some courts have responded by extending liability to one or more parent or successor corporations of the firms which originally caused the harm [19].

(4) *Proof of causation.* Proof of causation remains as the major obstacle for the toxic tort plaintiff in cases involving chronic health hazards. Causation is central to the tort system because it assures fairness and objectivity in determining responsibility and liability, and the courts have been more cautious in dealing with this obstacle.

The causation requirement can be, and has been, met in toxic tort cases, despite some difficulties:

> Causation could well be established in a tort action if, for example, the plaintiff provided proof of the presence of toxic chemicals, a plausible route of exposure, and expert testimony as to the cause of the injury. Moreover, the elements in the causal claim need not each be proved directly, but ... by circumstantial evidence. It is not necessary for the plaintiff to establish conclusive evidence of causation. As with other elements of a tort action, proof of causation may be by a preponderance of the evidence. (US National Science Foundation, 1983, pp 19, 20.)

In states like New Jersey, the plaintiff has the burden of proving causation in tort cases usually by a "preponderance of the evidence", definable in terms of a "reasonable probability" [20]. This is more liberal than the standards used in other types of cases (e.g. by "clear and convincing evidence", by evidence "beyond a reasonable doubt"). The adoption of new rules of evidence by many states (using the Federal rules as a model) has made it easier for plaintiffs to rely on witnesses who are experts in toxicology, epidemiology, animal studies, medicine, and statistics, and to have a single expert (usually a doctor with extensive knowledge of these other fields) draw inferences as to causation from the evidence introduced by these other experts (provided the evidence from these other fields of scientific expertise is admissible, i.e. it meets criteria, such as its "sufficient scientific basis to produce uniform and reasonably reliable results which contribute materially to ascertainment of the truth"; and further, that the evidence is found to be reliable, in that it meets other criteria, such as its "acceptance" in the scientific community) [21].

As more expertise in these various scientific fields is developed, and as research methods become more routine and reliable, the role of science as evidence in toxic tort actions will become more important to both plaintiffs and defendants. Of particular benefit to plaintiffs will be the new factual information that is now being made available by companies, as required by various federal and state regulations and laws. For example, the federal Occupational Safety and Health Administration (OSHA) now requires many employers to provide employees and their doctors and unions with access to company-compiled employee medical and exposure (to toxic substances) records [22].

OSHA has also recently enacted its "hazard communication" regulation which further requires certain employers (manufacturers) to disclose information to employees on the hazardous attributes of the chemicals they use (OSHA, 1983). Several states (some 24 states) have enacted laws which affirm this requirement and, in some instances, broaden its application to additional employers (e.g. Massachusetts extends the requirement to transportation, construction, and many other industries) (see Baram, 1984). And some ten states have recently enacted laws which mandate the disclosure of chemical hazard information by companies to various state and local officials and, in some instances, to community residents as well (Baram, 1984). Congress is now considering enacting a federal "community right to know" law, to extend the flow of hazard information to state and local officials and residents in all states to the public [23].

These new US requirements for the disclosure of company-held information on chemical hazards will provide workers, consumers, and community residents with new information which will be of use in establishing causation in toxic tort cases, and go far beyond the requirements of the EEC's "Seveso Directive" in forcing hazard information disclosure [24].

Finally, the causation issue will undoubtedly be influenced by growing company uses of biological monitoring and medical surveillance to protect workers (Baram, 1983). These new techniques for monitoring the workforce in hazardous industries provide new information profiles on human exposure, biological changes, and the evolution of latent disease. This information will, in turn, be available to workers, under the OSHA requirements discussed above (access to medical and exposure records), and will inevitably become available in various ways to community residents who seek the data. It will therefore represent an important body of information on disease causation, and will undoubtedly be used in toxic tort proceedings by plaintiffs to establish causation, and by defendants to refute causation (Baram and Field, 1984).

Thus, while causation remains an obstacle for plaintiffs, various legal and technical developments indicate that it will be diminished in the future (unless the new information increasingly establishes the multifactorial nature of human disease – e.g. that human life style, smoking, nutrition, and genetics play substantial roles in promoting diseases which may have been initiated by hazardous materials).

This brief review of toxic tort law in the USA provides information on some current strong trends which are enhancing the opportunities for private party plaintiffs to secure remedies and impose liability on industry.

In addition, other forces are at work which are already imposing great liability on industrial firms. These other forces, in the form of various federal and state laws for the abatement of risks from chemicals, pollutants and hazardous wastes, authorize federal and state agencies to establish detailed regulatory requirements for industrial activity, including testing, disposal practices, the filing of reports, and other disclosures of information for government use. Violations are subject to substantial civil penalties (e.g. up to $25 000 per day), to criminal penalties for violations which involved intent or fraud, and to civil actions in federal courts by citizens under certain limited circumstances [25].

To remedy existing and abandoned hazardous waste risks, federal law provides that the federal government can take whatever emergency measures are needed, and then secure compensation for the clean-up and other "response" costs, and for contaminated natural resources (e.g. groundwater), from the "responsible" parties (e.g. generators, transporters, disposers of the hazardous wastes) [26].

Under this potent law, the US Environmental Protection Agency (EPA) has identified hundreds of hazardous waste sites, and has launched a major clean-up program. For each site, it is identifying responsible parties and seeking, by negotiation, compensation from these parties. At many sites, the clean-up and other response costs exceed $10 million, and can run much higher. In addition, the costs of natural resource damage will be imposed in the near future, and these will also run into millions of dollars. Since hundreds of parties may be deemed "responsible" for each site, the EPA's negotiation efforts have been stalled, and it has resorted to litigation to secure the compensation authorized by federal law from the parties. Both negotiation and litigation efforts have run into factual complexities (e.g. allocation of responsibility for wastes) and have encountered the legal problem of what liability standards are to be used.

The responsible parties include past and present owners and operators of the site, transporters, and generators, defined as "any person who by contract agreement or otherwise arranged for disposal or treatment of hazardous substances owned or possessed by such person, by any other party ..." [27], thereby including producers and users of the toxic chemicals found at the site.

The issue of what liability standards to apply to the identified responsible parties has been addressed by numerous federal courts in the past few years, with most of the courts deciding that the strict liability theory applies [28], and finding the authorized statutory defenses (e.g. exercise of "due care") are unavailable in most cases involving generator disposal activities conducted before enactment of the statute [29].

These courts have also adopted a loose standard of causality. According to one court "The only required nexus between the defendant and the site is that the defendant has dumped his waste there and that the hazardous substance found in the defendant's waste is found at the site" [30]. This view has been extended by some courts, in their finding that the government can rely on documentary or circumstantial proof that the wastes were hauled to the site without proof that they were subsequently taken away, thereby allowing a minimal showing that the generator's wastes were at some time taken to the site [31].

Finally, the federal courts have grappled with the crucial issue of apportionment of liability among the numerous responsible parties involved at each site. Every court facing the issue of whether liability should be joint and several has construed the vague federal law and its ambiguous legislative history to find that Congress intended to permit joint and several liability, but did not require its imposition (Ollman, 1985, p 13).

Without joint and several liability, "the government would be able to recover from each generator only that portion of the total cleanup costs which it can prove was caused by the generator's wastes", a task of enormous technical difficulty and cost. Further, "the government would be precluded from recovering the costs of cleaning up wastes caused by unknown or insolvent parties ... [making] full recovery ... difficult, if not impossible..." (Ollman, 1985, pp 12–13).

The result has been a split among the federal courts on the issue of joint and several liability. All deem it legally permissible, and feel a uniform federal approach important and necessary. But the courts have divided into at least two camps, with some holding each defendant who "caused a single and indivisible harm subject to the liability for the entire harm unless there is a reasonable basis for dividing the harm according to the contribution of each defendant", and others holding that apportionment is a matter for the discretion of the court, when the defendant cannot prove his contribution according to certain criteria [32].

Thus, under federal law for hazardous waste clean-up, generators of the waste have generally been subject to strict liability, circumstantial evidence of causation, limited defenses, and the strong possibility of joint and several liability for toxic wastes they disposed of before the law was enacted. As noted earlier, various state laws have been enacted and are being used to secure additional awards for state agency clean-up costs and other remedies, imposing additional liability on parties responsible for hazardous waste problems.

Thus, two strong trends now converge on industry and create its new economic vulnerability: increased use of the common law of toxic torts by private parties in order to secure compensation and other remedies for harms to their interests, under increasingly favorable legal conditions; and increased use of new federal and state laws by government agencies to recover clean-up and other costs by means of negotiation or litigation, also under favorable legal conditions.

The economic impacts on industry have already proved substantial and threaten to become overwhelming. However, the industrial losses have been spread or diffused by insurance coverage. Thus, the insurance industry has borne many of the losses due to industrial liability for harms due to hazardous materials and is now lobbying to secure changes in tort liability rules. But various "positive measures" are also being urged and increasingly taken by industry and insurers *to prevent health and environmental risks and their associated losses*, and these may represent the most effective approach to overtaking and controlling economic vulnerability, to reviving the pollution insurance market, and to facilitating the public policy goal of preventing risk to health and environment:

> Companies are bolstering their risk assessment teams with increasingly qualified personnel, and are devising sophisticated policies that address more and more of the known pollution variables.... For all the efforts now underway, we are barely scratching the surface. (Borowski, 1984, p 35.)

These measures rely on increased use of *risk analysis*, the leadership of industrial and insurance officials, and the cooperation of government agencies; we shall now discuss them.

13.4. Using Risk Analysis to Meet the Challenge

13.4.1. Emerging uses of risk analysis in industry

Risk analysis can be defined in several ways. A general definition suitable for this discussion is that *risk analysis* is the process of identifying potential hazards for individuals and society, and estimating their magnitude and probability of occurrence with the aid of statistics, experiments, analytic methods of various fields of science and engineering, human experience, and judgment.

Industrial firms have traditionally relied on rudimentary forms of engineering risk analysis to identify potential accidents and other sudden hazards with economic consequences which may arise from their facility processes, sale of products, and other activities. Some of these analytic approaches have been supplemented and reinforced by safety engineering methods developed by insurers (as in the case of analyzing fire risks and workplace injury risks), private standard-setting groups [e.g. CONCAWE (Oil Industry Trade Association), the National Fire Protection Association, the American Society for Testing and Materials], and public officials (e.g. local fire marshals, state safety officials, federal transportation agencies).

Analytic procedures developed by these other parties have often been forced on industry, either directly or indirectly, by the law. Thus, compliance with government methods may be directly required for industry to secure permits to operate facilities or sell new chemicals; compliance with insurer methods may be a condition of the insurance contract and the basis for establishing the cost of insurance coverage; and compliance with the methods of private "expert" associations is forced by the need for industry to demonstrate it met "industry custom" and "state of the art" in the event of an accident and subsequent litigation by the injured parties, where this defense may be useful to the industrial defendant in avoiding liability [33].

The actual use of these analytic approaches in industrial decision-making constitutes ***risk management***, which can be simply outlined as:

> a logical procedure ... [which] entails:
>
> (1) identifying and analyzing a problem,
> (2) formulating alternative solutions,
> (3) choosing the best solution,
> (4) implementing the chosen alternative, and
> (5) monitoring the results to detect and adapt for any error or changes in conditions. (Head, 1978, p 9.)

But over the last decade of gradual pollution and latent disease losses, the prevailing risk analytic practices based on safety engineering have been shown to be inadequate for the task of identifying, measuring, and controlling the gradual and latent hazards arising from the routine activities of industrial firms, such as the continuing or repeated discharge of chemicals in small amounts into the environment. These "routine" activities, many of which received government approvals, create hazards which necessitate a greater reliance on new disciplines in risk analysis, such as the various fields of the health sciences, and biostatistics and probabilistic risk theory.

The new disciplines must also be integrated into subsequent risk management activities which rely on the analyses, in order to assure that managers act properly on the basis of analytic findings. For the new gradual or latent risks, the new disciplines are needed for establishing the full characterization or model of the risk problem, including exposure identification and analysis, and the efficacy of alternative solutions.

Reliance on the new health science disciplines and new methods of risk analysis is now being required for many chemical industry activities, particularly for the testing and disposal of chemicals, by recent government actions in the USA and the EEC [34]; and it is being promoted by the guidance of national and international expert groups [35], as well as by the need of corporate officials to avoid further losses from liability and compensation awards.

13.4.2. Basic premises of the new risk analysis for insurers

The task facing the insurance sector is threefold:

(1) *To define and model industrial risk contexts.*
(2) *To evaluate and choose appropriate methods of risk analysis* for use in establishing coverage.
(3) *To promote use of the methods* in the insurance sector and by insureds.

From the foregoing discussion of chemical industry hazards and the wide variety of injuries and losses, *two risk contexts* emerge most clearly:

(1) The context of the *facility* which produces, uses, or stores hazardous materials (e.g. chemical manufacturer, industrial or other user of chemicals who stores and uses them at a particular site).
(2) The less tangible context encompassing the *chemical life cycle which exists beyond any facility* context already considered (e.g. transport of chemicals to numerous interim storage and use sites, use of chemicals outside of processing facilities as in the case of residential and agricultural uses, disposal of the chemical).

For each of the two contexts, there are two types of *potential injurious occurrences* which must then be modeled: *nonroutine events*, such as explosions, spills, and other accidents; and *routine events*, such as emissions, discharges, and waste generation on a continuing or repetitive basis. This approach can be summarized by the matrix in *Figure 13.1*. In modeling risks of these four types (A, B, C, D), insurers must start with the assumption that the industrial activities in question may not have been preceded by careful, objective, and complete analyses, or by management decisions consistent with risk analysis findings.

Contexts	Injurious occurrences	
	Nonroutine	Routine
Facility	A	B
Life cycle outside facility	C	D

Figure 13.1. Risk scenarios.

For example, the industrial analysis may have been "synoptic" and not "sequential", i.e. it may have used a "once and for all" design engineering approach to risk, and not fully considered risk as an ongoing operational problem involving full consideration of personnel turnover, training, human factors, and other aspects of its capability for ongoing control [36]. Or the industrial approach may have deliberately excluded consideration of liability from gradual pollution or long-term illness (ten years and beyond) because the current managers would only be held accountable to directors and shareholders for liability incurred during their relatively brief tenure (commonly only one to five years in the USA) [37].

Further, many industrial firms have chosen to operate at risk levels which are higher and more prone to liability, rather than at achievable lower risk levels which are more costly in the short term, but which do not bear as high a loss potential. This may be due to one or more factors.

For example, some industrial officials look no further than achieving compliance with government regulations, and do not confront the more complex task of assessing risk and reducing it to even lower levels which may be achievable. This overlooks the fact that regulations are usually designed only to protect the general population from exposure to levels of risk which would produce significant numbers of injuries, and not to protect individuals who may be highly exposed or susceptible persons. In some cases, this industrial approach is due to overemphasizing regulatory compliance as the only goal for industry; in other cases, it is due to the industrial desire to avoid the short-term costs of greater risk reduction efforts, costs which may be outweighed by long-term tort liability and other losses [38].

Further, loss potentials are being increased by industrial efforts to reduce operating costs by deliberately avoiding full compliance with government regulations – efforts which will either be violations ignored by, or unknown to, government, or be allowed as permissible exceptions. Many firms realize the limitations of government monitoring and enforcement efforts, and deliberately operate with a planned number of violations knowing that the cost savings that accrue will outweigh any occasional government action and penalty. Firms also seek official exemptions from government regulatory requirements, often on the basis of economic hardship. Both circumstances reduce current operating costs but may increase long-term liability.

Finally, industrial firms which have been able to secure adequate and affordable insurance coverage for activities which have led to risks and losses borne over time by insurers have little incentive to conduct their activities at reduced risk levels. This may be due to pricing of insurance by private and government insurers that is inadequate – inadequate in that the costs to a particular firm with a record of losses may not be sufficiently differentiated from the costs to all other firms with better loss records.

These factors are depicted in *Figure 13.2* which is based on the theoretical notion that one can accurately predict risks and losses from a particular industrial activity. In the diagram, three sectors of risk are shown: the sector mandated by government regulation, the unreasonable risk sector prohibited by regulation, and the de minimis risk zone in which no risks of any significance will occur. Industry is legally required to confine its activity to the maximum risk level prescribed by government regulation. Thus, if perfect information and analytic capability to predict risk and loss for a particular activity are available, one can determine the losses associated with conducting industrial activity x at the maximum level of risk permitted by government $(L - x)$.

This can then be compared with the reduced losses $(L - y)$ that would follow from industrial activity y that is planned to be more stringent than government regulation and protect virtually any individual. Firms which deliberately exceed the maximum level of risk permitted by government by conducting their activity at risk level z, knowing that this will not be enforced as a violation or will be granted official exemption, will obviously incur the highest losses $(L - z)$.

Finally, *Figure 13.2* depicts the four loss coverage sectors provided by self-insurance, direct insurer coverage, reinsurance coverage, and, for certain types of risks, government insurance, hypothetically relating deliberate risk taking by industry to the availability of loss coverage, and opening up a number of options for insurers and government to use the availability and pricing of insurance to deter industrial risk taking [39].

Thus, some basic premises have been proposed for the development and use of risk analysis by insurers: the differentiation of risk

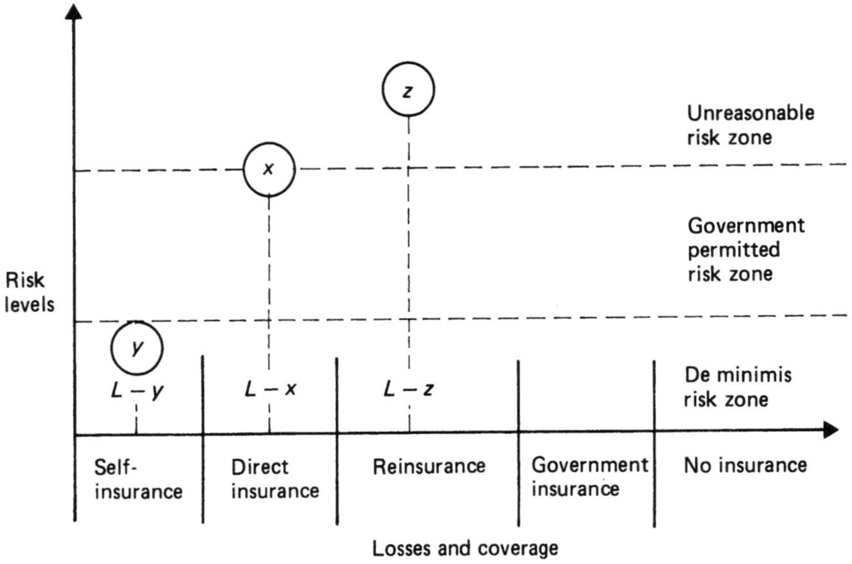

Figure 13.2. Risk, loss, and coverage.

contexts and events; the industrial failure to address risk control as an ongoing process problem; the industrial failure to deal with long-term losses in risk analysis because the accountability of industrial officials is short-term; and the deliberate conduct of industrial activities at risk levels which generate greater losses because of overreliance on government regulations in some instances, because of lack of government enforcement in other instances, and because of the availability of loss coverage without pricing for deterrent effect.

Finally, new laws and regulations in the EEC and the USA already mandate that industrial management conduct many of the elements of the risk analyses that have been discussed. For example, under the EEC's Seveso Directive (European Communities Council, 1982), the UK now requires the managers of each facility containing certain specified hazardous substances to prepare a "safety case" for the facility (see Advisory Committee on Major Hazards, Health and Safety Commission, 1984; Haigh, 1984).

The "safety case" is essentially a complete risk analysis for the nonroutine facility occurrence scenario described previously. It must contain an analysis of the installation; substances stored, used, and produced; processes involved; accident risks and internal and external accident triggering events (e.g. malfunction, flood); safeguards and control systems; and emergency response plans. The safety case contains much confidential information and its disclosure is mandated, but limited to designated government officials for evaluation.

After government review of its adequacy in protecting community residents from accidents, any inadequacies are to be cured by industrial management, and the government officials have the dual responsibilities of assuring that the final outcome is adequate to protect the public and of developing appropriate off-site emergency response plans for additional public protection (Advisory Committee on Major Hazards, Health and Safety Commission, 1984; Haigh, 1984; Lagadec, 1984). A model of the UK plan for implementing the Seveso Directive is provided in *Figure 13.3*.

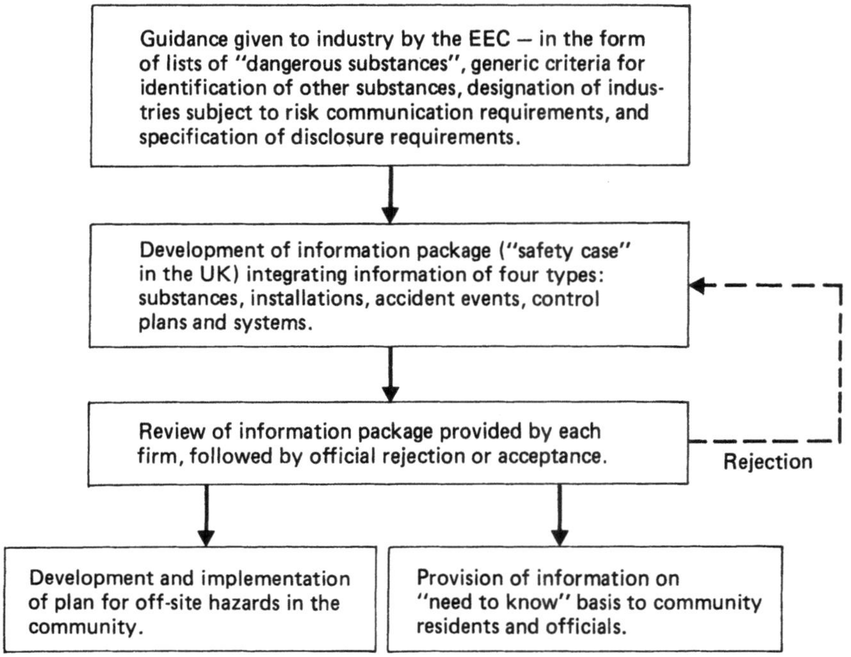

Figure 13.3. The UK system for controlling major accident hazards at facilities, following the EEC Seveso Directive.

Other EEC nations will similarly implement the Seveso Directive by 1989, and the EEC and various nations are undertaking a major research effort on many key attributes of facility accident hazards [40]. Although the USA currently lacks such requirements at the federal level, several proposals have been presented to Congress and enactment of a Seveso-like system is anticipated [41]. Anticipating government action, several major chemical firms in the USA have voluntarily adopted safety evaluation systems similar to those mandated by the Seveso Directive (e.g. the DuPont, Monsanto, and Rohm and Haas companies).

Thus, a major effort is underway in the countries of Western Europe and in the USA to have chemical firms conduct coherent and detailed risk analyses of their facilities to evaluate and prevent accident hazards, and to promote community response or damage control plans. The insurance industry could seize this opportunity, for example, and require as a condition for coverage, their access to and independent review of the safety case done by each industrial firm seeking insurance. Independent evaluation by insurers is advised, since the performance of companies and government officials may be of variable and even inadequate quality; and insurers may also want to use more stringent evaluation criteria for assuring that both risks and losses are reduced.

The "safety case" approach also presents an opportunity for insurers to force industrial risk analysis of routine facility occurrences, another of the four scenarios. Admittedly more complex because of the uncertainties about gradual pollution and chronic health hazards, the safety case can, nevertheless, be extended to deal with routine plant emissions and discharges.

13.5. Conclusion

The many new hazards to health and the environment that have been created by the chemical industry have been met by developments in the law which now assure that industry and its insurers will bear full financial responsibility for the harms.

The result is economic vulnerability that is now so pronounced for the industry and its insurers that it has caused the pollution insurance market to "collapse". Many firms are therefore relying on "defensive" strategies to try to restrict liability; but these do not prevent either risks or the development of new loss-spreading systems by government, both imposed on the private sector at considerable cost.

Industry and insurers also have positive options, however. They can improve their capability to prevent risks and losses, to reduce their economic vulnerability, and to preserve their autonomy from further government control. The foremost of the positive options involves striking at the root of the problem, the need to prevent risks to health and the environment by developing and using new methods of risk analysis and management.

Risk analysis is still in its infancy; and traditional risk management has been shown to be inadequate for the gradual pollution, latent diseases and other special hazards of the chemical industry. However, several factors indicate that new and powerful tools to overtake and control risks and losses and vulnerability are imminent.

Leading members of industry are supporting risk analysis and management initiatives. Government and academic efforts are being made to reduce theories to practice. And the advances in the health sciences, safety engineering, and stochastics that are being used to establish causation and responsibility for latent diseases, environmental degradation, and accidents that have occurred can also be used to act prospectively on potential causes of harm.

As these separate efforts converge, the tools for controlling risks and losses may become available for practice and systematic application in the private sector. Insurers should therefore join with industry, government, and academia in promoting this development, to protect their own interests and to protect human well-being.

Notes

[1] This chapter is based, in part, on research supported by the US National Science Foundation, under grant PRA 8212292, "Corporate Mangement of Health Risk". The author also wishes to acknowledge research assistance and information from the following: Mark Ollman and Deborah Novick, Boston University Law School; Philip Terrel, Liberty Mutual Insurance Co., Inc.; and J. Raymond Miyares, Bracken and Baram. A longer version of this paper was presented at the 12th General Assembly of the Geneva Assocation, Oslo, Norway, 24 June 1985.

[2] Comments by participants from insurance firms and reinsurance asociations at the Oslo and Vienna Workshops on Chemical and Hazardous Materials Risks, Geneva Association, Geneva, Switzerland, June and July 1984.

[3] Comments by participants at the Workshops on Chemical and Hazardous Materials Risks (see note [2]). See, also, "Liability insurance: the added burden of Bhopal", *Chemical Week*, 13 February 1985, 30–31.

[4] At the Workshops on Chemical and Hazardous Materials Risks (see note [2]).

[5] Comments by participants at the Workshops on Chemical and Hazardous Materials Risks (see note [2]). See also Sterling (1984).

[6] Nothstein (1984, pp 524). See also Walker (1984), which discusses recent decisions awarding punitive damages at multimillion dollar levels.

[7] *Ayer v. Jackson Township*, 189 NJ Super. 561 (1983).

[8] As reported in *The Times*, Trenton, NJ (5 June 1985, p 1). According to this report, the appeals court found that the plaintiffs had failed to establish, on a quantitative basis, any increased risk of illness, and had therefore failed to establish a factual basis for the medical surveillance remedy.

[9] See discussion in Nothstein (1984, pp 461–463).

[10] *Jackson Township v. Hartford Accident and Indemnity Co.* See discussion in Sterling (1984, p 222).

[11] For example, in addition to dozens of legal actions against industry for clean-up costs brought by the federal government in the USA, the Dutch and Swiss governments are also seeking millions of dollars in clean-up costs from various chemical companies: see the discussion in Sterling (1984, p 218). In the USA, many of these governmental actions are being taken under the authority provided by the Comprehensive Environmental Response, Compensation and Liability Act (CERCLA), 42 USC Sections 9601 et seq., which authorizes use of government funds for clean-up of hazardous waste sites under Sections 9604 and 9605, and the securing of compensation for these "response costs" from industry, by governmental action, under Section 9607. It also authorizes the government to seek injunctive relief under Section 9606 to require a private party to undertake the clean-up by itself.

[12] The use of cost–benefit analysis to set regulations has been mandated by Presidents Ford, Carter and Reagan, presumably to assure that the only regulations that will be enacted are those for which their benefits (e.g. lives saved) exceed their costs (e.g. industrial control costs): see *Executive Order 12291* (1981). This approach is often in opposition to many of the laws for risk regulation enacted by Congress, which authorize agency regulation on health and other nonmonetary grounds. Where such conflicts between Congressional authorization and Presidential order face an agency charged with regulatory responsibilities, the US Supreme Court has ruled that the Congressional mandate must prevail, and that a cost–benefit approach is not a permissible basis for regulatory action (*American Textile Manufacturers' Institute v. Donovan*, 101 S. Ct. 2478, 1981). Nevertheless, most agencies continue to use a cost–benefit approach wherever possible, because of their more immediate political accountability to the President and his Office of Management and Budget (OMB) (see Baram, 1980).

[13] The use of scientific findings as evidence for regulatory decision-making is now well established in the USA: see, for example, US Office of Science and Technology Policy (1985) and recent Risk Assessment Guidelines proposed by the US Environmental Protection Agency in 1984 and 1985. Industrial and private groups also promote science as evidence: see, for example, the American Industrial Health Council (1981) and recent reports by the Food Safety Council and the Chemical Manufacturers' Association.

[14] See, for example, *Lopez v. Sawyer*, 62 NJ 267 (1973); *O'Keefe v. Snyder*, 83 NJ 473 (1980); and *Tevis v. Tevis*, 79 NJ 422 (1979).

[15] *New Jersey Department of Environmental Protection v. Ventron*, 19 ERC 1505 (1983), p 1509.

[16] *Bridgeton v. BP Oil Co.*, 146 NJ Super. 169 (L. Div. 1976), in which an underground oil storage tank leak was involved.

[17] 90 NJ 191 (1982). See also *Fischer v. Johns Manville Corp.*, 193 NJ Super. 113 (1984), holding that punitive damages may be awarded in a strict liability action involving harms caused by asbestos products, in accordance with other "cases dealing with products ... when a manufacturer has knowledge, whether or not suppressed, that his product poses a grave risk to the health and safety of its users and fails to take any protective or remedial action" (p 132).

[18] *Ventron*, 19 ERC 1505 (1983), p 1510. See also *Perth Amboy v. Madison Industries*, NJ Super. Ct., App. Div. (21 April 1983), Nos. A-1127-8113, 1276-8173, 13 ELR 20554; and *Borel v. Fireboard Paper Products, Inc.*, 493 F.2d 1076 (5th Circ. 1973), *cert. denied*, 419 US 869 (1974).
[19] *Ventron*, 19 ERC 1505 (1983). See also *Ramirez v. Amstead Industries*, 408 A.2d 827; and *New Jersey Department of Transportation v. PSC Resources*, 419 A.2d 1151 (1980).
[20] *Germann v. Matriss*, 55 NJ 193, 221 (1970); *Kahalili v. Rosecliff Realty, Inc.*, 26 NJ 595 (1958).
[21] See Nothstein (1984, Ch 18), Epstein (1981), and recent court decisions including the following, decided under the New Jersey *Rules of Evidence*, NJSA 2A, 84A, Rule 56(2): *New Jersey v. Davis*, 96 NJ 611 (1984); *Bowen v. Bowen*, 96 NJ 36 (1984); *New Jersey v. Kelly*, 97 NJ 178 (1984); *Romano v. Kimmelman*, 96 NJ 66 (1984); *Procida v. McClaughlin*, 195 NJ Super. 396 (1984).
[22] 29 Code of Federal Regulations (CFR) Part 1910.20. See also other disclosure requirements by OSHA in 29 CFR Part 1904, and by the Environmental Protection Agency under the authority of the Toxic Substances Control Act, 15 USC, Sections 2601 et seq., at Section 2607 (reporting requirements imposed on industry).
[23] For example, the proposed Chemical Manufacturing Safety Act of 1985, introduced by Congressman James Florio.
[24] European Communities Council (1982). For an evaluation of the implementation of the Seveso Directive in the United Kingdom, see Haigh (1984).
[25] See the US Resource Conservation and Recovery Act, 42 USC Sections 6928 and 6972. See also US Toxic Substances Control Act, 15 USC, Sections 2601 et seq., which requires industrial testing of chemicals, and the reporting of health and safety studies and imminent hazard information, for example. Once again, violations are punishable by fines of up to $25 000 per day, criminal penalties, and citizen actions.
[26] See CERCLA, 42 USC, Sections 9601 et seq., particularly Sections 9607 and 9604. Note that the federal courts have been unanimous in holding that CERCLA creates a private right of action against responsible parties for the recovery of "necessary costs of response incurred by any other person" consistent with national regulatory plans: see *Wall v. Waste Resources Corp.*, 22 ERC 1785 (6th Cir., 6 May 1985).
[27] CERCLA ,42 USC Section 9607(b)(1).
[28] See, for example, *US v. Price*, 577 F. Supp. 1103 (DNJ, 1983); *US v. A&F Materials*, 578 F. Supp. 1249 (SD 811, 1984).
[29] See, for example, *US v. Wade*, 546 F. Supp. 785 (ED Pa, 1982); *US v. NEPACCO*, 579 F. Supp. 823 (WD Mo, 1984).
[30] *US v. Wade* (Wade II), 577 F. Supp. 1326 (ED Pa, 1983).
[31] Ollman, 1983, p 11, who cites and evaluates numerous cases.
[32] Ollman, 1985, pp 15–17; citing *US v. Chem-Dyne*, 572 F. Supp. 802 (SD Ohio, 1980) and *US v. A&F Materials*, 578 F. Supp. 1249 (SD 811, 1984).
[33] Compliance with custom or state of the art has been considerably weakened as a defense in tort actions over the last few decades. However, evidence of failure to comply by a defendant virtually assures that the plaintiff will prevail.

[34] For example, the chemical testing and other requirements of the US Toxic Substances Control Act, 15 USC Section 2601; and the EEC's *Sixth Amendment*.

[35] See, for example, US Office of Science and Technology Policy (1985); the Uniform Test Rules of the Organization for Economic Cooperation and Development, Paris; the Environmental Health Criteria reports of the World Health Organization (WHO), Geneva (e.g. WHO, 1984); and the recent useful report of the US National Science Foundation (Covello, 1985).

[36] Personal communication, Dr G. Volta, Commission of the European Communities Joint Research Centre, Ispra, Italy, 13 June 1985. This "Maginot Line" approach to risk, and its inability to contain risks, has been particularly acute in the nuclear power industry, and was cited by the Kemeney Commission as a major problem, following its evaluation of the Three Mile Island accident.

[37] This problem has emerged in the study of "Corporate Management of Health Risks", conducted at Boston University by the author with support from the National Science Foundation.

[38] See discussion in Richards and Silvers (1982).

[39] See Webb (1984), who warns that "the pricing structure of neither insurance nor reinsurance should be used to create intentional subsidies among buyers of any type of insurance" (pp 334–335).

[40] See Advisory Committee on Major Hazards, Health and Safety Commission (1984); Haigh (1984); Lagadec (1984). The research agenda includes future efforts on heavy gas cloud dispersion, the "domino effect" from multiple hazardous substances stored in proximity, human reliability, and the use of computerized data banks.

[41] For example, the proposed Chemical Manufacturing Safety Act of 1985, introduced by Congressman James Florio.

References

Advisory Committee on Major Hazards, Health and Safety Commission (1984), *The Control of Major Hazards, Third Report* (Health and Safety Commission, London).

American Industrial Health Council (1981), *Chronic Health Hazards: A Framework for Sound Science in Federal Decision-making* (American Industrial Health Council, Scarsdale, NY).

Baram, M., (1980), Cost–benefit analysis: an inadequate basis for health, safety, and environmental regulatory decision-making, *Ecology Law Quarterly*, **8** (3), 473–532.

Baram, M. (1981), *Alternatives to Regulation* (Lexington Books, D.C. Heath Co., Lexington, MA).

Baram, M. (1983), *Screening, Monitoring and Surveillance for Managing Occupational Disease Risks: Legal and Policy Considerations*, Paper presented at Conference on Biological Monitoring and Genetic Screening in the Industrial Workplace, Boston University Schools of Medicine, Public Health, and Law, held in Washington, DC 12–13 May.

Baram, M. (1984), The right to know and the duty to disclose hazard information, *American Journal of Public Health*, **74** (4), 385–390.

Baram, M. and Field, R. (1984), *Screening and Monitoring Data as Evidence in Legal Proceedings*, Paper presented at Conference on Medical Screening and Biological Monitoring for the Effects of Exposure in the Workplace, National Institute of Occupational Safety and Health, Cincinnati, OH, 12 July.

Borowski, P. (1984), The pollution liability market, *National Underwriter*, 20 July, p 35.

Cato Institute (1985), *Risk Management: Through Regulation or Insurance*, Conference materials, 16 May (Cato Institute, Washington, DC).

Covello, V. (ed) (1985), *The State of the Art of Risk Assessment* (US National Science Foundation, Washington, DC).

Epstein, S. (1981), Role of the scientist in toxic tort case preparation, *Trial*, **40**.

European Communities Council (1982), Directive on the major-accident hazards of certain industrial activities (82/501/EEC), *Official Journal of the European Communities*, **L230**, 8 May.

Haigh, N. (1984), *EEC Environmental Policy and Britain* (Environmental Data Services Ltd., London).

Head, G. (1978), *The Risk Management Process* (Risk and Insurance Management Society, New York).

Lagadec, P. (1984), *Major Technological Risks and Crisis Situations*, Paper presented at EPA Risk Communication Conference, Long Beach, CA, December.

Nothstein, G. (ed) (1984), *Toxic Torts* (Mcgraw-Hill, Colorado Springs, CO).

Ollman, M. (1985), *Generator Liability under CERCLA*, Unpublished paper, 10 May (Boston University Law School, Boston, MA).

OSHA (1983), "Hazard communication" rule, *Federal Register*, **48**, 25 November, 53 280.

Prosser, W. (1984), *Law of Torts*, 5th edn (West Publishing Co., St Paul, MN).

Richards, E. and Silvers, A. (1982), Risk management theory: reducing liability in corporate and medical environments, *Houston Law Review*, **19**, 251.

Rogers, W. (1977), *Environmental Law* (West Publishing Co., St Paul, MN).

Sterling, D. (1984), Environmental impairment liability: an insurance perspective, in J. Long and E. Randall (eds), *Issues in Insurance*, Vol 2, Ch 9, 217–277 (American Institute for Property and Liability Underwriters, Malvern, PA).

US National Science Foundation (1983), *Compensation for Victims of Toxic Pollution*, Report 83-6, July (National Science Foundation, Washington, DC).

US Office of Science and Technology Policy (1985), Chemical carcinogens: a review of the science and its associated principles, *Federal Register*, **50** (50), 10 371–10 442.

Walker, V. (1984), Punitive damages and management of environmental information, *Environmental Analyst*, **5** (9), 8–11.

Webb, B. (1984), Reinsurance as a social tool, in J. Long and E. Randall (eds), *Issues in Insurance*, Vol 2, Ch 10 (American Institute for Property and Liability Underwriters, Malvern, PA).

WHO (1984), ***Principles for Evaluating Health Risks to Progeny associated with Exposure to Chemicals during Pregnancy***, Environmental Health Criteria Report No. 30 (World Health Organization, Geneva, Switzerland).

13.D. Discussion

A.V. Cohen [1]

13.D.1. Introduction

My comments are necessarily those of a nonlawyer, but one with experience in administering risk regulation in the UK. Any views I express are personal, and do not necessarily coincide with the views of the UK Health and Safety Executive.

13.D.2. The role of insurance in encouraging safety

Insurance can bring home to owners of plant the existence of a risk, and spread any possible civil law liability. But insurance cannot be the full story in improving safety because:

(1) Liability for harm done will not always rest on the insured party. Much depends on the precise nature of the law.

(2) In any case, causation and fault may both be difficult to establish for "delayed effect" situations, like possible carcinogens. Even if causation can be established, and even with "joint and several liability", companies might well have gone out of business a generation after exposure.

(3) We need to examine the extent to which public liability policies properly cover the full effects of the "high-consequence, low-probability" situation.

(4) In some countries it has been found possible to charge differential premiums according to the level of risk. But, within a particular category of industry, the "spread" of financial risk can in effect act as a tax on the safe factory, to the benefit of the risky.

(5) Premiums tend to be low, partly because of the preceding items, partly because of competition, and partly because they are often part of a wider package. The premiums do not impact as sharply as they might on the occupier's awareness and only limited credit may be given for prevention measures.

13.D.3. The need for safety legislation and enforcement

Something more than civil law liability is therefore needed: regulatory and enforcement powers to ensure that occupiers reduce risks, to workpeople or to members of the public, to what proves a socially acceptable level. Even granting the increasing level of claims, insurance is unlikely ever to be an *alternative* to government regulation. The two must go together. What is regarded as socially acceptable regulation, or fair law, will vary from country to country. Typically, one can expect:

(1) A prescription for what is needed to achieve a certain level or standard of safety: zero risk is likely to be unobtainable.
(2) Active encouragement for occupiers to "self-inspect".
(3) Requirements to have general company safety policies, and disclosure of risks to "safety representatives" nominated by unions.
(4) Inspection to ensure that the prescriptions of the law are carried out.
(5) If the legal prescriptions are not carried out, legal (including criminal) sanctions.

The legal powers in the UK rest mainly with the Health and Safety at Work etc. Act 1974 [2]. The provisions of this Act are in general terms. Section 2 places a duty on "every employer to ensure, so far as is reasonably practicable, the health, safety and welfare at work of all his employees". Section 3 places a similar duty to "ensure, so far as is reasonably practicable, that persons not in his employment who may be affected thereby are not thereby exposed to risks to their health or safety". There is a Health and Safety Commission, appointed by the Secretary of State for Employment and including, among others, representatives of employers and trades unions, and a Health and Safety Executive, which employs the various inspectors and which has enforcement powers. Regulations can be made under the Act by the Secretary of State (Section 15), usually on the recommendation of the Commission and in any case after consulting them (Section 50).

These regulations can be considerably sharper than the general "reasonably practicable" requirement. Thus our way of controlling many dangerous substances, particularly carcinogens, and our proposed way of controlling all substances hazardous to health, is to require that exposure should not exceed a specified control limit, and in any case should be further reduced so far as is reasonably practicable. This latter point can mitigate many of the difficulties Baram has identified in regulations which merely set a control limit.

The regulations are true secondary legislation made by the Secretary of State and subject to parliamentary "negative approval". The recommendations for them arise in a forum representative both of employers and of trades unions. There is a wide range of regulations issued [3]. The Commission can also issue "approved codes of practice" and the Executive can issue guidance.

Before recommending a regulation, the Health and Safety Commission (Section 50.3) has to "consult any government department or other body that appears to the Commission to be appropriate". Consultative documents are usually published and invite comment. The Commission prepares its final recommendations in the light of consultation, but it does not have to justify every difference from every comment made to it. In principle it is possible for regulations to be ultra vires; but the grounds for legal contesting of a regulation are very much narrower in the UK than in the USA.

Delays are inevitable in preparing regulations, because of the need for a "legitimizing process" rooted in the country's legal and administrative practices. In the UK, this is consultation and is essentially slow. Legal contesting of a regulation is, I gather, more common in the USA, and will lead to "the law's delays". Either seems the inevitable price for forming legislation in a nonauthoritarian environment.

Enforcement of these regulations, and of the general duties of the Act, is the responsibility of the Executive and its inspectors, and, for some premises, of local authorities. All of these have power to prosecute, and to issue "improvement notices" and "prohibition notices" (Sections 21 and 22) which mean what they say: the recipient of such a notice may appeal (Section 24) but the onus of proof then rests on him. In a typical year there will be some 2000 prosecutions and some 15000 notices, of which notices approximately 200 will be appealed against, and perhaps ten appeals upheld.

Inspectors do most of their work informally by persuasion and informal advice rather than by using the arm of the law. Firms above a certain size are required to produce formal safety policies and, in many circumstances, to disclose information about risks to nominated "safety representatives" of their workforce.

Local authorities are responsible for enforcing the Act in certain premises and, among many other matters, for planning decisions, for emergency planning, and for licensing sites for the disposal of hazardous material. The Department of the Environment's Hazardous Waste Inspectorate can visit and advise local authorities, and has close connections with the Health and Safety Executive's general body of knowledge. The planning function, too, has an interface with the Health and Safety Executive: local authorities can and do seek our advice, and the famous Canvey Report (Health and Safety Executive, 1978, 1981) arose as a result of the invitation of the Secretary of State for the Environment, in connection with a planning decision.

13.D.4. Risk analysis

Recognizing that a hazard exists is relatively straightforward for normal industrial accident situations. It becomes much more difficult for situations with delayed effects on health, and even more so for the "high-consequence, low-probability" situation. Inevitably, many hazards have not yet been fully identified.

The next step must be the complex process of risk estimation. The process is considered at length in the Royal Society Report on Risk Assessment (Royal Society, 1983). It presents a familiar set of problems in estimating the effect of toxic substances: what weight to place on essentially limited animal experiments, how to apply results to the effect on man, etc. For potentially catastrophic situations, imaginative lateral thinking is needed, for all the ways in which things could go wrong.

In either case, the risk estimation is not an exact science: but an estimate that is correct to an order of magnitude is a good deal better than nothing. The inherent uncertainties must be kept in mind and do mean that:

(1) Tendentious criticisms of such estimates can and do arise.
(2) Risk analysis can estimate the size of the risk, and be very suggestive on how to reduce it.
(3) But the decision on how far to control a situation must rest with risk managers, overseen by a regulatory agency, and ultimately if necessary by government.
(4) Nevertheless, risk analysis, combined with consequential management, can be of real benefit in reducing risk. Canvey Island now presents a societal risk perhaps 20 times less than that at the beginning of the study.

13.D.5. The effect on industry

The growing liability to chemical firms arises from:

(1) The changing nature and scale of technology.
(2) Public expectations of safety and sensitivity to perceived danger, both of which may have increased in recent years.

The consequential unwillingness of insurers to cover large plant with potential for catastrophe, or for widespread harm, raises a number of problems, including:

(1) The fear of consequences, which may inhibit proper technical development. The employer cannot afford to be wrong and is reluctant to innovate.
(2) Once a major catastrophe has happened, the situation may well be beyond insurance. The real problem to a company may then be survival.

Risk estimation can put a scale on the hazard and thus help both insurers and regulators. But a technical process cannot by itself eliminate the above problems. Regulation (of which legitimization is an important component) is needed to reduce risk to acceptable levels. Such a process is unlikely to convince everyone: there will always be residual objectors and protestors.

For large works presenting "major hazards" the pattern, consistent with the Seveso Directive, is: notification, self-assessment under scrutiny from regulators, emergency planning, and inspection to follow up if necessary, with legal sanctions always held in reserve. A further element is that land use planning must take account of the "assessed residual risk".

Insurance, helped by risk assessment, thus has a role, but not an exclusive one, in the reduction of risk, particularly for firms that own large plants and thus have a strong financial interest in such reductions. But financial incentives may not be adequate to motivate small and medium-size firms. The need to control these must not be overlooked.

13.D.6. Conclusions

Insurance can well be of assistance in some areas in providing a financial incentive for reducing risk. This incentive is likely to be felt most sharply in large and sophisticated industrial undertakings, but hazards elsewhere should not be overlooked. In all areas legal powers for regulation and enforcement are also necessary, and the process of "risk assessment" can help both insurance and risk regulation.

The situation I have described is typically British in its style of agency interaction, of legitimization, and of enforcement. Other countries will have their own ways of handling such matters so as to make their decisions acceptable to those concerned.

Notes

[1] Crown Copyright Reserved – reproduction in whole or in part only with permission of Her Majesty's Stationary Office which permission to be obtained in the first instance from the author.

[2] 1974 c. 37.

[3] Some relevant regulations are:

The Dangerous Substances (Conveyance by Road in Road Tankers and Tank Containers) Regulations 1981 (SI 1981/1059).
The Notification of Installations Handling Hazardous Substances Regulations 1982 (SI 1982/1357).
The Control of Industrial Major Accident Hazards Regulations 1984 (SI/1984/1902).
The Classification, Packaging and Labelling of Dangerous Substances Regulations 1984 (SI/1984/1244).
Proposals for Dangerous Substances (Conveyance by Road in Packages etc.) Regulations 1984 (consultative document, Health and Safety Commission, HMSO, London, 1984).

References

Health and Safety Executive (1978), *Canvey: An Investigation of Potential Hazards from Operations in the Canvey/Thurrock Area* (HMSO, London).

Health and Safety Executive (1981), *Canvey: A Second Report* (HMSO, London).

Royal Society (1983), *Risk Assessment*, A Study Group Report (The Royal Society, London).

CHAPTER 14

Practical Aspects of Environmental Impairment Liability

A. Klaus

14.1. Introduction

Love Canal, Seveso, the *Torrey Canyon*, the *Amoco Cadiz*, and Bhopal are environmental catastrophes which will remain in our mind for decades. Such incidents have made us painfully aware that many by-products of industrial progress are threatening our environment.

Apart from the destruction of lives and irreplaceable values, such incidents often end up with billion dollar claims. Like shock waves, these monetary claims always shake the entire insurance industry around the globe. Surprisingly enough, however, the losses arising out of all these environmental disasters are low or even negligible compared with all fire losses. So, why are insurers making a fuss about these small indemnities?

The following may explain this clearly contradictory situation:

(1) Insurance of environmental impairment liability (EIL) is a branch with a small number of issued policies.
(2) The objects insured are biased toward the higher possibility of an incident occurring. This adverse selection is in flagrant contradiction to the basic rules of insurance theory.
(3) The total premium volume for the liability cover – under which such incidents are covered – is marginal compared with the sum for the insurance cover against fire/explosion perils.
(4) A fire occurs suddenly and accidentally, and the loss is normally quickly and precisely assessable. In contrast, in the case of many environmental losses, the situation is completely the reverse: the

date of loss and the cause of the losses are hard to assess, and it is very difficult to express the loss in terms of money.

(5) In the case of environmental impairment, loss investigation and settlement often take on a political aspect and the public at large is emotionally involved.

(6) The field of environmental losses is becoming more and more a milk cow for the enrichment of lawyers, clean-up contractors, excavators, waste incinerators, and others.

All in all, insurance of EIL on a full-cover basis is a branch with a very limited capacity, with a small number of issued policies, with the problem of antiselection, and with a high degree of incertitude.

However, despite these negative aspects, this type of insurance presents a challenging scope of work for the underwriters and the participating engineers.

14.2. How is EIL Organized?

The insurance industry's approach is relatively new and far from uniform. Under many standard forms of general comprehensive liability policies, damages arising out of a sudden and accidental pollution of the environment are also covered. The Bhopal incident is a typical example of such a sudden and accidental type. In contrast, the main target of EIL insurance born in 1974 is to cover also gradual events. Incidents like a leaking tank are normally not covered under the general comprehensive liability policies: these incidents are typically covered by an EIL policy.

A gradual leak or discharge of effluent may go unnoticed for a long time before its harmful properties are realized. This makes it almost impossible to identify an occurrence within any given policy period. The term "occurrence" must be avoided. Therefore, as a necessity, such an EIL policy is always issued on a claims-made basis.

One of the main problems of an EIL insurance consists in the calculation of premiums. Insurance rates are normally calculated on a statistical basis. They depend on historical information about size of claims and frequency. Unfortunately, only insufficient information on this type of loss is available to provide a basis for rating these pollution risks. Furthermore, the historical claims situation may not be a reliable guide for the future. In recent years there has been a growing awareness of environmental deterioration and loss of amenity. Improved scientific techniques give victims a greater chance of locating the source of a specific environmental impairment.

In setting tariffs for EIL risks, insurance companies have followed a highly subjective approach: instead of collecting information about claims, they have considered the likelihood of such claims. In principle, the following approach has been adopted.

Processes and substances are listed according to their possibility of causing harm to water, land, and air. At this stage we are solely concerned with the likelihood of environmental damage, not with whether or not a claim will arise nor with what the extent of such a claim will be. The final result is a score-table for different industrial activities. It gives an indication of the pure risk and is the main element of the EIL tariff system. Other elements are: country, history, degree of technical development, turnover of the client's company, claims consciousness of the population, etc. Together with experience in the field of liability, we finally end up with our tariff system.

To cover a particular risk with an EIL policy, an essential part is a careful risk inspection of the premises of each client. Such a survey serves to identify risks and to assess the likelihood and size of possible loss scenarios. Fundamentally important is the assessment whether the estimated loss scenarios give rise to actionable claims. Surveyors and underwriters must discuss these points together. Underwriting a misjudged risk could end up disastrously for the underwriter.

Analysis, evaluation, and treatment of risks, so-called "risk management", has become of utmost importance not only in the field of fire insurance (where these techniques have been developed) but also in the field of liability insurance.

14.3. Experience with EIL Insurance

Much of our know-how has been collected so far by trial and error. I am pleased to have the opportunity to share our experiences at Swiss Re with you. Selected areas that may be of interest to you are now discussed.

14.3.1. Clients

Our clients mainly belong to the group of "carriers" and "disposers" of toxic or nontoxic waste. Another group of clients is involved in transportation, storage, handling, and processing of oil products. These clients are well aware of the high risk potential they carry with their business. However, the vast majority of plant managers, when making a cost–benefit comparison, still decide against an EIL cover.

The high publicity surrounding EIL insurance provoked the public to believe that thousands of clients and a tremendous premium income must be involved. In fact, Swiss Re, as a reinsurance company, is involved through the ceding insurance companies in a disappointingly low number: there are only a few dozen clients in the portfolio.

14.3.2. US shock

A few years ago, because of the very difficult situation of the liability insurance market in the USA, we decided (also through fear of the courts) to turn away from the US EIL market. In our opinion, this was a necessary and vital decision: just look at North American newspapers and read about the compensation that insurance companies have to pay.

14.3.3. Worldwide EIL insurance capacities

Today, as far as we know, there exists a national EIL pool in Italy with a capacity of approximately US $3 million. Another national pool is in France, with a capacity of French Francs 30 million. A most interesting experiment has recently been started in the Netherlands, where 53 insurance companies bring in their EIL risks on a voluntary basis. The pool has a capacity of approximately US $1.5 million. Our own international underwriting capacity is US $3 million. There may be further insurance capacities floating around. But, all in all, we guess that the maximum combined limit of liability available may hardly reach US $10 million for a single risk in Europe.

Compared with the possible scenarios which may occur, this maximum limit of liability is in some cases not more than a drop on a hot stone. Compared with the late 1970s when a client easily obtained a limit of US $30 million, this is a considerably different situation.

In the USA, a dramatic EIL capacity slash has been observed during the last few months. There, most buyers will now only be able to purchase a maximum of US $10 million in EIL coverage, down from limits of more than US $65 million in early 1984. This slash in EIL capacity is caused by the lack of reinsurance capacity. An important characteristic of this business is that it is heavily reinsured. The reinsurers are obviously suffering from severe losses and have a very negative feeling towards the US courts. To some extent the situation applies also to Europe. The reinsurers' distaste for pollution risks is likely to continue in the near future.

14.3.4. Loss experience

The total number of losses and claims (so far) against our clients is small. However, the average amount paid per loss is remarkably high. This is not astonishing, since the character of the EIL cover is to indemnify an insured against a catastrophic event. This includes his having to bear a considerable deductible; therefore only the costly accidents are filed.

All the losses can be divided into two classes:

(1) Storage and transportation of oil and oil products.
(2) Waste disposals.

The loss handling philosophy is remarkable. What we learnt in the first place is that the reinsurers are too far away from the risk to be directly involved in the investigation and settlement of a loss. Secondly, the insurance companies reinsured such risks to a large extent, in some cases up to 99%. Consequently, they are not very interested in exposing their highly qualified expensive technical staff for such a case when most of the risk is reinsured. In the end, both insurer and reinsurer pay without a proper loss investigation, and the technical staff remains in the office. This situation calls for improvement in the immediate future.

14.4. Perspectives of EIL Insurance

The future of EIL coverage with retroactive cover on a claims-made basis is not secured. Sooner or later the following actions should be taken.

14.4.1. Reorganization of insurance technique

In our opinion, the most appropriate insurance technique for such a business is the formation of a pool (national or international). Such business can be done on a much more economic basis when the administrative expense is in relation to the number of policies and premium income, respectively. Such a pool could easily get a considerable volume and, as an additional benefit, the technical advisers could acquire considerable experience in a short time.

Within a pool, the above-described risk management technique and, possibly, loss investigation could be carried out much more efficiently and with greater benefit to every party involved.

14.4.2. Improvement of claims/loss handling technique

From the viewpoint of insurance, the main target in the claims handling and adjusting process is the compensation of the damages caused to the victims and to back unqualified claims in order to satisfy the victims and to keep the loss figures in acceptable proportions.

Pollution of the environment belongs to the most complex and delicate group of losses. Each and every case is a political issue. Whereas catastrophic fire losses - even those with casualties - are mentally accepted as an inevitable blow, the simplest pollution case produces highly emotional reactions. In such situations, it is rather difficult for the loss adjusters to keep their feet on the floor. In many cases, those responsible are confronted with a flood of claims. To separate the chaff from the wheat it is essential that underwriters, loss adjusters, and the risk engineers work shoulder to shoulder. This may sound very simple, but it is not at all obvious. There is more than one case where these three categories of highly qualified people of one and the same insurance company had no communication. Everyone is pottering about in his own garden. Some new possible organizational methods are required.

EIL business calls for teamwork between commercial and engineering specialists. It requires an interdisciplinary approach in order to be able to cope with every accelerating change. The aspects of cause, consequences, and special circumstances can never be considered in isolation. With the elimination of the growing pains mentioned and the improvement discussed, EIL should finally get off the ground.

CHAPTER 15

Recent Developments Concerning the Legal Regime and Insurance Problems related to the Transportation of Hazardous Materials by Sea

E. Orlando

Transportation of hazardous goods and materials, by any means of conveyance, is obviously a rather risky activity. In this short chapter I will only refer to transportation by sea, but this is certainly not to say that transit by other means (road, rail, inland waterways, or air) may not imply hazards or involve lesser legal and insurance implications, albeit different ones.

My restricting this exposé to maritime transport is meant to limit a subject that is otherwise too vast, and follows the specific commitment of marine insurers – and of the International Union of Marine Insurance (IUMI) – to this singular problem, leaving any other field open to their colleagues in the various sections of the insurance industry (motor, general third party liability, and aviation respectively).

In maritime transport, however, hazards are more evident and obvious: not for nothing was marine insurance historically by far the first form of insurance industry. Marine underwriters find little problem in providing protection for the property (ship and cargo), for the freight, and in respect of collision liability; at the same time, "Protection and Indemnity (P and I) Clubs" specialize in affording coverage for all other aspects of shipowners' liability.

In contrast, no established market exists at present for any Third Party Liability (TPL) insurance related to cargo, since to date cargo by itself has hardly been considered to be a target for any recovery actions, in tort or otherwise.

Nowadays, the ideas of ecological and environmental protection are developing rapidly, and there is a growing feeling that general and collective interests should somehow be entitled to full compensation when damaged: thus, the issue of liability insurance in connection with maritime transport of hazardous substances is an open problem. Even before insurance is arranged, a number of legal questions need to be answered: Where are the responsible parties to be found? What is the nature of such liability? What limits (if any) are appropriate in respect of that liability?

These three questions are strictly interconnected, and perhaps the easiest one to answer is the second. According to the philosophy presently prevailing, it is possible to predict that the issue will be governed by the principle of strict liability – following, for example, the liability of air carriers, or that of nuclear plant operators. I will take this for granted here, without discussing the merits or otherwise of this system, which would divert us from our main subject.

Once this assumption is accepted, which party shall bear the burden of such liability? The various indications (jointly or alternatively) could be: the carrier, the shipper (or receiver), the owner of the goods, and maybe even the producer. I am inclined to believe that here the wealth of different possibilities is far from helpful with regard to finding a universally acceptable solution.

It is worth mentioning that in one specialized area – that of ocean transport of oil and oil products – the solution was arrived at by mutual agreement and cooperation among all interested parties (which include carriers and industries): I am referring to the schemes known as CRISTAL and TOVALOP, backed at government level by the Civil Liability Convention 1969 and the Fund Convention 1971.

This complex system has been positively tested on many occasions. However, as it refers to a single specific and homogeneous trade, it would appear to be of little help as a model for rules to be applied to the transport as general cargo of the greatest variety of goods and materials.

Yet, when in 1984 the International Maritime Organization (IMO) held a preliminary discussion on a draft "Convention on Liability and Compensation in connection with the Carriage of Hazardous and Noxious Substances (other than oil) by Sea" (the "HNS Convention"), at least one aspect of the agreement governing the liabilities arising out of the maritime carriage of oil and oil products was taken as a model by the drafting panel – namely, the involvement of both the carrier and the carried goods. A two-tier liability system was suggested, with liability falling firstly upon the shipowner up to a certain limit, and subsequently upon the cargo once that limit had been exhausted.

This mechanism has many merits, but I see one major disadvantage in that at the first level the newly defined liability would coexist with all other shipowners' liabilities: quite understandably, shipowners and P and I Clubs strongly dislike this solution involving a separate exposure with a limit of its own. In the case of a major accident, this could imply a number of problems and delays in view of the need to previously establish the amounts claimed by all the damaged parties involved, and their respective rights of priority should the statutory limit prove insufficient to compensate all of them.

Many problems arise in connection with the second liability layer, especially as to identification of the party having to accept liability for the cargo: the easiest solution would appear to be to single out the shipper, who is also the most clearly identified (both the owner and the receiver may commonly change and change again during the course of a sea crossing).

Marine insurers have expressed in principle their readiness to provide the insurance capacity required, since their service in some ways parallels cargo insurance, though there are major differences in that both the nature of the guarantee and the amounts at stake are very different. With regard to the latter difference in particular, attention should be paid to the fact that goods of relatively modest value could be required to bear extremely high liability limits.

It is common knowledge that the 1984 IMO discussions afforded few positive results: the differing national and trade interests came into collision on the points I have mentioned, as on many more (level of the limitations, direct recourse of the claimants against liability insurers, safeguard of the ultimate recourse against any party eventually at fault, inclusion in the Convention of the liability in respect of parcel goods as against bulk cargo only), and the issue was referred back to the drafting committee – which, I am sorry to believe is tantamount to saying "ad Kalendas graecas" (indefinitely postponed)! I fear that, if this subject is taken up again under the shock of a major accident, the solutions could be sought for by other bodies (e.g. the United Nations Conference on Trade and Development) and be influenced more by political than technical and commercial considerations. Pending all this, marine insurers may do no more than to confirm their willingness to contribute to the finding of the appropriate solutions in the best interest of all parties concerned, making these solutions the least expensive for the trade and the shipowners' industry.

I shall add a brief reference to another collateral problem which is being currently debated. Based on a study by the Comité Maritime International, the IMO is discussing a revision of the 1910 Brussels Convention on Salvage and Assistance at Sea. This issue is closely connected with the "HNS" problem, since it envisages going beyond the traditional "no cure, no pay" principle which has been followed whenever the salvaged ship has been carrying hazardous substances (other than oil and oil products as cargo): thus, the salvage operation would simultaneously be aimed at preventing (or minimizing) pollution hazards.

Here again a model exists – namely, the private instrument known as the "Funding Agreement" which was arrived at after the coming into force of the Lloyds' Open Form 1980: this is the new standard salvage form now in use in connection with salvage operations concerning loaded oil tankers.

On the one hand, there is a generally felt need to encourage salvors to take up such often tricky and dangerous operations, by an appropriate enhancement of the salvage award; on the other hand, we are presently short of a designated counterpart to share the burden of such enhancement along with the salved property and interests – the ship, cargo, and freight.

In view of the urgent need to maintain an active and well-equipped salvage industry in the best general interest of the shipping community, the various parties concerned – shipowners, shippers, P and I Clubs, and marine insurers – are endeavoring to finalize an agreement for the allocation of the increased salvage awards that are anticipated. Such an understanding, once it has been arrived at, should, in my opinion, be only temporary: it is an anomaly to agree that the burden of the extra cost (the amount of which, in any case, is never precisely identified) which is referrable to the safeguard from pollution should be borne by insurers who have not undertaken to afford any direct coverage for the liability arising out of the possible pollution incident.

CHAPTER 18

Insuring Environmental Liabilities

R.M. Aickin

I shall begin by saying what I am not going to do. I am not going to discuss how to underwrite environmental impairment and the mechanisms of risk assessment; I am not going to address the problems of the waste disposal site issue; nor am I going to discuss claims-made coverages. I have written on these subjects elsewhere (Aickin, 1984a,b, 1985). In this short chapter, I am going to discuss insurance in general terms – what it can do, what it is doing, and what its limitations are. The basic tenet of insurance is to spread the costs of the misfortunes of the few among the many who run the same risks. In providing this service, an insurer's overriding concern is to know his maximum exposure to a loss. In other words, he needs to know that in the event of some misfortune he will not have to spend more than a certain fixed sum of money.

There are restrictions to the scope of insurance. It can only deal with pure risks. It cannot deal with risks which are so large that they become society's problem and should be dealt with by society or the state directly. The traditional example of this is the "war risk" which, except in special circumstances, is uninsurable. War damages (e.g. during the Blitz in London in the last war) have been met by the state.

The other thing which it is important to distinguish are the classes of property and liability insurance: these things are separate and distinct.

In the USA, liability serves three purposes: first, to compensate victims; second, to provide a punishment and deterrent for wrongful acts; and, third, to provide retribution. Insurance is designed for and can deal well with the first aspect of compensation. It deals badly with the retribution and the deterrent aspects. Indeed, if these are dealt with by insurance, the deterrent effect and the punishment effect are removed from the perpetrator.

Looking at environmental liabilities, we need to consider the potential size of the liabilities; although I said I would not deal with the waste problem, I must briefly refer to it here. I have talked to a number of officials about the cost of the waste site problems over the past several years and they admit that the size of Superfund at $1.6 billion is totally inadequate to deal with the problem. They have privately given me numbers between $125 and $250 billion. The US Congress Office of Technology Assessment has recently released a report identifying the cost of the waste site problem in the USA as in excess of $100 billion. This is a significant sum of money. It is between two and three times the annual premium volume of the US property casualty industry and is several times that industry's total surplus, i.e. the amount of money it has in the bank. If society or the courts are intent on forcing this cost onto the insurance industry, they will bankrupt the insurance industry because it does not have the money to pay for the loss or the wherewithal to collect it.

The insurance industry provides a vital function to millions of individual citizens who require fire insurance so that if their houses burn down in the night they will still be able to have a roof over their heads. It seems to me inconceivable that society will let the insurance industry become bankrupt; and, therefore, being an optimist, I think it is inevitable that this problem will be solved in some other way than forcing the cost onto the insurance industry.

I believe that the size of the potential exposure is such that it becomes a societal risk, that somewhere a ceiling must be placed upon the liability of individual companies, and that, if liability exceeds this level, it will have to be borne by the state as a societal risk.

There has been a lot of talk of the insurance industry exercising the role of policeman. There is some historical precedent for this, in boiler machinery and lift insurance. I should emphasize that this is essentially a first party property insurance: a lift is a discrete piece of machinery, and it becomes fairly easy to examine this and to develop a checklist for the examination. This is a very different proposition from that of liability insurance.

With respect to Massachusetts, waste generators are reluctant to come forward and say that they need waste disposal sites. I must say that there are good reasons for this. I am sure that we all have our own conceptions of what a waste generator is. I would also venture to suggest that, whatever our individual perceptions, they do not include soft drinks manufacturers, people making dairy products, or people manufacturing soap.

Yet it is precisely these three industries that headed the list of contributors to one of the major nasty septic waste disposal sites in the USA. These industries have enormous incentives not to be associated with hazardous waste. If the public thought that yogurt was a major contributor to hazardous waste, they would be very likely to stop eating it. This would be bound to be against the dairy product manufacturers' interest. Over and above this, there is a concern for secrecy with regard to environmental matters. My company, ERAS (International), had a client, who must remain nameless, with whom we had a serious disagreement. I have to say that they were more than cooperative and free in providing us with information. Some of their facilities were heavily contaminated. In a number of them there was 7 m of product floating on the water table beneath the facility. The company was aware of this problem and did not wish insurance for it. We were also aware of it and did not wish to provide insurance.

The traditional way of dealing with this is to draft an exclusion which says that "this policy shall not apply to the following liability". This created enormous problems. The insured or the potential insured refused to accept such an exclusion – not because they wanted coverage but because they wanted the existence of the problems suppressed. After some consideration, we countered by saying, "well, if you won't accept this exclusion which we have drafted, we will draft an exclusion dealing with the problems described in pages 7–10 of the report." This produced an even worse reaction of horror. This revised exclusion was even more unacceptable because it admitted the existence of a report which third parties might ask to read. Eventually we amicably agreed that we would not provide insurance and they would not buy it because of this problem of disclosure of sensitive information. Other insureds have reacted to this problem in a different fashion: they have simply refused to provide information.

It is not feasible to expect an insurance industry to act as a policeman. Insureds are unwilling to provide information or wish to hedge it around in secrecy to such an extent that it is dealt with under a stringent confidentiality agreement where it is a private transfer of information between an insured and an insurer. How much more will they object and have reservations about providing this information when the insurer is in the position to blow the whistle and to close them down (and may have a statutory obligation to do so)?

The second point which I think is of great importance is that, if I am lucky in dealing with one of the major companies represented at this Conference on Transportation, Storage, and Disposal of Hazardous Materials, I may have one day to evaluate the risks of that major corporation. Certainly I will have had people go out to do inspections and to gather information – they may even have spent a week on the corporation concerned – but it is quite unrealistic economically to expect a report of the scale or of the detail of the Canvey Island report (Health and Safety Executive, 1978, 1981), which cost 300 000 pounds sterling, to be provided on which I can base my decision. Therefore, as an underwriter I base my decision upon personal skills. It is obvious to me that this is troublesome and that both the social scientists and the systems analysts shy away from personal skills.

They both consider that human intervention and human skills make life difficult for them and they cannot account for it in their modeling; yet, an underwriter is totally dependent upon these personal skills. This means that it is very difficult to be put in the role of the policeman and to say that I have closed someone down because I have a personal feeling, perhaps on the basis that I do not like the color of someone's hair.

One may ask how the insurance industry manages to get away with this. The insurance industry manages quite simply because it is a competitive free market. We were the first people providing environmental insurance. Because we did so and were perceived to do so profitably, other people came into the field. Some of those were better than we were at evaluating risks; most of them were less good. However, this allows for competition. You come to me and I decline to insure you because I do not like the color of your hair, so you go to someone else who may not have this aversion and obtain coverage. This means that the insurance industry is cyclical: when we make losses as an industry, some of those people who came in late and have made greater losses because they are less good will leave. At that stage, competition is smaller, rates become more expensive, and insurance becomes more difficult to obtain.

One must take into account the method by which insurance is regulated. Insurance is regulated because it provides valuable service to individual citizens like you and me, and it is important for our long-term security that an insurance company should not go bankrupt and leave us without coverage when it does so. As a result, the regulators and legislators deal with insurance in its own terms. They take premium volume as a measure of the risk and they say that, all things being equal, an insurance company will strive to balance its income and expenditure and therefore will expect a total costs to income ratio of about 100%. They then go further and say that, in the worst case, one might make a 30% loss. As a result, one must show sufficient financial strength for a third of one's premium volume.

While this is acceptable in personal lines business and other business where rates do not fluctuate greatly, it is not acceptable or a sensible method for regulation in commercial lines. If I were foolish enough to write 10% of, let us say, EXXON's program at a premium of 0.1%, and in the following year when rates had hardened tenfold I could obtain a 1% premium, then, in maintaining the same premium volume, I would only be able to write 1% of EXXON's liability program.

For those of us who worked in the market, the signs that the market had changed were clear between the last week of May and the first week of July in 1984. While it is obvious that Bhopal did not help matters, it is quite clear that it did not have a great impact on change in the market, as the symptoms were evident well before the Bhopal accident.

According to Smets (Chapter 4, note [21]), Union Carbide had $200 million of capacity for liability insurance available to them at the time of the Bhopal accident in December 1984. It is also fairly common knowledge that Union Carbide are supposed to have turned down an offer for a further $150 million which would have brought the total capacity of liability insurance up to $350 million.

There were other corporations whose philosophy was to buy the maximum insurance coverage available and who last year purchased the $350 million or so of coverage which was available to them. One such company that I know of, when it came to renew its coverage in 1985, was only able to place, through its brokers, $47 million of capacity. It is therefore conceivable that, had Bhopal occurred six weeks later, in January 1985, there would only have been perhaps $50 million of coverage available, as it is probable that Union Carbide have a 1 January renewal date. This is not a well-known or publicized fact but it seems to me that it is a strong support for the argument that there should be a ceiling above which the state must take over responsibility for the liabilities as a societal problem.

This shrinkage of capacity has been universal. The result is that insurers do not have the capacity to write the traditional business that they would like to. In the circumstances where they cannot write as much business in the traditional areas that they think they understand, it is very difficult to expect them to divert that capacity to other areas where they need to talk to scientists in order to have the risks explained to them. I do not think that we should view the shrinkage of environmental impairment liability capacity as a running away from environmental risks so much as a preference to devote the scarce resources of the insurance industry to more traditional areas of activity.

In conclusion, let me say that I believe that the insurance industry has things of value to contribute in this area but that it cannot be expected to take on the entire costs of environmental problems, nor can it be expected to take on the role of policeman because it does not have the resources to do so. The insurance industry cannot afford to expend more than, say, 10–20% of its premium income in acquisition and analysis of risk data. If, in the case of Canvey Island, we are dealing with risks of the order of 1 in a 1000 and are therefore only receiving thousands of pounds in premium for the risk, we simply cannot afford to spend 300000 pounds sterling on evaluation.

References

Aickin, R.M. (1984a), Environmental impairment liability – an adolescent class of insurance, *Post Magazine and Insurance Monitor*, **CXLV** (27 September), 2305–2306.

Aickin, R.M. (1984b), Claims first made wordings: the lessons of EIL, *Post Magazine and Insurance Monitor*, **CXLV** (8 November), 2674–2675.

Aickin, R.M. (1985), *Environmental Impairment Liability Underwriting*, Paper presented at RIMS Conference.

Health and Safety Executive (1978), *Canvey: An Investigation of Potential Hazards from Operations in the Canvey/Thurrock Area* (HMSO, London).

Health and Safety Executive (1981), *Canvey: A Second Report* (HMSO, London).

CHAPTER 17

The Role of Insurance in Risk Spreading and Risk Bearing

W. Pfennigstorf

17.1. Introduction

To describe the role of insurance in risk spreading and risk bearing is to describe insurance – risk spreading *is* insurance. Insurers have been pioneers in the application of probability theory, risk theory, and statistical methods; using these tools, they have developed highly sophisticated techniques for relieving the burden of most of the risks associated with modern life and economic activities. This chapter reviews the major factors that determine and limit the actual and potential extent of insurance coverages applicable to the risks associated with transporting, storing, or disposing of hazardous materials.

Hazardousness is a dynamic concept: it depends on information, and therefore changes over time. Although dynamite is capable of causing great damage, its dangerous properties are known so well that it is not considered too dangerous to work with, give adequate precautions, and insurers have no difficulty providing coverage for the risk that remains. The greatest difficulties arise, for insurers no less than for engineers and businessmen, where the potential of a substance to cause harm is not known or cannot be measured. Liability insurers also have to accept the perceptions of hazardousness under which courts and legislatures determine liability, even if these perceptions are not consistent with available medical evidence.

17.2. Technical Limitations of Insurance

Insurance does not eliminate risk: it does not keep harmful events from occurring (although the existence or availability of insurance coverage and the operating practices of insurers and insureds may have some effect, in one or the other direction, on the probability of occurrence). All that insurance can do is to provide compensation, usually in the form of payment of money, for the *consequences* of harmful events, and only to the extent that they can be expressed in economic or monetary terms.

This is particularly significant for harms caused by hazardous materials. No amount of money – and, for that matter, no insurance policy – can provide adequate compensation for the emotional suffering of those afflicted with a disabling disease or a genetic defect, or even the irritation and inconvenience associated with minor everyday forms of pollution, let alone the destruction of natural resources and the persistent contamination of the environment.

Therefore, compensation, whether financed through insurance or otherwise, can never be a substitute for preventive control. Nor can insurers be relied on to control the risk in a manner that would be equivalent to administrative control. Proposals, much discussed in recent years, to employ insurers as "surrogate regulators" (Pfennigstorf 1982), would create extremely difficult problems of cost allocation, supervision, enforcement, and conflicts of interests. As Aickin points out in Chapter 16 of this volume, even the initial investigation and evaluation of risks burdens insurers and insureds with considerable expenses and other problems that may be out of proportion to the amount of the premium that the insured can afford to pay in view of the volume of its business.

What insurance can do, and can do very well, is to soften the impact of the economic consequences of random harmful events, by pooling risks and averaging the cost of compensating losses. Most insurers provide coverage for a fixed premium for a specified period. The advantage for the insured is obvious; for the insurer, however, fixed premiums involve the risk of accurately predicting the total losses that will have to be paid from the total premiums collected. This is much easier in some lines of insurance (e.g. life insurance) than in others (e.g. general liability insurance). Insurers have learned to soften the impact of large-scale loss fluctuations and of unexpectedly large losses by establishing reserves, by maintaining a security margin of capital and surplus funds, and especially by obtaining reinsurance coverage.

Despite these safeguards, it remains one of the most urgent concerns of any insurer to be able to predict future losses as accurately as possible and to obtain the widest possible spread of risks for reliable statistical averaging. These goals determine to a large extent what risks an insurer will solicit or accept, and on what terms; and they are also the reason why insurers find it impossible to provide coverage for extended periods for fixed premiums.

Insurers often call the risks that they decline for such technical reasons "uninsurable" – a term that suggests some objective standards. Actually, perceptions and definitions of insurability are vague and subjective, and differ widely among insurers, depending on, among other things, each individual insurer's basic attitude towards risk, its financial conditions, its appraisal of market conditions, and its officers' and underwriters' personal evaluation of the risks assumed.

The insurers' usage of the term "risk", incidentally, is anything but precise. They use it in the meaning that it has in common language and also in the special meaning of risk theory. In the daily practice of the insurance business, however, the word is most frequently used to indicate the person or object or entity in whom or in which the risk may or may not materialize in the form of an injury, damage, or loss. In identifying coverages, insurers usually refer to perils or hazards, which represent commonly perceived potential causes of loss, such as fire, collision, or exposure to a specified substance or condition.

Despite their efforts and achievements in analyzing risk factors, insurers must make most of their decisions on the basis of incomplete information and largely subjective evaluations, and subject to political and market pressures.

To some extent, concerns about insurability can be overcome by higher premiums. Indeed, some insurers like to say that every risk can be insured provided the price is right. Even though slightly exaggerated, this statement reflects a basic truth of the insurance business, but it also serves as a reminder that the dividing line between insurance and gambling, though clear in theory, is difficult to define in practice. A case approaching the borderline is satellite launching insurance, where premiums have reached 20% of the insured value and are expected to increase to 30%.

Insurers are sensitive to the need for precise terms defining the risks to be covered under a specific policy, and the extent to which they are to be covered. They are also sensitive to the special risk inherent in changes in the original risk situation on which the loss projection and premium calculation were based, and they have developed a variety of rules and techniques to minimize the potential effect of change. These concerns are particularly relevant for the risks of hazardous materials.

17.3. Special Problems Posed by Hazardous Materials

17.3.1. Relevant types of insurance

Although liability insurance has received primary attention recently, losses caused by hazardous materials are among the risks covered by several types of insurance. Most life and disability (accident and health) insurance policies are all-risks policies – their benefits are payable if the insured person dies, is injured, or becomes sick, regardless of the cause, subject to a few precisely defined exceptions, which normally do not include exposure to a hazardous substance. Likewise, all-risks property insurance policies cover damage caused by hazardous materials unless this is specifically excluded. The only standard exclusion in property insurance is that for exposure to radioactive materials or radiation.

There can be problems of coordination between different types of coverage. Chemical spills on the insured's own premises are a matter of property insurance, not liability insurance. However, if the chemicals reach the groundwater or otherwise pose a risk to third parties or the environment, the question of liability and liability insurance coverage comes into play. Liability insurers may voluntarily pay for clean-up in such cases to avert or minimize losses, but so far they have refused, with few exceptions, to make such payments part of the regular coverage.

First party accident and health insurance can have a substantial indirect effect on liability insurance claims. Where health care costs and loss of income due to disability are covered under a comprehensive social insurance system, there is less incentive to sue someone else for damages than where tort liability is the only or an important additional source of compensation.

17.3.2. Dimensions of the risk

Insurers face uncertainties that are not limited to the random incidence and distribution of losses but relate to the nature of the risk and the extent of the losses that can be caused. Past experience, which ordinarily provides the basis for at least a rough estimate of future losses, is of no help with many types of hazardous materials.

In addition to chemical, physical, and biological factors, liability insurers must also consider uncertainties of a legal nature – the confusing, conflicting, and rapidly changing rules that determine who is liable to whom under what circumstances and in what amount. This appears to be a greater problem in the USA than in Europe, for a variety of reasons that can only be listed here but cannot be discussed in detail. They include the absence of a comprehensive national health insurance system, resulting in greater reliance and pressure on the liability system as a source of compensation for everyday accidents. Other factors are: the collateral source rule, which permits a claimant to recover more than his actual economic loss; the multitude of jurisdictions; the adversary style of proceedings in civil matters; the contingent fees that attorneys customarily charge for representing a plaintiff; the American Rule, under which each party bears the cost of its own attorney, regardless of the outcome; and the availability of highly skilled and very aggressive attorneys specializing in representing plaintiffs in personal injury claims.

Hazardous materials pose several special problems for which the traditional rules of tort law offer no satisfactory solution and for which therefore new rules more favorable to injured persons have been considered necessary. These problems are:

(1) The large number of identical or similar injuries that can result from one incident, producing a large number of claims, which, if litigated individually and independently, could exhaust the physical resources of the judicial system, and could lead to long delays in the disposition of claims and to unequal compensation for similar injuries.
(2) The difficulty of tracing the causal connection from the polluter's culpable conduct through the release and the claimant's exposure to the eventual injury and economic loss, which places extra weight on the plaintiff's burden of proof.
(3) The fact that in many cases a causal relationship between exposure to a given chemical and a specific type of health damage (e.g. cancer) cannot be established individually but only by inference from statistical correlations for large groups of individuals.

Liability has been expanded by court decisions and by state and federal legislation, and it has been expanded in many directions and in many respects. Some of the changes that have occurred recently have already made it exceedingly difficult for insurers to estimate potential aggregate future losses. The tendency to expand the scope of liability is continuing, and the nature and extent of future changes are unpredictable (Pfennigstorf, 1979).

The difficulty of anticipating changes in the risk situation, as well as the impact of the changes on the insurer's risk, increases with the time considered. For this reason, insurers generally issue policies only for short periods of time, or subject to termination upon notice, which gives them an opportunity to review their underwriting decisions and to adjust premiums and contract terms.

The difficulties are further compounded by the fact that losses from a single incident involving hazardous materials can reach catastrophic proportions. The disaster potential of hazardous materials is even greater than that of a natural disaster since losses are not spread among many property insurers but rather are transformed into liability claims and in that form are concentrated on a small number of defendants and their insurers. This means that insurance policies issued for individual risks would have to respond to the full catastrophic amount of the total loss – with a much smaller population of insured risks available to supply the necessary premium volume and spread of risk.

17.3.3. Inherent limits of liability insurance

The general rule is that losses caused deliberately by the insured are outside the concept of risk, which implies an event whose occurrence is uncertain and beyond the control of either party. Where in life and health insurance coverage is provided for the consequences of intentional acts, such as suicide, it does not constitute an exception in the strict sense but reflects a different risk. In liability insurance, coverage of claims arising from the insured's intentional acts has even been declared to be contrary to public policy (Pfennigstorf, 1979). Similar concerns about the randomness of losses have kept insurers from providing coverage for the consequences of decisions made or actions taken in the regular course of business. And further, insurers have generally found that they are unable to provide security against events such as inflation or depression, which affect not individuals randomly but society as a whole.

From the beginning, liability insurance has been subject to conflicting demands in the market and in public policy, which have tended to interfere with the development of a sound and consistent theoretical basis and effective business practices. Furthermore, the legal complexity and the conflicts of interest inherent in the three-party relationship linking insurer, insured, and third party claimant have been responsible for many misunderstandings and misinterpretations in the courts.

In the USA, general liability policies traditionally limited the coverage to claims based on bodily injury or property damage caused by an accident. The term "accident" was later replaced by the term "occurrence", which was in turn defined, awkwardly, as an accident (including some nonaccidental exposures). This change was in part a reaction to adverse interpretations of the term "accident" by the courts and in part an effort to accommodate the demand for liability coverage against claims arising from nonsudden (though unintended and unexpected) events, including certain releases of hazardous materials (e.g. from underground storage tanks) that occurred and continued without the insured's knowledge (Pfennigstorf, 1979).

This was followed by a clause excluding all pollution claims except those based on sudden and accidental releases. The courts, however, have tended to disregard the exclusion except where the release was clearly intended by the insured.

European insurers also generally base liability insurance coverage on an occurrence or event causing bodily injury or property damage, and they generally exclude from the coverage all claims based on the gradual effect of exposure to substances or conditions. They have avoided most of the problems experienced by the US insurers by leaving the traditional exclusion undisturbed and satisfying the demand for coverage of nonaccidental releases or exposures through an endorsement or a separate policy (Boediker, 1980; Spiller, 1981).

Most recently, special policies have been developed for pollution liability, or environmental impairment liability, which typically cover both sudden and nonsudden releases (Pfennigstorf, 1982). In turn, future standard general liability policies will, beginning in 1986, exclude all pollution liability claims (Malecki and Flitner, 1985). This will resolve the problem of distinguishing between sudden and nonsudden releases. It will not, however, provide a definitive answer to the more difficult question of how to distinguish between intended or expected releases (or injuries?) on one side and unintended and unexpected releases on the other side, or, more generally, between the situations and claims that the insurer intends to cover and those that are not meant to be covered.

European and US insurers have found it equally difficult to define precisely the event that has to occur within the period covered by the insurance to trigger the coverage. There are few problems in the case of an accident in the conventional sense – a violent conspicuous event in which all of the essential elements of a covered claim occur simultaneously or in short sequence. In the case of exposure to some hazardous materials, however, there may be intervals of many years between the incidence of these elements – the insured's act or omission that forms the basis for his liability under tort law, the release of the materials into the environment, the third party's exposure, the injury, the incidence of economic loss, and eventually the making of a formal demand for payment.

Policies covering claims for bodily injury or property damage generally specify the incidence of the injury or damage as the crucial event. That event itself is not easy to define or to identify in nonaccidental situations, however. That has become painfully clear to US insurers in the recent controversy about asbestos-related claims. Even among the insurers themselves there was disagreement about the meaning of their policy terms. The courts, not surprisingly, have tended to resolve the disagreements in favor of the insured.

Further, the insurers found that the long intervals between occurrence and eventual claim settlement all but defeated the purpose of limiting policy periods to one year at a time. US insurers are now being asked to pay claims for asbestos-related diseases and for the costs of cleaning up hazardous waste sites on the basis of policies written 50 or more years ago. There is no doubt that at the time when the policies were written no one contemplated or anticipated, or should have anticipated, claims of this nature and amount. Nevertheless, the occurrence formula would seem to imply coverage for all claims, however late brought, that are based on an occurrence during the policy period. All that is needed for the coverage to be activated is for a court to recognize the claim as well founded and timely made.

The insurers have responded by replacing the occurrence formula with a policy that limits coverage to claims that are made within the period specified in the policy (Malecki and Flitner, 1985). Claims-made policies are singularly effective in dividing up the risk of long-latency claims and thus making it, in shorter intervals, calculable and indeed insurable. In the process, the risk of long-term changes in the risk situation, which the insurers found to be beyond their capabilities, has been returned to the insureds. That means that in order to be fully protected against the risk of liability claims arising some time in the future from a present condition or activity, an operator must be prepared to continue a liability insurance policy in force for as long as there is a possibility that claims may be made, and to pay potentially increasing premiums for that coverage, even after the activity is terminated and no longer produces any revenues.

17.3.4. Limitation of compensable claims and damages

Liability insurance was designed for the risk of liability arising under the rules of tort law or, in the countries of the civil law system, the body of statutory law known as "the law of delicts". These rules, and the case law relating to their interpretation and application, have for years provided relatively reliable guidance to insurers with respect to the measure of damages payable for different types of injuries, and with respect to the calculation of damage to property.

US insurers always had to operate with a greater degree of uncertainty in this respect than European insurers, owing to the multitude of jurisdictions and to the role of the jury in determining the amount of damages to be paid. General or nonpecuniary damages awarded for pain and suffering, emotional distress, and similar effects have constituted especially volatile elements in the evaluation of potential claims.

Recently, the tendency of courts to award punitive damages in large amounts has added a new dimension of uncertainty.

The uncertainty produces, and is in turn increased by, a high rate of litigation. Consequently, litigation expenses have become an alarming burden for liability insurers.

Yet another problem is presented by claims relating to expenses incurred by government agencies or by the insured himself for clean-up or remedial measures and for the restoration of natural resources. These expenses can be part of the loss resulting from damage to property within the accepted meaning of that term, as when materials released by the insured have contaminated the property of a third person, and in such cases coverage under a liability insurance policy is usually not in doubt.

Liability insurance is not meant to cover the cost of cleaning up spills on the insured's own premises, however. They are supposed to be covered by property insurance. Where spills of this kind have presented a risk of harm to others, insurers have voluntarily assumed the cost of certain clean-up measures taken with their approval to avert or minimize liability claims.

During the past several years, as public awareness of environmental pollution has increased and the legal and technological bases for effective clean-up have been expanded, there has been increasing pressure on insurers to provide coverage for such expenses. Indeed, clean-up claims are about to exceed damage claims by individuals in frequency and severity. The insurers have recognized the risk and the need for coverage, and some of them have offered to cover it to some extent.

There is, however, great uncertainty about the amount of the expenses that are necessary in such cases. Among other things, the insurers are concerned about the difficulties involved in defining what is "clean", about the fast pace of development in monitoring and clean-up technology, and about the vast range of discretion that environmental agencies have in selecting the targets and methods of clean-up and remedial actions. Under some Acts, the agencies have the power to act when there is only a threat of a release. The risk of such actions and the resulting expenses is extremely difficult to reconcile with traditional notions of liability insurance.

17.3.5. Restraints of the insurance market

The US insurance market for commercial coverages in general is characterized by wide fluctuations, or cycles, moving from periods where coverages are offered freely and at low premiums to periods of high premiums and restrictive underwriting practices. The effect of normal market movements is amplified by the practice of some insurers to follow the general trend of the market not by raising or reducing their premiums for the volatile classes of business but rather by entering the market or withdrawing from it on short notice. At times, the reaction of the market to a series of large losses is best described as panic. The normal delays and difficulties involved in adjustment combined with sudden contractions of the market can produce temporary shortages.

For several years in the early 1980s, there was excessive price competition among insurers who were primarily interested in cash flow to benefit from the prevailing high interest rates. Eventually, as rising losses could no longer be compensated with investment returns but started to affect the insurers' surplus, the market did turn around, with the effect that commercial coverages of all kinds have become more expensive and more difficult to obtain.

In the European insurance market, the fluctuations appear to be less violent. Generally it seems that both insurers and insurance buyers value continuity and think in broader perspectives and longer terms in setting their pricing and underwriting policies (Boediker, 1980; Spiller, 1981). The marine insurance market also appears to be relatively stable. This seems to be the case especially for pollution liability coverages, which are provided either by pools or by Protection and Indemnity Clubs or similar mutual organizations (Treasury Department, 1982).

The market for pollution liability insurance for land-based facilities in the USA has been affected not only by the general cycle in the commercial insurance market but also by extraordinary losses in other types of liability insurance, as well as by persistent concerns about the tendency among courts and legislatures to extend the range of polluters' liability.

17.4. Outlook

Since the limitations just discussed affect only certain situations, they suggest a discriminating approach. It should be possible to provide insurance protection subject to appropriate definitions and limitations relating to situations, substances, types of occurrence, type of injury, type and amount of remedy, etc. To some extent, also, the availability of insurance coverage is influenced by traditional differences in the nature and attitudes of the insurers serving a specific market area.

Thus, the fewest problems seem to exist in the field of marine transportation, where releases are almost always the result of accidents or at least of incidents of short duration, and where international agreements provide a basic framework of liability rules and insurance requirements (Treasury Department, 1982, 1983).

Land transportation (or transportation on inland waterways or through the air over land) presents more difficult problems. Releases in densely settled areas can expose large numbers of individuals to hazardous materials directly or through contamination of surface waters or groundwater, the soil, or the air. On the other hand, releases of materials in the course of transportation are almost always accidental, which means that action can be taken promptly to contain the release and to minimize exposure. Even though the loss potential can reach disaster proportions, the large number of risk units should make it possible to spread the risk, provided universal or near universal participation can be obtained.

The problems tend to be most serious where hazardous materials are stored or disposed of. The risk burden of existing sites containing materials accumulated over a longer period will have to be shared by all. Even if attempts are made to force liability for cleaning up these sites on the generators of the substances (provided, of course, that they can be identified, that they still exist, and that they are solvent), the costs will have to be borne indirectly by the community. The temptation is great to force these costs on insurers on the basis of liability policies written many years ago. However, since this particular risk was not considered at the time when those policies were designed and their premiums established, and since no reserves exist to cover the claims, the necessary funds must be taken from surplus, exposing present policyholders in unrelated lines of business to an increased risk of insolvency and creating a need for general premium increases for present policyholders to make up the loss.

In the case of most disposal facilities, the risk of liability continues even when the facility is closed. Many of the materials retain their hazardous properties indefinitely. That they will eventually escape is certain; it is only uncertain when. No insurance scheme has been designed or proposed that would provide the necessary long-term coverage. Apart from the length of the time involved and the difficulty of predicting losses and calculating premiums, the fact that closed facilities have no revenues from which to pay premiums creates doubts about the economic viability of any such scheme. In the USA, the Superfund Act of 1980 has created a special fund for postclosure claims. It is financed through contributions paid by operating facilities (Treasury Department, 1982).

It has even proved quite difficult to secure the availability of the funds that are needed to pay for the work involved in closing and sealing a disposal site, and for the monitoring that is required under the environmental regulations for a certain period after closure. US regulations require the necessary amount to be estimated and to be accumulated in a trust fund over the active period of the facility. If for some reason the facility must be closed prematurely, the amount in the trust fund may still be inadequate. This risk of premature closure is one that could be covered by insurance. Indeed, such insurance is anticipated in the relevant regulations, and an appropriate policy has been offered in the market. This coverage, of course, is unlike any of the conventional property or liability insurance coverages but resembles in many respects a life insurance policy.

Finally, special insurance problems are presented by the contractors that perform clean-up and remedial work in connection with releases, removing hazardous materials and contaminated soil from one site, transporting and separating them, and depositing them again at some other site. These contractors may become subject to liability like anyone else who handles hazardous materials. The difference is that, if they are unable to obtain adequate insurance coverage and consequently are unable to continue their operations, the entire system of environmental control could be disrupted.

It should be pointed out that most of what has been said so far about availability of insurance applies to small and medium-size operations. Large manufacturers, because of the volume of their operations, are in a position to absorb (or self-insure, or retain) the cost of all but the very large claims, and they need to obtain insurance coverage only for the amount by which a catastrophic loss may exceed the amount of the retention. Also because of their volume, they can negotiate comprehensive insurance agreements covering all of their (excess) risks on terms and at premiums that are not available to ordinary policyholders. They are also likely to have their own captive insurance companies, which give them direct access to the international reinsurance market. Like everyone else, however, they feel the effects of general movements in the insurance and reinsurance markets.

17.5. Challenges

17.5.1. Market prospects

The insurance market, nervous though it is, has the capacity to respond quickly to a strong and broad demand for coverage. The withdrawal of some commercial insurers and reinsurers from the pollution liability insurance market has caused concern that there may not be enough insurance coverage available for all existing risks. Paradoxically, however, it is also true that the demand for the coverages that are offered has been less than expected. All the pools have complained about the small volume of business that they have been able to attract.

Pool managers and insurers agree that coverage amounts could be increased substantially if the volume of business increased.

If, for whatever reason, the response of the existing commercial insurance market should be too slow or too timid, the demand, if sufficiently strong, can be expected to turn quickly to alternative sources of supply. The most obvious reaction to an unresponsive established market is the formation of cooperative or mutual insurance organizations. This approach has been used many times in the history of the insurance business, most recently for medical malpractice liability, municipal risks, and lawyers' professional liability. In the current discussion, mutuals are rarely mentioned, but there has been much talk about association captives and risk retention groups, both of which are essentially just different names for the old concept of mutual insurance.

The experience of mutual insurance organizations for professional liability risks suggests that their development tends to be slower in times of a soft commercial market, when coverages are readily available at low premiums. That, in addition to the relatively low level of risk awareness, may explain why so far there seems to have been little interest in mutual insurance in the hazardous materials industry. The interest can be expected to increase as coverage in the commercial market becomes more expensive and more difficult to obtain.

Mutuals are voluntary organizations of businessmen. To be effective, they must operate in most technical respects like commercial insurers, and over time they tend to be assimilated by the commercial market.

Eventually, if a perceived shortage of insurance coverage is not filled either by the existing market or by newly created mutuals, it may become necessary for the government to step in. Government action may take the form of providing encouragement for the formation of mutual insurance organizations, or of requiring their formation, or of requiring insurers to establish a supplementary system for making insurance available for those facilities that cannot find an insurer in the market. If the need for insurance coverage is considered crucial for public policy, the government may even have to assume the role of an insurer of last resort.

17.5.2. Remaining limitations

Even the most responsive commercial market and even the most comprehensive mutual insurance plan cannot solve all the problems of insuring the risks of hazardous materials, however. We can expect coverage to be available, but it will have its price, which will be higher than what insureds used to pay for liability insurance in the times of cash-flow underwriting, and it will not provide the long-term security to which insureds had become accustomed under the traditional occurrence-style policy.

Rather, the coverages that are available now and will be available in the future are limited in several respects:

(1) Their amounts are fairly modest compared to the catastrophic dimensions of potential liability.
(2) They are claims-made policies and thus provide coverage only for claims made within the period specified in the policy.
(3) They do not cover the cost of clean-up measures on the insured's own premises except under exceptional circumstances.
(4) Many policies exclude specific substances (e.g. asbestos) by name, or exclude certain types of injury or damage (e.g. genetic damage).

In addition, insurers are extremely selective in their underwriting, and consequently, coverage will be unavailable, or available only at a prohibitive price, for high-risk facilities.

No insurer can be expected to assume the risk of change of risk factors associated with long-latency claims, at least not until the pace of development in technology, science, medicine, the law, and the economy stabilizes sufficiently to permit reliable long-range loss projections and corresponding financial planning. Therefore, claims-made policies will be the rule rather than the exception. One possible alternative, suggested by Spühler in his Discussion of this chapter, would be to provide for automatic termination of coverage upon exhaustion of the policy limits.

The claims-made policy is by no means an ideal contract instrument. For one thing, by designating the making of a claim as the crucial event for triggering the insurer's liability, the policy changes the conditions governing interpretation and application of the rules concerning notice of loss and cooperation, creating uncertainty and danger of abuse. On the one hand, the insurer will have to insist, as always, on being informed promptly of any event that is likely to result in a claim. Ordinarily, this information is used for efforts to avert or minimize the loss. Under a claims-made policy, however, the insurer may use the information to cancel the policy or to refuse to renew it before the claim is actually filed, and thus to escape liability. On the other hand, the insured, who is bound by the policy to cooperate with the insurer in trying to avert or minimize the loss, is under the temptation, once he knows facts that could lead to a claim, to precipitate the claim before the policy expires and before the insurer acquires the knowledge that would induce it to cancel. The task of reconciling the conflicting interpretations and interests will fall on the courts, which will be tempted, as usual, to favor the insured. In the meantime, there will be considerable uncertainty about the extent and effect of the information that the insured must provide to the insurer.

Second, the separation of the traditional bond between the range of covered claims and the activity from which they arise has serious and far-ranging consequences for the protection that can be provided to injured persons and the public through mandatory liability insurance. The effectiveness of liability insurance requirements as a device to protect injured persons rests on the fact that ordinarily they can be enforced conveniently in conjunction with licensing and other regulatory controls at the time when the liability-prone activity takes place. If, however, in order to protect potential claimants, a policy must be renewed for many years after the activity has ended, the insurance requirement becomes for practical purposes unenforceable.

In other words, claims-made policies are useless as means to satisfy a statutory insurance requirement, except with respect to claims involving the short-term effects of an accidental (conspicuous) incident. And, since it appears that for the foreseeable future claims-made policies will be the only commercially available type of liability insurance, this means that mandatory or compulsory insurance can no longer be considered a viable option for securing payment of third party compensation or clean-up costs.

In fairness it must be pointed out that even an occurrence-based liability insurance policy could not provide full protection for long-latency claims over several decades. To begin, the amounts of coverage would become inadequate as a result of general currency depreciation. In addition, every system of compulsory insurance has to deal with a number of other problems related to definition, enforcement, availability, premium control, and protection of third party claimants against defenses based on violations of the insurance contract.

To criticize commercial insurers for their reluctance to provide long-term coverage for liability risks would be not only futile but indeed unjustified. The real culprits are the uncertainty and the confusion that exist with respect to causation and liability over long periods of time. Indeed, it appears that the problem is not so much with the insurers' inability to provide coverage for an existing liability risk but rather with the inability of the liability system to deal with situations for which it was not designed. Widespread dissatisfaction with the performance of the conventional liability system has been the reason why the rules have been changed and are likely to be changed further with a view to making it easier for victims to receive compensation for losses that according to widely held perceptions deserve to be compensated. This state of affairs reduces considerably the range of potential alternatives.

17.5.3. Potential actions and alternatives

The US insurers have recently concentrated their efforts on a campaign to halt or even to reverse the trend toward expansion of liability. This raises several questions. First, one might ask why the insurers, which are only indirectly affected and could presumably protect their own interests adequately by increasing premiums and limiting coverages, have chosen to fight the battle of those who are affected immediately by increased risks and higher insurance costs and who therefore would presumably be able to present a much more convincing argument in the political arena. Second, if the insurers succeeded in eliminating some claims, that would not prevent the underlying release, exposure, injury, or loss from occurring and consequently would not eliminate but only redirect the pressure for some kind of compensation, except perhaps in the case of some types of nonpecuniary or punitive damages.

In most respects, those trying to restrict tort liability can expect to succeed only if they can offer a preferable or at least acceptable alternative source of compensation. The insurers' own interests are affected not by the liability rules as such but by the tendency of the courts to misinterpret policies to expand coverage retroactively to include new risks and changed liability rules. This problem calls for different remedies.

Legislatures have tended to respond to the generally perceived need to make compensation available to those who suffer injury or loss due to exposure to hazardous materials, without requiring the victims, or the government with respect to clean-up expenses, to incur the cost, delay, and uncertainty associated with a tort liability claim, by establishing compensation funds of various kinds. The scope of these funds is usually limited to specific types of pollution or specific classes of materials, and most of the funds existing in the USA are in fact intended primarily or exclusively for financing government clean-up operations. Some details are provided by Pfennigstorf (1979).

All specialized pollution compensation funds suffer from two weaknesses. First, because they are specialized, they require claimants to show not only that they suffer from a specified health defect but also that they were exposed to a specified material or type of material under qualifying circumstances that may serve as a basis for a presumption of causation. Even with very broad and generous presumptions, the burden is a heavy one for many claimants considering the long periods of time involved. Because of the inherent complexity of the factual situations and the need to establish a causal connection between exposure and injury, the procedures cannot be much less complex, protracted, and costly than those involving liability under tort law, and they cannot even offer much more certainty with respect to the outcome.

The second weakness of special compensation funds is that they do not, as a rule, preclude the injured person from pursuing a liability claim under tort law. Since the range of compensable damages and the amount of compensation provided by the funds are limited, there is a continued incentive for claimants to try to obtain a larger award, including nonpecuniary damages and perhaps even punitive damages, through a court action. Only very feeble attempts have been made so far to discourage such actions, and it is unlikely that they will ever be precluded completely.

It may be time for insurers and insurance buyers to reconsider the basic principles on which the business of liability insurance has been conducted. Where the liability risk is of a long-term nature, it calls for long-term coverage. Indeed, the business of liability insurance has to some extent already been transformed from a short-term business into a long-term business. Insurers have had to adopt some of the techniques usually associated with such long-term coverages as life or disability insurance. The next logical move would have to be toward longer contract terms, or contracts written for indefinite terms, as are customary in Europe, with restricted termination and ample periods of notice, or some other means of assuring continuity of coverage. The amounts of coverage would have to be adjustable, as would the premiums. In addition, there could be sizable premium deposits at the beginning of the contract, as well as special reserves to spread out the long-term risk. Conversely, there would have to be a provision to let the policyholder share in investment income and in savings from lower than expected claims. A long-term adjustable liability policy along these lines would assure a measure of continuity and perhaps a more equitable allocation of risks. Otherwise, however, it would still leave much to be desired. It would not unburden the insured of the risk of change in the risk situation, it would not solve all problems inherent in the liability system, and it would most certainly create some new problems of its own.

Indeed, the scope of available insurance coverage is not likely to improve significantly as long as compensation is legally tied to specific substances, exposures, or conditions. The problems of causation and proof that have been associated with the conventional tort liability system can be eliminated only by a scheme that relies neither on individually established liability nor even on individually established causality but is instead based on the principles of all-risks, first party insurance.

This approach would not make it necessary to abandon or weaken the "polluter pays" principle. On the contrary, it, and it alone, would make it possible to allocate the costs of pollution commensurately with the available evidence of causation, i.e. on a collective rather than individual case-by-case basis. Under a tort system of liability, depending on the weight that is given to the statistical evidence, the claimant receives full compensation or nothing, or perhaps partial compensation reflecting the statistical data for his group. None of the results is satisfactory. The recent tendency has been to award full compensation, resulting in a cost burden on the polluter that is inconsistent with the available evidence in that it disregards the effect of other actual or potential contributing factors or alternative causes.

Under a general all-risks first party compensation system, all injured persons would be compensated fully in accordance with the standards of the system, regardless of the cause of their injury or disease. However, for all the diseases for which the available evidence indicates that exposure to a specific hazardous substance or condition is a contributing factor, an appropriate portion of the total cost of the benefits would be allocated to those determined to be responsible for the release or the exposure, as a group. The collective cost allocation would thus treat both injured persons and responsible polluters as groups rather than individuals and would thereby avoid the problems of unclear or multiple causation at both ends of the chain.

There are precedents for this approach. Specifically, article L 213-1 of the French *Code des assurances* imposes on all automobile owners subject to obligatory liability insurance a special charge, to be collected by the insurer along with the premium, to reimburse the carriers of social sickness insurance for a part of the medical costs attributable to automobile accidents but not compensated through the liability system.

Even the most effective system of general compensation as just outlined will not make liability insurance obsolete. Individual liability remains a viable option in all cases where causation and blame can be determined with little difficulty and conventional means. For this and other reasons, the tort liability system is unlikely to lose its attractiveness for injured persons and their attorneys, at least as long as it offers rewards that, although more chancy and more difficult to obtain, are potentially much larger than the benefits that can be provided by even the most generous first party compensation system.

From this perspective, any change in the law that would limit the amount of damages recoverable under tort law to the actual economic loss and to certain maximum allowances for pain and suffering and other nonpecuniary damages would not only limit the size of individual claims but would at the same time reduce the incentive for injured persons to resort to tort actions in search of large awards.

17.5.4. Challenges for insurers

The discussion so far has focused on the limits of the risk protection that can be provided by commercial insurers, and on possible alternatives. It should be clear, though, that these limits and problems affect only a relatively small part of the universe of risks associated with hazardous materials. They constitute exceptions. Insurability, with appropriate caution, remains the rule. What needs to be done is to identify more precisely those risks that can be covered, the conditions under which they can be covered, and the premiums at which they can be covered. This is primarily a matter of technical research, claims analysis, and actuarial studies. In the second place, it is a matter of coverage design and policy drafting.

The US insurers have taken the bold and unpopular steps of changing to claims-made coverage and of excluding all pollution liability from the general liability insurance policy. They have not yet, however, found the courage to exclude coverage for punitive damages and to limit coverage for general (nonpecuniary) damages. An effort by the Insurance Services Office to exclude punitive damages from general liability coverage failed for lack of support at a time when irresponsible competition in the commercial insurance market was at its peak. Coverage of punitive damages under a policy of liability insurance has always been questionable and is indeed considered contrary to public policy in several states. The same reservations apply to general damages to the extent that they are imposed to punish or deter. As long as the insurers do not take these actions, which are clearly within their power and responsibility, their efforts to have liability rules changed by legislation will continue to lack persuasiveness and indeed credibility.

Elimination or limitation of coverage for punitive and general damages would be beneficial in several respects. Apart from being able to better predict future claims, the insurers would eliminate the major incentive for judges and juries to escalate awards in reliance on the presumed "deep pockets" of the insurers.

Much remains to be done by insurers, especially in the USA, with respect to the definition of the events that they intend to cover as distinguished from those that they do not want to cover. The struggle that began with the accident formula and continued with the occurrence definition and the limited pollution exclusion has not ended with the introduction of the claims-made policy and the total pollution exclusion in the general liability policy.

Now, as always, the insurers that agree to cover the risk of liability for pollution or for the effects of exposure to hazardous materials, even if they do not limit the coverage to "sudden and accidental" releases, do not intend to cover the consequences of every release of whatever nature. It is obvious, for instance, that the insurer does not intend, and the insured does not expect, the policy to cover claims based on events that are part of the normal everyday operation of the insured plant or facility. Although, as noted earlier, claims of this kind still retain some elements of randomness in incidence and amount of resulting loss, most insurers and business managers regard them not as a risk to be insured but as a regular cost of doing business. US insurers now rely primarily on the exclusion of "bodily injury, property damage, or environmental damage which is expected or intended from the standpoint of the insured". This formula has its own shortcomings, however. They have been discussed extensively by Pfennigstorf (1979). It may become necessary in the future to spell out in more detail which situations are and which are not meant to be covered. Pfennigstorf (1979) and Spühler in his Discussion of this chapter have outlined the different factors that insurers may have to consider in developing more detailed risk classifications and more precise definitions of coverages.

Some environmental impairment liability insurers hope to achieve the desired limitation of coverage by providing that there will be no coverage if the insured violates applicable laws or regulations in the area of environmental protection.

Faron (1986) has suggested that even clean-up expenses can be covered in the case of precisely described substances, situations, and operations (such as gasoline tanks at service stations), where loss patterns show great similarity, where there is sufficient loss experience to make the risk calculable, and where there is a sufficient number of units of risk exposure to permit an adequate spreading of the risk.

The insurers' efforts to define the limits of their coverages have often been frustrated by the courts. In many cases, there was indeed an ambiguity, for which the insurers had to accept the blame and the economic consequences. In many other cases, however, the courts have distorted the meaning of a clearly written clause to make the insurer pay for losses that were not intended to be covered. Most recently, in the context of losses caused by hazardous materials, US insurers have appealed to the Congress for help in their struggle against unsympathetic courts, and the Congress appears to be ready to provide that help in the form of rules governing the interpretation of certain words or clauses in existing policies (such as "sudden and accidental" or "expected or intended").

17.6. Conclusions

The role of insurance in dealing with the risks of hazardous materials is a complex one, shaped by conflicting interests and public policy goals. Those working with hazardous materials want the broadest coverage for all the risks that they face, including those that no one is yet aware of, at the lowest price. In the context of environmental policy, however, the role of insurance is ambivalent: if preventing harm to the environment and to individuals and deterrence of pollution are perceived as the primary goals, any kind of insurance must be viewed with suspicion, as a potential cause of distortion in the allocation of social costs. Where the compensation of victims of pollution and the financing of clean-up and rehabilitation measures are concerned, however, insurance is an ally.

If a potential polluter is unable to obtain insurance and consequently goes out of business voluntarily or is bankrupted by uninsured claims, this might be welcomed, from a strictly environmentalist point of view, as the expected and intended result of deterrence through effective cost allocation. Different public policy goals may, however, require the polluter to continue its operations, either because its products are considered essential for the commonweal or because its termination would have unacceptable economic or social consequences.

About 30 years ago, the deterrent effect of potentially ruinous uninsurable liability caused US power companies to refuse to commit themselves to nuclear energy when it became available for private use. They changed their minds only after the Congress, through the Price–Anderson Act of 1957, had limited their liability, prescribed the details of the necessary insurance coverage, and introduced the federal government as an insurer of last resort. In some quarters there are now serious misgivings about that decision.

Nevertheless, we may soon face a similar situation in the hazardous materials area, with one important difference: nuclear power in the 1950s was merely considered desirable as an alternative to conventional power generation relying on shrinking supplies of fossil fuels. The present need for hazardous waste disposal sites and clean-up contractors, in contrast, is immediate and urgent. If they cannot obtain insurance, they must be allowed and indeed be required to operate without it, irrespective of their economic viability, and some way must, and will, be found to provide adequate compensation to the persons who may be injured as a result of the operation, and to pay for necessary clean-up in the event of a release.

Moreover, persons who are injured by exposure to a hazardous substance, or who suffer from symptoms that are believed to be caused by such exposure, are popularly perceived as deserving compensation. It is this perception that has been primarily responsible for the recent expansion of liability by courts and legislatures, and thus indirectly for the present restriction of available liability insurance. For the sentiment favoring compensation, it does not matter much whether the responsible polluter is insured, whether liability can be established under tort law, whether a polluter can be identified, or even whether there is medical evidence linking the injury to the exposure.

The introduction of the claims-made policy may become the final proof that the popular demand for broad and uncomplicated compensation can no longer be satisfied through liability insurance or through liability claims, but requires an all-risks-type first party compensation system under which persons suffering from injuries and diseases receive compensation without regard to the cause or causes. Political factors are likely to make this conclusion difficult to accept, and it will be even more difficult to decide on the details of an appropriate alternative system. There is an important role to play for private insurers in such a system. What that role will be will largely depend on the insurers' ability and willingness to recognize the demand and to cooperate with legislatures and regulators in designing and implementing the system.

In conclusion, it is worth repeating two statements made earlier in this paper. First, in the vast majority of activities or situations involving hazardous materials, insurance performs its function as a risk-spreading mechanism without difficulty. Limited or unavailable insurance coverage is the exception, not the rule. Second, in a system of environmental control the role of insurance, although important, can only be ancillary, as a backup mechanism for compensating economic losses and financing clean-up expenses. Neither compensation nor insurance, however, can be an acceptable substitute for prevention, removal, or neutralization.

References

Boediker, T. (1980), *Umweltverschmutzung und Haftpflichtversicherung: Eine Untersuchung über bestehende Deckungskonzepte unter besonderer Berücksichtigung risikotheoretischer Aspekte*, Doctorate Dissertation (Department of Economics, University of Hamburg, FRG).

Faron, R.S. (1986), Insuring pollution liability in the United States, in W. Pfennigsdorf (ed), *AIDA Studies in Pollution Liability and Insurance*, pp 178–184 (AIDA, Budapest).

Malecki, D.S. and Flitner, A.L. (1985), *Commercial General Liability: The New Claims-Made and Occurrence Forms* (National Underwriter Co., Cincinnati, OH).

Pfennigstorf, W. (1979), Environment, damages, and compensation, *American Bar Foundation Research Journal*, 349–448.

Pfennigstorf, W. (1982), Insurance of environmental risks: recent developments, in *ABA Environmental Law Symposium*, pp 57–81 (American Bar Association, Chicago, IL).

Spiller, J. (1981), *Umweltproblem und Versicherung*, I-VS Schriftenreihe, Band 8 (Institut für Versicherungswirtschaft, St Gallen, Switzerland).

Treasury Department (1982), *Hazardous Substance Liability Insurance*, A Report in Compliance with Section 301(b) and Section 107(k)(4)(A) of PL 96–510 (Treasury Department, Washington, DC).

Treasury Department (1983), *The Adequacy of Private Insurance Protection under Section 107 of the Comprehensive Environmental Response, Compensation and Liability Act of 1980*, A Report in Compliance with Section 301(b) of PL 96–510 (Treasury Department, Washington, DC).

17.D1. Discussion

J.G. Cowell [1]

17.D1.1. Introduction

Hazardous waste is rightly being seen as the key problem of the present and future generations of mankind. My concern, as an executive of the Comité Européen des Assurances, is with "concepts and policies" – in particular with problems of compensating injury or damage caused by the generation, movement, and disposal of hazardous waste.

The *Global 2000 Report to the President* of the USA (Barney, 1981) on entering the twenty-first century reminds us that "environmental problems do not stop at national boundaries". The Report continues:

> If present trends continue, the world in 2000 will be more crowded, more polluted, less stable ecologically, and more vulnerable to disruption than the world we live in now. Serious stresses involving populations, resources and environment are clearly visible ahead. Despite greater material output, the world's people will be poorer in many ways than they are today.

Environmental issues have become beloved of the media. Hardly a day goes by without the appearance of some item or other concerning the environment [2].

Product liability was the issue of the 1970s. Environmental impairment liability (EIL) is the issue of the 1980s and beyond. And legal developments in the field of defective products are now being applied to environmental impairment. What we are seeing today is the snowball effect of heightened public awareness of the pollution problem – not least in the light of the Mexico City (19 November 1984) and Bhopal (2 December 1984) disasters.

And, inevitably, everyone wants to get on the bandwagon – lawyers, legislators, regulators, consumers, scientists, engineers, and even insurers. There is a growing feeling among insurers and reinsurers that it is only within the last two or three years that we have even begun to perceive the true dimensions of the environmental impairment risk. And we still have a long way to go before we fully understand the insurance implications of this increased awareness of the problem.

17.D1.2. Key problems

I should like to address the question of risk spreading and risk bearing under four heads:

(1) The problem of underwriting separate EIL cover.
(2) The problem of mobilizing capacity and coverage.
(3) The problem of legal liability and insurance cover.
(4) The problem of anticipating future developments.

First, however, I shall briefly consider risk management.

There is a fundamental misconception about the risk management discipline which still rears its head whenever risk management is under discussion. Risk management is not just a new way of looking at the old loss prevention long practiced by insurance carriers. It is rather a new approach to an old problem – the problem of risk rendered more acute than ever before by the accelerating pace of technological development.

There is a basic difference in this new approach. The risk manager is concerned with the identification and treatment of risk which may, but need not, include consideration of risk transfer, usually to an insurer. The insurer is concerned with the insurability of risk, and his risk engineering services are geared to improving the quality of risk from the point of view of insurance. Insurance has (and will continue to have) an important role to play in dealing with risk. But it is not the only answer to risk. Traditionally, insurers have limited themselves to – or at least concentrated on – the risk transfer option. Today, insurers are having to become involved in what may be called the "whole risk" approach – seeing risk in its entirety, not solely in terms of whole or partial transfer of risk to insurers. Other risk financing options are available (self-insurance, captives, mutual funds, etc.) and need to be assessed objectively. It is with this in mind that we should address the problems raised earlier.

The Problem of Underwriting Separate EIL Cover

The problem of underwriting EIL is complicated by the fact that the notion of suddenness has lost much of its meaning in cases of gradual pollution involving leakage or seepage. Here the settlement of claims is complicated by numerous factors such as:

(1) Delays before the occurrence or manifestation of injury or damage (e.g. harmful effects developing over an extended period of time).
(2) Delays after the occurrence or manifestation of injury or damage (e.g. identification of the cause of injury or damage, identification of the person liable).

Separate EIL insurance for land-based risks is unlikely to take off in Western Europe in the immediate future. First, the insurance buyer in Western Europe is not at all convinced that he needs separate EIL cover. And this for a very good reason. In most countries the commercial liability policy usually covers sudden and unexpected events. By implication it excludes events which are neither sudden nor unexpected. Unfortunately, however, there is no clearly established cutoff point between sudden and nonsudden events – and the courts are well known for finding against insurers in case of doubt. Consequently, the insurance buyer sees no reason why he should buy (necessarily expensive) EIL insurance when he can already obtain cover under the commercial liability policy. In the circumstances the outlook for separate EIL insurance looks bleak.

Liability insurers can be forgiven for envying their opposite numbers in property insurance who have very largely succeeded in mastering the problems of risk analysis and risk control. Property insurance (damage to own property) is characterized, above all, by knowability and calculability. Property insurers know the limits of insurability in the sense of being able to quantify their exposure to loss. The same cannot necessarily be said of liability insurers.

In effect, the boundaries of liability insurance (damage to third parties) are constantly being rolled back beyond the horizon. This is particularly true of environmental impairment where liability insurers are finding themselves increasingly obliged to cover risks which, quite simply, are unknowable and therefore incalculable. Roger Anderson has suggested:

> What Bhopal has demonstrated is that while the actual chance of something happening might be remote, the losses that can be incurred in today's technological environment are huge. Claims incidence becomes irrelevant, as does the mathematical probability of a catastrophe. (Editorial, *Insurance Week*, 18 January 1985.)

Does this mean that such risks are uninsurable? Not necessarily – although underwriters differ on the extent to which you can put a price on risks which are unknowable. After all, it is much easier to be wise after something has gone wrong than it is before. It also means that we have to recognize the implications for liability insurers of being obliged to pay claims today in circumstances which quite simply could not have been conceived when first going on risk. Bhopal could, in fact, happen anywhere. And the final cost of such a disaster will be measured not in hundreds but in thousands of millions of dollars.

The Problem of Mobilizing Capacity and Coverage

The problem of mobilizing capacity and coverage to deal with EIL exposures is a very real one. The US market for EIL has virtually dried up, and European markets are becoming distinctly nervous. Insurers are equally worried about the sort of losses they may be called upon to compensate under the commercial liability policy, and are not too sure what to do about it.

One reason is the very diversity of insurance markets, which is both their strength and their weakness: their strength because the competitive spirit among insurers generally ensures the availability of some sort of cover provided you are ready to pay the price; their weakness because that same competitive spirit makes them more than reluctant to pool their resources or adopt a uniform approach to underwriting. Furthermore, insurers and insurance buyers rarely agree on what is the right cover at the right price. There is more than a grain of truth in the couplet that "those who can afford insurance don't need it, and those who need insurance can't afford it".

Again, there is something to be said for the view that "exceptional risks require exceptional measures". Nuclear risks are a case in point. Pools were (and still are) seen as the only way of effectively mobilizing maximum capacity to provide the necessary coverage for what were (and still are) exceptional risks. But we would be wrong to strain the comparison between nuclear risks in particular and environmental risks in general. We can, however, draw a number of useful lessons from experience in the nuclear field. First, nuclear liabilities are excluded from the commercial liability policy and are written exclusively by the nuclear pools – except in the case of marine risks written on the open market in London. Second, governments recognize that there are limits to what insurers can offer by providing backup guarantees to supplement cover given by the pools. Third, liability insurance in respect of nuclear exposures is compulsory. This is not the case in respect of (other) environmental exposures in Western Europe.

Pooling experiences outside the nuclear field in Europe have hardly been satisfactory so far, very largely because the pools (in France, Italy, and The Netherlands) are having to compete on unequal terms with cover given under the commercial liability policy. True, the US Insurance Services Office recommendation on excluding all pollution cover from the Commercial General Liability (CGL) policy comes into force on 1 January 1986. But the lesson does not yet appear to have been learnt in Europe.

The Problem of Liability and Cover

There is frequent confusion between legal liability and insurance cover. This is true whether or not we are speaking about compulsory insurance. Any decision whether or not liability should be limited in amount is a matter for the legislator but there is no question whatsoever of insurers being in a position to provide unlimited cover for high-exposure risks. Insurers must be in a position to calculate, no matter how imperfectly, their exposure to potentially catastrophic loss, particularly where such calculation is complicated by low frequency of loss.

This contrasts with experience in the field of motor insurance where high frequency of loss allows insurers to calculate their exposure with greater (but not absolute) certainty. The fact that most motor insurers presently provide unlimited cover in respect of bodily injury should not allow us to overlook the fact that the idea of unlimited liability (and unlimited insurance cover) was born at a time when it was impossible to imagine any one accident placing the future of the motor insurer in jeopardy: insistence today on unlimited cover in high-exposure areas could well lead to insurers quite simply withdrawing from the market.

The Problem of Future Developments

It is becoming increasingly clear that insurers face even greater problems in the field of environmental impairment than elsewhere in civil liability insurance. What concerns insurers most is lack of clarity, especially concerning the system of liability. How do you put a price on future claims caused by factors which could not have been known to the underwriter when he went on risk? Whether or not we shall find a satisfactory "insurance solution" depends very much on how far we go in applying the "polluter pays" principle and "cradle-to-grave" rules of liability [3].

How, in fact, do you determine the moment of loss? Was it at the design or production stage? Was it at the moment of first exposure (e.g. first exposure to asbestos dust)? Was it at the moment of first appearance of symptoms of injury? Was it at the moment of medical diagnosis?

These questions are not academic. They are vital to determining which insurance policy should cover the injury or damage. Historically, insurers have tried to address these questions in two ways: first, by providing cover under the policy in force at the time the act giving rise to injury or damage was *committed*; second, by providing cover under the policy in force at the time the injury or damage *occurred*. Both the *act committed basis* (only used to any extent in Switzerland) and the *occurrence basis* (widely used in Europe and North America) have distinct drawbacks.

Cover under a policy often written 20 or more years ago is likely to be wholly inadequate in today's conditions – not least in the light of monetary inflation. Furthermore, it may prove impossible to trace the policy in force at the time the act was committed or the injury or damage occurred. Difficulties are compounded in toxic waste cases involving a large number of generators using the same pits, ponds, landfills, etc. US courts have typically driven a bulldozer through insurers' defences (e.g. by holding all insurers liable to compensate victims from, say, the moment of first exposure to the moment of medical diagnosis). Europe has been comparatively immune from the toxic waste litigation fever which has gripped the USA. But this may not always be the case, particularly in the light of increased public awareness of the long-tail problem where injury or damage can only become apparent many years after first exposure.

One way of dealing with the problem has been to write policies on a *claims-made basis*. This is the method favored by reinsurers and EIL underwriters, though liability insurers in general are reluctant to abandon practices established over many years. Cover in this case is provided under the policy in force at the time the claim was *notified* to the insurer. Both insurer and insurance buyer are relatively secure in the knowledge that the level of indemnity is (more or less) adequate to meet today's level of claims. In addition, liability of the insurer is cut off at the end of the policy period: the only real problem occurs where there is no insurance in force to cover future claims (e.g. in case of bankruptcy or cessation of activity).

Finally, there is the question of liability itself. Should liability be channelled? And, if so, who should be liable? Alternatively, should liability fall on the holder of the waste – the person having care, custody, and control of the waste at the time of injury or damage?

These are issues on which insurers will have to think very carefully in the light of future demands for EIL cover (under a separate policy or under the commercial liability policy). Insurance can do much to compensate for damage to persons and property. It can never be a substitute for a properly developed program of loss prevention and loss control. And it certainly does nothing to reduce the need for tighter administrative supervision over hazardous waste movements.

17.D1.3. Conclusion

Insurers have mixed views on compulsory insurance. In the end it may be the only way of dealing with selection against insurers, and mobilizing capacity and coverage in the EIL field. Compulsory insurance does raise serious problems of enforcement and identification of the polluter – not to mention the desirability of further state intervention. Insurers should not be expected to act as policemen or to bear the burden of unidentified or uninsured polluters. Appropriate arrangements to back up insurance cover will have to be investigated.

It is anybody's guess whether cover will continue to be offered under the commercial liability policy or will be provided under separate cover. Certainly, many EIL underwriters favor writing and rating the EIL exposure separately. Others see a pooling of resources as the only realistic solution to a problem which can only get worse before it gets better.

Litigation attitudes must (and will continue to) influence developments on both sides of the Atlantic. But we should always keep in mind the fact that the prevailing wind in liability blows from West to East: what happens in America today could happen in Europe tomorrow – despite real differences in law and practice including, of course, heightened litigation consciousness, contingency fees, punitive damages, and jury system.

The English statesman Canning once remarked that he had called the new world into existence to redress the balance in the old. When we look at some of the excesses in the field of litigation in the USA we can be forgiven for thinking that it is we in the old world who need to redress the balance in the new.

Notes

[1] The views expressed here are those of the author alone and do not necessarily represent those of the Comité Européen des Assurances or its member associations.

[2] See, for example, the *Wall Street Journal* on waste clean-up (31 May 1985) and waste reduction (4 June 1985).

[3] "Von der Wiege bis zur Bahre" (Goethe) ["From Cradle to Grave"].

Reference

Barney, G.O. (1981), *The Global 2000 Report to the President* (Council on Environmental Quality and Department of State, Washington, DC).

17.D2. Discussion

J. Spühler

17.D2.1. Preliminary remarks

In contrast to Pfennigstorf's approach in outlining the role of insurance in risk spreading and risk bearing in connection with hazardous materials and the problems involved from the standpoint of a lawyer, in the following the issues are viewed from the viewpoint of an underwriter. It is therefore necessary to broaden the range of the discussion in the following manner:

(1) It is not only the situation during transportation, storage, or disposal of hazardous materials that creates all the problems, but each and every phase during the lifetime of such materials: production, storage, transportation, handling, use, consumption, disposal, and redisposal.
(2) The problems around hazardous materials cannot just be discussed and analyzed with regard to such materials in their "waste" state, because it is an impossible task to distinguish clearly between the state of "usefulness" of a product and that of its being a "burden" which somehow has to be got rid of properly.
(3) The kind of insurance cover for injuries, damages, economic losses, and nonpecuniary losses is not of primary importance to uncover the problems arising from hazardous materials. This is to be observed despite the fact mentioned by Pfennigstorf that liability insurance has been focused primarily on such materials and harmful occurrences initiated by them.

In addition to this broadening of the discussion, the following central question must be taken into account: To what extent do the fundamental goal of insurance and the basic principles attached to it correspond to the specific situations arising from the unique qualities of hazardous materials by gaining, producing, transporting, storing, handling, using, and disposing of them? In this context, the fundamental goal of insurance consists in the transfer of uncertain burdens resulting from unforeseen events hitting a firm or an individual person to well-defined and limited costs within a well-defined period of time so that as a residual there are no, or only restricted, imponderabilities left to a firm or an individual.

Now, before being in a position to outline an acceptable solution for insurance coverage in respect of hazardous materials, we have to consider the primary qualities of such materials as far as they are related to primary insurance principles.

17.D2.2. Insurance-related unique qualities of hazardous materials

Hazardous materials may at any time and anywhere cause bodily injuries, property damage, nonpecuniary claims, clean-up, and removal to a very large number of firms and individuals within a very large area. The very high monetary extent may be created by an accident or from nonsudden pollution that occurs gradually over an extended period of time, with the causal initiation of the relevant events very often a long time in the past. Within this general situation, the following general items have to be observed: the instability of the risk situation, the kinds of consequences arising from the risk, the long-range character of the risk, the extent of compensation arising from the risk, and the determination of the event turning the risk into an actual loss. Analysis of these items discloses the following.

Instability of the Risk Situation

The calculation of the premium for an insurance coverage valid over a medium period of time requires a certain minimum degree of stability of the risk situation. This prerequisite is not given in connection with hazardous materials for the following reasons:

(1) New technologies facilitate the discovery and tracing of toxic effects of products which were previously considered as absolutely safe. Or an unforeseen mixing of products which themselves are completely nontoxic may bring about a highly toxic conglomerate. This perpetual and very often drastic change in the situation calls – under fixed premiums and policy conditions – for a drastic limitation of the underwriter's engagement in respect of the period of time at risk. This means that the policy period is shortened or that policy wordings contain a clause that gives to the underwriter the right to change at least the premium rate charged within short intervals.

(2) Furthermore, the abrupt and unforeseeable changes on the legal scene bring about a highly uncertain state for the underwriter. This is particularly true in the field of liability insurance. New legislation introducing strict liability rules or imposing specific costs on the insured, as well as court decisions, lead to unanticipated aggravation of the risk borne by the underwriter.

Kinds of Consequences Resulting from Hazardous Materials

The consequences of hazardous materials consist primarily of bodily injuries to people who are in contact with such materials in the course of their daily work or to people who find themselves "outside" the actual handling area. The range is practically unlimited, as the air and flows of water very often act as carrying media for a toxic release leading to the harmful impact on human beings over a very large area. The same effect can be noted with respect to property damage. Many production sites have to stand idle for a long time owing to their contamination by accidently released hazardous materials used there, and this leads to very significant loss of use. In addition to bodily injuries and property damage, the underwriter is increasingly faced with non-pecuniary losses resulting from the release of hazardous materials. Furthermore, various kinds of costs for the prevention of expected losses and for the limitation of losses that have already occurred as well as clean-up costs and costs for restoration may be incurred.

These facts lead to the following conclusions:

(1) The kinds of consequences show that it is not only liability insurance but all lines of insurance that may be impacted by hazardous materials. This is very significant, as most insurers today do business in all lines of nonlife insurance.

(2) The buyer of insurance coverage is not primarily concerned about the type of insurance he needs. This is illustrated by the fact that it is immaterial to him whether he is financially hit by loss or damage to his own property or by his own bodily injuries on the one hand, or by those of others due to his failures, negligence, or undue activities for which he has to stand good on the other hand.

17.D2.3. Long-range character of risks inherent in hazardous materials

Hazardous materials do not necessarily cause negative effects on the spot when being gained, produced, and so on. And, very often, their negative effects cannot be discovered at the moment when they arise. This causes the high degree of uncertainty about the state of risk at the time of granting insurance coverage. Such a situation is highly undesirable for the underwriter, as he aims at limiting his engagement in respect of the period of time as set forth in the policy.

In the light of these conditions, the so-called "claims-made" policies were created and offered on a large-scale basis. The negative effects of this type of policy are far smaller in practice than stated by Pfennigstorf. If the relationship between insurer and insured is based on a common understanding and on the principle of mutually honoring each other's interest, the claims-made policy must be considered as a well-based instrument for solving the present problems in the best possible way.

High Extent of Compensation Arising from Hazardous Materials

The consequences of negative effects are of an extent that cannot be compared with anything else. This is primarily seen in the field of liability insurance. The present demand in respect of the extent of insurance cannot be met by the insurers. Their bad experience in product liability and environmental impairment liability brought about severe restrictions in order to secure a sound balance between their overall engagements and their financial standing. The question arises as to whether the situation could be improved by transferring the coverage for environmental impairments from the general liability policy into a special environmental impairment liability policy on a larger scale than today. This question must be answered in the negative. The underwriter's engagements cannot be increased by creating separate policies for risks that have been part of another kind of liability policy hitherto. His engagements in each of the two types of policies have to sum up within a well-defined period of time.

Absence of Well-determined Events Resulting from Hazardous Materials

The absence of well-determined events resulting from hazardous materials places quite a restriction on the insurability of negative effects due to such materials. All kinds of wordings – be they accident-oriented or based on the term "occurrence" – in fact show very serious failings. On the other hand, sound underwriting is not possible by offering coverage to each and every origin of claims. No coverage can be given to willful acts or omissions. The same is true for all planned and intended operations which are connected with negative effects arising from hazardous materials. The effects arising from situations of this kind have to be balanced out financially by including their consequences in the production and handling costs. This attitude concerning willful and random acts and omissions very clearly shows where we have to draw the line between the transfer of risks to the insurer on the one hand and the financing of losses by funds to be accumulated by the polluters on the other. Thus we come to the conclusion that insurance is aimed at covering the negative consequences of nonroutine and irregular operations in the course of any activity.

17.D2.4. Outline of an acceptable solution

Considering the general goal of insurance as well as the basic principles of underwriting associated with that goal, we can outline the framework for an acceptable solution to be realized by insurance in relation to hazardous materials as follows:

(1) The negative effects must be uncertain at least as to the moment when they may arise. And they must not have occurred prior to the inception of an insurance cover. Furthermore, they must be free from any willful act or omission.

(2) In order to safeguard the balance between prefixed premiums and the extent of the engagement of the insurer, it must be possible for the insurer to alter the premium rate and the extent of coverage during the course of the policy to match the major alterations in the assumptions previously taken into account when establishing the insurance coverage.

(3) The consequences covered must be of an objective and real nature. Nonpecuniary and noneconomic losses must be excluded.

(4) Consequences resulting from normal gaining, producing, handling, using, transporting, storing, and disposing cannot be subject to an insurance coverage. They represent part of the ordinary well-expected costs of the business activities.

(5) High amounts of insurance asked for must be limited and they must be supported by an adequate deductible to be borne by the insured.

(6) In order to cope with the unrealistic interpretation of policy wordings by courts, in order to escape the task of setting a strict line between "accidental events" or "occurrences" and "gradual developments and emissions", and in order to bring about a clear assignment of these to a specific moment of time, a very reasonable kind of coverage in respect of the amount available to cover losses consists in the following scheme.

A prefixed amount – say, US $50 million – agreed upon by insurer and insured sets the limit up to which the total of all losses and costs resulting from hazardous materials within a specific period of time and being claimed against the insured are paid by the insurer. At the moment when the said amount is exhausted, the policy lapses automatically; otherwise the policy ends at the end of the period agreed upon between insured and insurer. Such a setup has the advantage of avoiding misinterpretation of the policy wording in respect of the parts mentioned earlier and therefore results in a high degree of determinateness for both parties involved.

17.D2.5. Concluding remarks

Insurers are constantly studying ways and means to improve the insurance coverage of hazardous materials in order to meet the needs of insureds in accordance with the constant developments in technology, economic situations, medical know-how, as well as in law and its interpretation by courts. However, legal and court-related exaggerations do not solve the problems. Part of the burden arising from hazardous materials can be transferred to insurance setups in all lines underwritten today. And this part will be taken by the insurers in fulfilling their task in respect of their set goal. But the rest of the burden has to be met by other means.

Epilogue

From Seveso to Bhopal and Beyond

P. Kleindorfer and H. Kunreuther

A principal objective of the International Institute for Applied Systems Analysis (IIASA) Conference was to develop a set of recommendations for future studies between researchers and practitioners that would improve the management of hazardous materials. To this end we invited six practitioners to participate in a panel discussion followed by small group meetings [1]. Each of the small groups was asked to outline a set of research needs in their area with a concern for linking theory with practice. The research recommendations developed by the small groups are presented below by linking them to the following themes highlighted by the conference:

(1) *Problem context.* There is a need to increase our understanding of the problems and opportunities facing firms which manufacture products that create toxic waste, the alternatives open to transporters of hazardous materials, and the challenges facing interested parties involved in the siting of storage and disposal facilities.

(2) *Risk analysis.* There is a need to document the potential benefits and inherent limitations of risk analysis both at the assessment level and at the level of how data are communicated to the different interested parties (e.g. the public, industry). In particular, we need to understand how bargaining and negotiation can facilitate the decision process and enable interested parties to reach compromise solutions.

(3) *Risk management and insurance.* There is a need to understand the role that legal institutions and regulation can play in facilitating the production, transport, and storage of hazardous

materials. What is the appropriate role of insurance in dealing with these problems?

We now summarize a set of key ideas raised by the panel members, open discussions at the Conference, and research recommendations from the small groups by specifying a set of topics related to the above three areas.

E.1. Problem Context

E.1.1. Production of hazardous materials

In his panel presentation, Perry Hopkins provided a set of principles for firms to follow in preventing and dealing with Bhopal-like incidents. In particular, he emphasized the importance of firms recognizing the need for expertise and safety in technology when they deal with highly hazardous materials which are also essential for meeting society's needs. Manufacturing line management must be full participants in the development and implementation of all business planning involving hazardous materials, and they need to participate actively in developing industry standards, government regulations and laws for the management of hazardous materials worldwide. Hopkins also pointed out the need for improved methods of detection of hazardous materials discharged into the atmosphere from devices that relieve pressure in vessels that might otherwise rupture. Currently, there are no means to neutralize, disperse or otherwise protect the environment from these discharged materials.

Key Research Areas for Production

(1) How do we reconcile differences between allowable concentrations or doses in the workplace and those in the external environment?
(2) How can industrial firms remain competitive while still addressing the hazard and risk concerns of the public?
(3) How does one introduce industrial practice and compliance techniques into small firms for problems involving chronic and gradual pollution, as well as possible explosive accidents?
(4) Determine the effects of various organizational structures and managerial behavior on the levels of risks in a firm.
(5) Undertake a state-of-the-art survey of current safety apparatus used in industry and its effects on reduction of risk (e.g. the use of dated or modern equipment).
(6) Undertake case studies of management practices in dealing with risks for firms of different sizes, including a survey of risk levels

accepted by safety managers in different types of industrial facilities.

(7) What are effective procedures for developing trust between the public and firms? How do successful companies deal with this issue?

(8) How do different firms deal with mismanagement issues? Can risk scenarios play a creative role in this process?

E.1.2. Transportation of hazardous materials

In the panel discussion, Frederik Bjørkman stressed the importance of the carrier of hazardous materials as an identifiable party in the hazardous materials process. He indicated that the public is frequently unaware of the sender or the receiver of the goods, but can normally identify and will demand compensation from the transporter. For this reason, local authorities frequently place a number of restrictions on carriers to protect the general public.

Key Research Areas for Transportation

(1) What is the linkage between the transportation of hazardous materials and the siting of disposal facilities from the point of view of managing risk and dealing with the costs of an accident?

(2) Can one determine minimal acceptable standards for transporting hazardous goods? What is the evolution and rationale of international conventions regarding these standards?

(3) Can one develop regulations for dealing with transportation of goods that have an opportunity of being appropriately monitored and controlled? What is the past experience with these types of regulations in different countries?

(4) What are appropriate liability and insurance mechanisms for covering transport of hazardous materials? How easily can these be enforced in practice?

E.1.3. Siting of storage and disposal facilities

Research in this area needs to be designed so that consideration is given to the technical, social, political, and economic aspects of the siting process. As in the other areas, there is a need to investigate implementation problems associated with siting a new facility and ways of enforcing any rules and regulations. As was pointed out by a number of participants in the meeting, the siting issue encompasses a wide variety of problems at all levels, including legal issues, public participation, as well as the role of policy tools such as compensation and insurance for facilitating the process.

Key Research Areas for Siting

(1) What are the trade-offs between equity and efficiency considerations in making siting decisions?
(2) What role can risk assessments play in siting decisions? Can one develop a set of criteria for evaluation of the risks resulting from siting in one place versus another?
(3) Does insurance availability influence siting? What type of insurance would be most useful in this connection?
(4) How can one bring the public effectively into the siting process? In what ways can technical assistance be useful in enhancing public participation?
(5) How can one enhance public trust in institutions and facilitate the siting of hazardous facilities?
(6) What role can compensation and benefit sharing play in conjunction with other policy instruments such as regulations for facilitating the siting of hazardous facilities?

E.2. Risk Analysis and Decision Processes

E.2.1. Risk assessment

Orio Giarini, in his panel discussion comments, indicated that in the nineteenth century it was much easier to determine risks from industrial plants such as a textile mill by undertaking a specific inspection. In the last 20 years, technology has changed so rapidly that it is difficult for those in the plant to fully understand technological risks and even more difficult for outsiders, such as insurance inspectors, to monitor and understand these risks. This may create moral hazard problems if those inside the firm have a better conception of the risk than those outside (e.g. insurers). This is one reason that insurance for many types of risk is not available today. To improve the situation with respect to risk assessment, the following research needs have been outlined.

Key Research Areas for Risk Assessment

(1) Creation of new data bases and coordination of existing ones on the toxicity of chemicals in different environments and on their physical effects.

(2) Development of simple assessment models that contain only key variables and can be applied to a number of different situations without large expenditures of time or money.
(3) Conducting of a survey of risk assessment activities of international and national institutions.
(4) Development of worst case scenario methodologies for use in a variety of industrial settings.
(5) Impact of scientific uncertainty in undertaking risk assessments and settling differences between experts. What role can science courts play in adjudicating the process?

E.2.2. Risk perception and communication

There was considerable discussion at the Conference on the differences in perception between experts undertaking risk assessment and the lay public. There was general consensus on a need to study ways of improving the communication of information on the nature of the risk to the general public as well as the costs and benefits of alternative policy strategies.

Key Research Areas for Risk Perception and Communication

(1) How can one make risk information more relevant to specific interested parties (e.g. top-level officials, lower-level executives, government agencies)? How do different political regulatory statutes and cultural factors affect these communication needs?
(2) How can one better communicate uncertainties with respect to probabilities, consequences, and trade-offs between different alternatives? What is the role of computerized decision support systems in improving the way individuals process this information?
(3) What information about risks do people feel they need in contrast to information that the government feels may be "good for them"? Do people want to know about the chances of potentially catastrophic accidents in the future (e.g. an earthquake in California) if they are living in the area? Does framing of data in different forms (e.g. gains versus losses) change people's cognitions or just their responses?
(4) What do people feel is a fair political process and how does that affect the way information is/should be presented to them? Is it useful to consider compensation or benefit sharing as a way of facilitating communication?

E.2.3. Research on bargaining and negotiations

In his panel discussion, Károly Bárd indicated the need to focus on the different interested parties affected by hazardous materials and to determine the appropriate role of insurance and compensation as policy tools for facilitating bargaining and negotiation. He indicated that one needed to link these alternative mechanisms to the objectives of society. For example, there are differences between socialist economic planning, where insurance is compulsory, and the available coverage in other countries where a voluntary system is in place. Political, psychological, social, and economic conditions may set up different atmospheres for bargaining and negotiating on environmental matters.

Key Research Areas for Bargaining and Negotiation

(1) What is and should be the role of experts and expert knowledge in the negotiation process? Should expert knowledge be used as a tool for bargaining or is it perceived as a constraint on bargaining?
(2) What is the best way to prepare people for negotiation? Are existing training programs helpful in resolving conflicts on environmental problems?
(3) At what stage in the policy process – policy formation, standard setting, implementation – do bargaining and negotiation occur? What opportunities exist for bargaining and negotiation that are not presently being exploited?
(4) How do bargaining and negotiation processes differ among political cultures? Are there specific lessons that can be transferred from one country to another?

E.3. Risk Management

E.3.1. Legal institutions

In his panel discussion, Ludwig Krämer indicated that in the European Economic Community Treaty there are over 100 binding legal instruments which relate to risk and the environment. He pointed to the Seveso Directive as a model for dealing with plant safety practices and one that is likely to be exported from Europe to the USA. Since conflicts over values, facts, and policy actions will continue to be adjudicated by law, it is clear that legal institutions and practices will be a cornerstone of risk management. The following areas were deemed especially important research topics here.

Key Research Areas for Legal Institutions

(1) What are the potential reforms to the legal system in the USA with respect to toxic tort and compensation for damages? Are there any lessons from European countries which may be helpful in this regard?
(2) Evaluate the joint and several liability system in the context of hazardous materials. What are the costs and benefits of continuing with this type of arrangement?
(3) What are the economic incentives of toxic tort law and liability with respect to product developments (e.g. pharmaceuticals) and health and safety procedures within firms?
(4) How does the legal framework influence the availability of insurance? What reforms would be helpful in providing increased coverage against environmental pollution damage?

E.3.2. Regulation

Several of the panelists discussed the importance of regulations, with appropriate monitoring and control procedures, as risk management tools for the hazardous materials problem. The question of when to utilize regulations produces a wide range of responses. In some cultures, there is a reluctance to impose regulations unless there is a clear failure of market-like mechanisms such as effluent-charge incentive systems. In other countries, regulations are a way of life. The hazardous materials problem is viewed by most countries as one that needs to be at least partly remedied through regulating activities of plants, transporters and those who operate storage and disposal facilities. In addition, the public is extremely reluctant to sanction new facilities that have the potential of causing damage to health and safety without assurance that strict regulatory and control procedures will be enforced.

Key Research Areas on Regulation

(1) What are the interrelationships between standard setting, monitoring, and enforcement of regulations in the responses of firms producing hazardous materials as by-products?
(2) What is the relationship between self-regulation by firms and externally imposed regulations by government agencies?
(3) What are the appropriate regulatory authorities and enforcement mechanisms associated with hazardous materials storage and disposal facilities? How will regulations facilitate the siting process?

(4) What types of regulations can assist the process of bargaining and negotiations for transport, storage, and disposal of hazardous materials?
(5) What role can regulation play for dealing with hazardous materials problems when the causality of certain health effects cannot be ascertained?
(6) What lessons can be learned from international comparative research on hazardous materials regulations for model regulatory legislation?

E.3.3. Research on insurance

In his commentary, Michael Stradley pointed out that the market for environmental impairment liability has collapsed in the USA. Reinsurers throughout the world are reluctant to provide coverage in the USA because of the uncertainty as to the magnitude of claim settlements in court. This has created a lack of capacity in the industry and raised questions as to alternative mechanisms for insuring interested parties against potentially catastrophic losses.

Key Research Areas for Insurance

(1) How can uninsurable events be made insurable for protecting against damages from hazardous materials? What actions are needed to increase the capacity of insurers and reinsurers? When can claims-made rather than occurrence-based policies be helpful in this regard?
(2) Is there a need for government involvement for dealing with catastrophic losses through some type of reinsurance program? Is the Price–Anderson Act or the Black Lung Program a useful model for some type of government–private system in the USA?
(3) What type of self-insurance plans by industry are likely to be successful in filling the gap in insurance protection?
(4) What are the incentive effects of insurance in increasing the safety level and protective activities of industrial firms?

In order to undertake research on these issues, there needs to be an open dialogue between the academic community and real world practitioners. The IIASA Conference on Transport, Storage, and Disposal of Hazardous Materials was designed as a first step in this direction. Hopefully, the process will be accelerated in the coming years.

Note

[1] The members of the panel were: Károly Bárd, Board of Insurance Enterprise, Hungary; Frederic Bjørkman, Director of the Swedish Regulatory Authority for Transport of Dangerous Goods; Orio Giarini, Secretary General of Geneva Association; Perry Hopkins, Director of Manufacturing Services at Du Pont; Ludwig Krämer, Commission of the European Communities, Belgium; Michael Stradley, Engineering Consultant, American Nuclear Insurers' Association, USA.

List of Conference Participants and Contributors to the Proceedings [1]

Dr AICKIN, R. Malcolm, ERAS (International) Ltd., Environmental Risk Analysis Systems, Ibex House, Minories, London EC3M 1HJ, UK.

Prof BARAM, Michael, Boston University School of Public Health, School of Medicine, 80 East Concord Street, Boston, MA 02118, USA.

Dr BÁRD, Károly, Managing Director, Állami Biztositó, Ulloi ut 1, 1813 Budapest IX, Hungary.

Mr BAUMANN, Ernst, Risk Management Service Centre, Swiss Reinsurance Company, 50/60 Mythenquai, Zurich, Switzerland.

Ms BEAUCHAMP, Martha, Managing Attorney, Office of General Counsel, American Petroleum Institute, 1220 L Street, NW, Washington, DC 20005, USA.

Mag BERGER, Heinz, Head of Department IV/4, Transport of Dangerous Goods by Road, Bundesministerium für Verkehr, Sektion IV, Karlsplatz 1, 1010 Vienna, Austria.

Mr BJORKMAN, Fredrik, Director, Competent Authority for Dangerous Goods, The National Administration of Shipping and Navigation, 60178 Norrkøping, Sweden.

Mr BOHNENBLUST, Hans, Ernst Basler & Partners, Consulting Engineers and Planners, Forchstrasse 395, 8092 Zurich, Switzerland.

Prof BRODECKI, Zdzislaw, Department of Maritime Law, Faculty of Law, University of Danzig, ul Swierczewskiego 27, 81-831 Sopot, Poland.

Mr COHEN, Adrian V., Health and Safety Executive, Baynards House, 1 Chepstow Place, London W2 4TF, UK.

Mr COLLART, Jean E.G., Director, Royale Belge, 25, Bd du Souverain, 1170 Brussels, Belgium.

Dr COVELLO, Vincent T., Program Manager, Risk Analysis Program, National Science Foundation, 1800 G Street, NW, Room 1233, Washington, DC 20550, USA.

Mr COWELL, John G., Deputy Secretary General, Comité Européen des Assurances, 3, Rue Meyerbeer, 75009 Paris, France.

Dr DAVIES, J. Clarence, Executive Vice President, The Conservative Foundation, 1717 Massachusetts Avenue, NW, Washington, DC 20036, USA.

Dr DAVIES, Gary A., Attorney at Law, 602 Gay Street, Suite 507, Burwell Building, Knoxville, TN 37902, USA.

Dr DINGLINGER, Friedemann, Ministerium für Ernährung, Landwirtschaft, Umwelt and Forsten Baden-Würtemberg, Postfach 491, 7000 Stuttgart 1, FRG.

Dr FEDRA, Kurt, IIASA, Schloss Laxenburg, 2361 Laxenburg, Austria.

Dr GIARINI, Orio, Secretary General, Association Internationale pour l'Etude de l'Economie de l'Assurance ("Geneva Association"), 18, chemin Rieu, 1208 Geneva, Switzerland.

Mr GROSJEAN, Christian, Association Internationale pour l'Etude de l'Economie de l'Assurance ("Geneva Association"), 18, chemin Rieu, 1208 Geneva, Switzerland.

Prof HARKER, Patrick, Department of Decision Sciences, The Wharton School, University of Pennsylvania, Philadelphia, PA 19104, USA.

Mr HAWKSLEY, James, L., ICI Group Safety Services, Bessemer Road, Welwyn Garden City, Herts AL7 1HD, UK.

Mr HOPKINS, Perry, Director of International Manufacturing, E.I. Du Pont de Nemours and Company, Room 4020, The Du Pont Building, Wilmington, DE 19898, USA.

Mr JAIN, D.K., First Secretary, Indian Embassy, Kärntner Ring 2, 1010 Vienna, Austria.

Mr JONES, T. Jeffrey, President, Delta Management Group, 1414 Fenwick Lane, Silver Springs, MD 20910, USA.

Prof KASPERSON, Roger E., Center for Technology, Enviroment and Development (CENTED), Clark University, Worcester, MA 01610, USA.

Dr KLAUS, Alfred, Risk Management, Swiss Reinsurance Company, 50/60 Mythenquai, Zurich, Switzerland.

Prof KLEINDORFER, Paul, R., IIASA, Schloss Laxenburg, 2361 Laxenburg, Austria.

Ms KNAP, Zsuzanna, Állami Bizositó, Ulloi ut 1, 1813 Budapest IX, Hungary.

Mr KRAMER, Ludwig, DG - XI - Environment, Commission of the European Communities, 200, rue de la Loi, 1049 Brussels, Belgium.

Prof KUNREUTHER, Howard, Director, Center for Risk and Decision Processes, The Wharton School, University of Pennsylvania, 3620 Locust Walk/CC, Philadelphia, PA 19104, USA.

Prof LAGADEC, Patrick, Laboratoire d'Econométrie, l'Ecole Polytechnique, 1, rue Descartes, 75230 Paris, France.

Prof LEE, Terence R., Department of Psychology, University of Surrey, Guildford GU2 5XH, UK.

Prof LIND, Niels C., Director, Institute for Risk Research, University of Waterloo, Waterloo, Ontario N2L 3G1, Canada.

Dr LINNEROOTH, Joanne, IIASA, Schloss Laxenburg, 2361 Laxenburg, Austria.

Ms MATTHEWS, Marjorie K., Chief, Evaluation and Analysis, Transport Dangerous Goods Directorate, Transport Canada, Twr. 'B', Place de Ville, Ottawa, Ontario K1A 0N5, Canada.

Dr MORICI, Alessandro, Nuclear Regulatory Directorate, Comitato Nazionale per la ricerca e per lo sviluppo dell'Energia Nucleare e delle Energie Alternative (ENEA), Via Vitaliano Brancati, 48, 00144 Rome, Italy.

Prof MUNN, Robert E., IIASA, Schloss Laxenburg, 2361 Laxenburg, Austria.

Dr NASCHI, Giovanni, Executive Director, Nuclear Regulatory Directorate, Comitato Nazionale per la ricerca e per lo sviluppo dell'Energia Nucleare e delle Energie Alternative (ENEA), Via Vitaliano Brancati, 48, 00144 Rome, Italy.

Dr NICHOLS, Joanne, Massachusetts Department of Environmental Management, Commonwealth of Massachusetts, 19th Floor, 100 Cambridge Street, Boston, MA 02202, USA.

Prof O'HARE, Michael, John F. Kennedy School of Public Policy, Harvard University, Cambridge, MA 02138, USA.

Prof O'RIORDAN, Timothy, School of Environmental Sciences, University of East Anglia, Norwich NR4 7TJ, UK.

Dr ORLANDO, Enrico, Central Manager, Riunione Adriatica di Sicurità, Corso Italia, 23, 20122 Milan, Italy.

Dr OTWAY, Harry, Sector Head, Technology Assessment, Commission of the European Communities, Joint Research Centre, 21020 Ispra (Varese), Italy.

Mr PIASECKI, Bruce, Science, Technology and Society Program, Cornell University, Ithaca, NY 14850, USA.

Dr PFENNIGSTORF, Werner, Research Attorney, American Bar Foundation, 750 North Lake Shore Drive, Chicago, IL 60611, USA.

Prof POCCHIARI, F., Director General, Istituto Superiore di Sanità, Viale Regina Elena 299, 00161 Rome, Italy.

Dr RENN, Ortwin, KFA-TUG, Kernforschungsanlage Jülich, Postfach 1913, 517 Jülich, FRG.

Dr ROSENTHAL, Isadore, Director, Safety, Health and Environmental Affairs, Engineering Division, Rohm and Haas, Box 584, Bristol, PA 19007, USA.

Mr SCHROEDER, P., Zurich Insurance Company, 2 Mythenquai, 8002 Zurich, Switzerland.

Prof SLOVIC, Paul, Research Associate, DECISION RESEARCH, 1201 Oak Street, Eugene, OR 97401, USA.

Prof SMETS, Henri, Principal Adminstrator, Environment Directorate, OECD, 15 avenue de l'Amiral Bruix, 75016 Paris, France.

Dr SPÜHLER, Jürgen, Management Committee, Zurich Insurance Company, 2 Mythenquai, 8002 Zurich, Switzerland.

Dr STALLEN, Pieter-Jan, Netherlands Organization for Applied Scientific Research (TNO), Centre for Technology and Policy Studies, PO Box 541, 7300 AM Apeldoorn, The Netherlands.

Mr STRADLEY, Michael W., American Nuclear Insurer's Association, 270 Farmington Avenue, Farmington, CT 06032, USA.

Dr von WINTERFELDT, Detlof, Institute of Safety and Systems Management, University of Southern California, University Park, Los Angeles, CA 90089-0021, USA.

Dr WYNNE, Brian, School of Independent Studies, University of Lancaster, Bailrigg, Lancaster LA1 4YN, UK.

Dkfm Dr ZIDEK, Heinz, Director, Marine Department, Erste Allgemeine Versicherungs-A.G., Brandstätte 7–9, 1010 Vienna, Austria.

Note

[1] The following contributors to these *Proceedings* did not attend the Conference:

V. Silano (Istituto Superiore di Sanità, Rome), coauthor of Chapter 3.
G. Zapponi (Istituto Superiore di Sanità, Rome), coauthor of Chapter 3.

J Deprimoz, author of the Discussion of Chapter 4.
Miley Merkhofer (Applied Decision Analysis, Menlo Park, CA), coauthor of Chapter 8.
Steven M. Swanson (American Petroleum Institute), coauthor of the Discussion of Chapter 12.

Index

MIX
Papier aus verantwortungsvollen Quellen
Paper from responsible sources
FSC® C105338

If you have any concerns about our products,
you can contact us on
ProductSafety@springernature.com

In case Publisher is established outside the EU,
the EU authorized representative is:
Springer Nature Customer Service Center GmbH
Europaplatz 3, 69115 Heidelberg, Germany

Printed by Libri Plureos GmbH
in Hamburg, Germany